The Pesticide Detox

The Pesticide Detox

Towards a More Sustainable Agriculture

Edited by

Jules Pretty

London • Sterling, VA

First published by Earthscan in the UK and USA in 2005

ISBN: 1-84407-142-1 paperback
       1-84407-141-3 hardback

Typesetting by JS Typesetting Ltd, Porthcawl, Mid Glamorgan
Printed and bound in the UK by Cromwell Press Ltd, Trowbridge
Cover design by Andrew Corbett

For a full list of publications please contact:

Earthscan
8–12 Camden High Street, London, NW1 0JH, UK
Tel: +44 (0)20 7387 8558
Fax: +44 (0)20 7387 8998
Email: earthinfo@earthscan.co.uk
Web: **www.earthscan.co.uk**

22883 Quicksilver Drive, Sterling, VA 20166-2012, USA

Earthscan is an imprint of James & James (Science Publishers) Ltd and publishes
in association with the International Institute for Environment and Development

A catalogue record for this book is available from the British Library

Library of Congress Cataloging-in-Publication Data

Pretty, Jules N.
     The pesticide detox : towards a more sustainable agriculture / Jules Pretty.
          p. cm.
     Includes bibliographical references and index.
     ISBN 1-84407-142-1 (pbk.) – ISBN 1-84407-141-3 (hardback)
     1. Agricultural pests–Biological control. 2. Pesticides–Environmental aspects. 3.
     Organic farming. I. Title.

SB975.P75 2005
632'.96–dc22

                                                                        2004024815

Printed on elemental chlorine-free paper

Contents

List of Contributors

Emer Borromeo works at the Philippine Rice Research Institute, and formerly worked at the Plant Protection Department of the International Rice Research Institute, Los Banos, The Philippines.

David Buffin is former coordinator of the UK and Europe programmes at Pesticide Action Network, UK, and is now at City University.

Donald C. Cole is associate professor in epidemiology and community medicine with the Department of Public Health Sciences of the University of Toronto, Canada. His research has included the health risks associated with pesticide exposure among Nicaraguan and Ecuadorian farm families.

Charles Crissman is regional representative for sub-Saharan Africa at the International Potato Center (CIP), Nairobi, Kenya. His research has included the economic, health and environment tradeoffs of agricultural technologies.

David Dent is director of CABI Bioscience, a Division of CAB International, UK.

Barbara Dinham is director of Pesticide Action Network (PAN-UK).

Kevin Gallagher is IPM specialist at the Global IPM Facility of the UN Food and Agriculture Organization (FAO) Plant Protection Service, Rome.

Tara Pisani Gareau is at the University of California Santa Cruz, where she is studying the role of hedgerows in vegetable cropping systems in the Central Coast region of California for improving biological control of insect pests, and the sociopolitical incentives and disincentives for habitat enhancement in farms.

Hans Herren is director general of the International Centre of Insect Physiology and Ecology, Nairobi. He was awarded the World Food Prize in 1995.

Rachel Hine is research officer at the Centre for Environment and Society, University of Essex, UK.

Peter Kenmore is coordinator at the Global IPM Facility of the FAO Plant Protection Service, Rome.

Jan-Willem Ketelaar is coordinator at the FAO Vegetable IPM programme for South East Asia, Bangkok.

Misa Kishi is senior environmental health specialist at the JSI Research and Training Institute, US, and visiting researcher at the Center for International Forestry Research (CIFOR), Indonesia.

Markus Knapp is a scientist and project coordinator for the Integrated Control of Red Spider Mites project at the International Centre of Insect Physiology and Ecology, Nairobi.

Catrin Meir is a freelance consultant who has worked in farmer participatory research and training to reduce pesticide use, and the evaluation of IPM programmes in Central America since 1992.

Tom Mew is head of the Plant Protection Department of the International Rice Research Institute, Los Banos, The Philippines.

Peter Ooi is coordinator of the Regional Cotton IPM programme at the FAO Regional Office, Bangkok.

Sam Page is organics and soil health specialist at CABI Bioscience, UK Centre.

Myriam Paredes is a consultant based in Quito, Ecuador. Her PhD research at Wageningen University and Research Centre, the Netherlands, is on the interfaces between development interventions, technology and rural people.

Jules Pretty is professor of environment and society in the Department of Biological Sciences at the University of Essex, UK. His previous books include *Agri-Culture* (2002, Earthscan), Guide to a Green Planet (editor, 2002, University of Essex), *The Living Land* (1998, Earthscan), *Regenerating Agriculture* (1995, Earthscan), and *Unwelcome Harvest* (with Gordon Conway, 1991, Earthscan).

Niels Röling is emeritus professor of agricultural knowledge systems in developing countries at Wageningen University, the Netherlands. His previous publications include *Facilitating Sustainable Agriculture* (editor with A. Wagemakers, 1998, Cambridge University Press) and *Extension Science: Information Systems in Agricultural Development* (1988, Cambridge University Press).

Fritz Schulthess is principle scientist and project coordinator for the Biological Control of Cereal Stemborers project at the International Centre of Insect Physiology and Ecology, Nairobi.

Carol Shennan is director of the Center for Agroecology and Sustainable Food Systems at the University of California Santa Cruz.

Stephen Sherwood is area representative for the Andes Region at World Neighbors and is based in Quito, Ecuador. His graduate research at Wageningen University and Research Centre, the Netherlands, is on the sociobiological developments that led to the modern-day pesticide dependency and ecosystem crisis in the Northern Andes.

J. Robert Sirrine is at the University of California Santa Cruz, where he is studying northern Michigan cherry farmers in a network of economic institutions in which they are losing power over their decisions, and integrating agroecological theory into the evaluation of practices designed for sustainable production.

Harry van der Wulp is senior IPM policy specialist at the Global IPM Facility of the FAO Plant Protection Service, Rome.

Janny Vos is farmer participatory IPM specialist at CABI Bioscience, UK Centre.

Hermann Waibel is professor of agricultural development economics at the University of Hanover, Germany.

Stephanie Williamson coordinates Pesticide Action Network Europe and works in the international and UK programmes at PAN-UK, where she focuses on IPM.

Preface

There was once a town where all life seemed to live in harmony with its surroundings. The town lay in the midst of a checkerboard of prosperous farms, with fields of grain and hillsides of orchards. . . Then a strange blight crept over the area and everything began to change. . . There was a strange stillness. . . It was a spring without voices. . . The people had done it themselves.

With these words Rachel Carson's fable of a *Silent Spring* (1963) became famous worldwide. She painted a picture of a healthy community in town and country-side. This idyll, which could be anywhere in the past, delights visitors and locals alike. But it falls into a mysterious silence, *'which lay over fields and woods and marsh'*. The community had withered and died, and apparently all because of the widespread use of pesticides. This simple story is so compelling that more than 2 million copies of the book have been sold, and it continues to sell well. This is impressive for any book, let alone one mainly documenting the ills of the world.

Of course, the truth behind the fable plays out rather differently in real life, as no town has died solely because of agricultural pesticides, and neither has all the wildlife been eliminated. But there is something in what she says that remains significant more than 40 years later. Since the early 1960s, the world population has more than doubled, and agricultural production per person has increased by a third. Over the same period, the use of modern inputs for farming has grown dramatically, and they have been very effective in helping to increase agricultural yields. Pesticides are now available in the remotest regions of the world. Farmers can see their short-term effect – killing insects, weeds and diseases, and leaving the crops and animals to flourish. Yet there has been a hidden cost to pay. Harm to environments and human health has accompanied some of these fundamental changes in food production systems. For far too long we have accepted these costs as the unfortunate but necessary side-effects of progress.

Yet in the last decade of the 20th century, many communities around the world have begun to see some remarkable revivals. The pesticides that harm environments and human health are increasingly being identified, and alternative, cheaper and safer management methods have been developed and now adopted by several million farmers. Food production by these farmers has not been compromised, which is a surprise to many. Something is happening. The spring may have been silent, but the prospects for the 21st century are now changing. In a small Asian village a rice farmer says *'my fields have been silent for*

30 years, now they are singing again'. Pesticides had eliminated the unnecessary wildlife, but now the frogs are back. What brought about these changes? When Asian rice farmers first began to learn about the beneficial effects of predators and parasites in field schools, and about how to grow rice with limited or no pesticides, they changed their practices by the tens of thousands. Yields were maintained or improved, and costs cut substantially – good for both families and the environment. This time, the people have done the right thing for themselves.

Remarkably, this story is beginning to be played out in different ecological and social settings around the world. But progress towards safer agriculture is still relatively rare. Each year, pesticide use in agriculture amounts to some 2.5 billion kg – about 400g for every person on the planet. Yet we still have limited knowledge about the causal relationships between harmful products and adverse health and environmental problems in the field and at home. Some people say these costs simply have to be accepted, as sustainable alternatives cannot work for both the environment and food security. Despite great progress, the world's agricultural and food systems are still not always ready to take on board the principles of sustainability.

This book seeks to address some of these difficulties and set out some new solutions. Pests, diseases and weeds eat, infiltrate and smother crops and grab their nutrients. If farmers stood back and let nature take its course, there would be insufficient food. They must do something. Pesticides are easy to use, although often costly for farmers. In addition, they frequently involve considerable costs to society in the form of public health and environmental costs. Alternatives often appear more difficult to implement, but are more sustainable in the long term. Their broad introduction, however, continues to face many challenges.

There is, perhaps, less of a choice than many may like to think. Recent food scares have underscored the importance of food safety. Contamination of water resources with pesticide residues is increasingly becoming an important issue in a growing number of countries. And recent studies are indicating that the poisoning of farmers and their families in developing countries is far worse than previously thought.

Governments are now beginning to tighten their pest and pesticide management policies, supported by a growing body of evidence to show that food can be produced in more sustainable ways. There is enormous scope for further reductions in pesticide use, and where pesticide use remains justified, there are often less hazardous alternatives to the products currently being used. This book describes the problems associated with pesticide use and highlights a range of initiatives that provide viable alternatives, with special attention given to integrated pest management (IPM).

The International Code of Conduct on the Distribution and Use of Pesticides defines IPM in this way:

> *IPM means the careful consideration of all available pest control techniques and subsequent integration of appropriate measures that discourage the development of pest populations and keep pesticides and other interventions to levels that are economically justified and reduce or minimize risks to*

human health and the environment. IPM emphasizes the growth of a healthy
crop with the least possible disruption to agro-ecosystems and encourages
natural pest control mechanisms.

In this book this approach to IPM is sometimes called community-IPM, low-toxicity IPM, ecological IPM or even just ecological pest management (EPM), implying that the approach is something more than just a reduction in pesticide use. Despite many positive national and international intentions and commitments, and even though less hazardous alternatives are often readily available, large quantities of undesirable pesticides continue to be used in many parts of the world. These include products with acute toxicity hazards or chronic health hazards. Some are persistent in the environment and/or disrupt ecosystem functioning.

This book explores the potential for the phasing out of hazardous pesticides and the phasing in of cost-effective alternatives already on the market. The priority criterion for phasing out is acute mammalian toxicity in view of the high incidence of farmer poisoning, especially in the tropics where protective clothing is not available or is too costly or uncomfortable to use. Other criteria include chronic health hazards and hazards to ecosystems. But such phasing out of undesirable products and the phasing in of new ones will need to be accompanied by supportive policy measures. Policy changes may include: the removal of subsidies on products scheduled for phase-out; taxation of products with high social costs; financial incentives to encourage local development and the production of new products; incentives to encourage partnerships between local producers in developing countries and producers of non-toxic products in Organisation of Economic Co-operation and Development (OECD) countries; a review of lists of registered pesticides; the establishment, monitoring and enforcement of maximum residue limits; and investment in farmer training through farmer field schools.

There has been promising progress, with many of these policy measures now implemented in various countries. But what is still missing is a comprehensive and integrated approach by all countries, in which the idea of agricultural sustainability is placed centre-stage. What would happen if this occurred? Would there be sufficient food to meet growing demand? Would the rural towns come to life? Would the birds and frogs sing again? The answer could be a resounding yes, if we come to appreciate that fundamental changes in pest management in agriculture are beneficial for farmers, consumers and the environment. Such collective successes are clearly very hard to achieve, but this book sets out some of the opportunities to make progress.

This book is a compilation of chapters on selected subjects that together constitute a larger picture about the changes necessary for pest and pesticide management. It describes the current concerns about the side-effects of pesticides, and demonstrates the feasibility of change on the basis of a number of concrete cases from both developing and industrialized countries.

In Chapter 1, Jules Pretty and Rachel Hine review pesticide use and the environment. Pesticides are now widely used in food production systems across the world, and increasingly, in some countries, in the home and garden. Some 2.5 billion kg of active ingredients are applied each year, amounting to an annual market value of some US$25–30 billion in the 1990s and 2000s. Just over a fifth of all pesticides are used in the US. However, most pesticide markets in industrialized countries are no longer expanding, and companies are looking to developing countries to increase sales. More than 800 products are in regular use worldwide. Pesticides have become ubiquitous in environments worldwide, some reaching hazardous levels for humans. Pest resistance has become increasingly common, with 2645 cases of resistance in insects and spiders recorded in the late 1990s. The problem for regulators is that causality is very difficult to establish. This is graphically shown by the amount of scientific effort required to understand the effects of pesticides on wild bird populations. A further reason to be cautious now comes from concerns about the endocrine disrupting properties of some pesticide products.

In Chapter 2, Misa Kishi questions what we know about the health impacts of pesticides, and shows that pesticides do harm human health, although their effects are not widely recognized and their full extent remains unknown. This is true in both industrialized and developing countries, and for both their acute and long-term effects. However, the extent of the problem is far greater in developing countries. In industrialized countries, the focus has shifted from occupational exposure to the effects of long-term low-level exposure to the general population. While the problems of acute effects in developing countries have been recognized to a certain extent, the perception promoted by the pesticide industry is that the number of acute pesticide poisonings due to suicides is greater than occupational poisonings. However, this is not supported by the evidence. For a variety of reasons – including the underutilization of health facilities by agricultural workers, the inability of health personnel to diagnose pesticide poisoning, and the lack of understanding of the importance of reporting – the underreporting of occupational poisoning is very common. This in turn misleads policy-makers. Furthermore, even when no data exist on the adverse effects of pesticides, it cannot necessarily be assumed that there are no problems.

In Chapter 3, Jules Pretty and Hermann Waibel provide a comprehensive analysis of the full cost of pesticides. Unfortunately, the external environmental and health costs of pesticides are rarely addressed when calculating whether or not pesticides should be used in agriculture. Data from four countries is incorporated into a new framework for pesticide externalities, and this shows that total annual externalities are US$166 million in Germany, US$257 million in the UK, US$1398 million in China (for rice only) and US$1492 million in the US. These externalities amount to between US$8.8 and $47.2 per hectare of arable and permanent crops in the four countries – an average of US$4.28 per kg of active ingredient applied. This indicates that the 2.5 billion kg of pesticides used annually currently impose substantial environmental and human health costs, and that any agricultural programmes that successfully reduce the use of pesticides that cause adverse effects create a public benefit by avoiding such costs. A total of 62 IPM initiatives from 26 countries are analysed to illustrate the trajectories that yields and pesticide use have taken. There is promising evidence that pesticide use can be reduced without yield penalties, with 54 crop combinations seeing an increase in yields while pesticide use fell. A further 16 crop combinations saw small reductions in yield with large reductions in pesticide use, and 10 saw increases in yields accompanied by increases in herbicide use.

In Chapter 4, Barbara Dinham discusses the role of corporations in shaping modern agricultural production. The products of their research and development dominate the agricultural input market, and the industry is now highly concentrated into six research-based companies, with a large number of generic companies seeking to gain a greater foothold on sales. The health and environmental side-effects of many of these products have been acknowledged, and some have been removed from the market as a result. Nevertheless, many hazardous pesticides, and others associated with chronic health concerns, continue to be freely available in developing countries. Workers and farmers who are not able to protect themselves are still using these products under inappropriate conditions. The major companies have signed up to the FAO code of conduct, and its implementation is crucial to reduce the adverse effects of pesticides. More assertive action may be needed in developing countries to find less hazardous and more sustainable pest management solutions for poor farmers. The most important step companies could make would be to remove the most toxic pesticides from the market, particularly in countries where conditions are unsuitable for their use, and introduce less hazardous products and technologies.

In Chapter 5, David Dent provides an overview of agrobiologicals and other alternatives to synthetic pesticides. Some attempts have been made to substitute pesticides with agrobiologicals, the biological equivalents of synthetic pesticides. These include biopesticides based on bacteria, fungi, viruses and entomopathogenic nematodes and a range of other off-farm inputs, including pheromones and macrobiological agents such as predators and parasistoids. Many agrobiologicals represent safe and effective alternatives to pesticides, but systems of registration and regulation tend not to favour them. IPM requires the availability of a range of options to farmers so as to ensure the long-term control of pests, diseases and weeds. Pest management can be made safer by eliminating the most hazardous products, substituting them with safer biocontrol agents and biopesticide products, implementing administrative controls that emphasize training and education in the safe use of existing products and improved agroecological knowledge, and making available personal protective equipment only as a measure of last resort.

In Chapter 6, Catrin Meir and Stephanie Williamson analyse farmer decision-making for ecological pest management. Farmers in both developing and industrialized countries are increasingly faced with rapid and profound changes in production technologies, processing and purchasing systems, and market requirements. These changes require new management skills and knowledge if farmers are to remain competitive in global markets. Sound decision-making about pest management strategies and pesticide use is critical, even for those farmers growing mainly subsistence crops for local consumption, since most farmers face rising production costs, increased competition and growing consumer concerns about food quality and safety. This chapter reviews what is known about farmer decision-making for pest management and why it is important if farmers are to be motivated to reassess their approaches to pest management, as well as to make them more aware of alternatives to pesticides. Farmer perceptions are described, together with external influences on farmer decision-making, and the training and agricultural extension methods that aim to influence farmers' pest management knowledge and practices.

In Chapter 7, Niels Röling sets out a radical vision for the human and social dimensions of pest management. This chapter presents an approach based not on causes but on human reasons. In trying to explain sustainability, the aim is not to look for causes and effects in the physical world but for human reasons in terms of people's 'gets', 'wants', 'knowing' and 'doing'. This translates as an exercise in reinterpreting the perfectly valid instrumental discourse about agricultural sustainability into a totally different discourse based on cognition and learning. The chapter is based on the assumption that we live not in an epoch of change, but in a change of epoch. We have successfully built technology and an economy that allowed a sizeable proportion of humanity to escape much of the misery of previous generations. However, in the process of co-evolving our aspirations and technologies, we have transformed the surface of the earth. This chapter reiterates the indispensability of a constructivist perspective for mobilizing the reflexivity and resilience required during a change of epoch. It provides a theoretical underpinning for the human predicament of having to juggle coherence and correspondence, and further analyses pressure in terms of the nature of human knowledge and its inadequacy. The challenge is not in dealing with land but in how people use land.

In Chapter 8, Kevin Gallagher, Peter Ooi, Tom Mew, Emer Borromeo, Peter Kenmore and Jan-Willem Ketelaar provide a detailed analysis of low-toxicity IPM for rice and vegetables in Asia. The powerful forces that drive these two systems could not differ more. Rice production is a highly political national security interest that has often justified heavy-handed methods to link high yielding varieties, fertilizers and pesticides to credit or mandatory production packages and led to high direct or indirect subsidies for these inputs. Research to produce new varieties and basic agronomic and biological data was well funded. Vegetable production, on the other hand, has been led primarily by private sector interests and local markets. Little support for credit, training or research has been provided. The high use of pesticides on vegetables has been the norm, due to a lack of good knowledge about the crop, poorly adapted varieties and a private sector push for inputs at local kiosks to tackle exotic pests on exotic varieties in the absence of well-developed management systems. However, other pressures are now driving change to lower pesticide inputs on both crops. Farmers are more aware of the dangers of some pesticides to their own health. The rise of Asian incomes has led to a rise in vegetable consumption that has made consumers more aware of food safety. More farmers are producing vegetables for urban markets, so driving competition to lower input costs as well. Integrated pest management programmes in both crops aim to reduce the use of toxic pesticide inputs and the average toxicity of pest management products that are still needed whilst improving the profitability of production.

In Chapter 9, Hans Herren, Fritz Schulthess and Markus Knapp analyse a variety of approaches for low to zero pesticide use in tropical agroecosystems, particularly in Africa. Agricultural production in tropical agroecosystems is greatly affected by pests with the result that synthetic pesticide use has been rising. This is particularly true for cash and horticultural crops that have a significant economic return. Recently, however, the use of pesticides is being restricted on crops destined for export, following the introduction of new maximum residue levels in industrialized countries. Six key issues for pest management decision-making are identified. These are: (i) education and information availability; (ii) economic environment and imperatives; (iii) agricultural production systems; (iv) availability and affordability of alternative pest management tools and implementation strategies; (v) market requirements, consumer education; and (vi) policy environment. Two detailed case studies are analysed: lepidopteran cereal stemborer management, and biological control in vegetables, and conclusions drawn on the practicalities of eliminating synthetic pesticides from the 'ecological' IPM toolbox without jeopardizing the quality and quantity of food production, whilst at the same time improving farmers' revenues and the sustainability of their production systems.

In Chapter 10, Stephen Sherwood, Donald Cole, Charles Crissman and Myriam Paredes focus on improving ecosystem and human health in the northern Andes by revealing problems and solutions in Ecuador's Carchi province. Over 60 per cent of the rural population were found to have had their nervous system functions affected by pesticides. Very high rates of human poisoning were discovered: 171 per 100,000 population, with mortality at 21 per 100,000, the highest recorded rates anywhere in the world. This high incidence may not be because the situation is particularly bad in Carchi, but because researchers sought systematically to record and document it. Meanwhile, the principal position of the national pesticide industry continues to be farmer education through 'Safe Use' campaigns. This continues, despite misgivings that the notion of the safe use of highly toxic chemicals under the social and environmental conditions of developing countries is almost impossible. The project team worked with interested stakeholders to inform the policy debate on pesticide use at both the provincial and national level. Its position has evolved to include the reduction of pesticide exposure risk through a combination of hazard removal, the development of alternative practices and ecological education. Their experience led them to conclude that more knowledge-based and socially oriented interventions are needed. These should be aimed at building farmer capacities, promoting more regenerative agricultural practices, and improving markets and policies.

In Chapter 11, Stephanie Williamson provides new evidence from Benin, Ethiopia, Ghana and Senegal on pesticide use and the opportunities for implementation of IPM in a variety of crops. Pesticide use in Africa is the lowest of all the continents, accounting for only 2–3 per cent of world sales, and averaging in the 1990s, 1.23 kg ha^{-1} compared with 7.17 kg in Latin America and 3.12 kg in Asia. This low use appears to suggest correspondingly low level health and environmental hazards. Regrettably, this assumption is wrong, as African farmers currently use many WHO Hazard Class Ia and Ib products, and few users take precautionary measures to prevent harm. Once again, alarmingly high rates of pesticide poisoning were recorded. The research on eight cropping systems in four countries revealed increasing interest in IPM training. Integrated pest management and agroecological concepts need to be brought into the mainstream curricula in agricultural colleges and schools, with practical educational materials adapted for African cropping systems. Persuading more decision-makers and other important stakeholders to accept the IPM concept and its practical implementation is a vital priority in the transformation of African farming systems for the benefit of rural communities and their consumers.

In Chapter 12, Janny Vos and Sam Page analyse the case of cocoa management in West Africa. Concern is expressed in this area about the impact that the sudden phase-out of toxic pesticides could have on smallholder farmers. Cocoa originated in South America and is now cultivated in West Africa (Côte D'Ivoire, Ghana, Nigeria, Cameroon), South America (Brazil and Ecuador) and Asia (Indonesia and Malaysia). Up to 90 per cent of the world's cocoa is produced by smallholder farmers, cultivating on average less than 3 hectares each. As cocoa is an exotic plant in West Africa, it has contracted a number of serious new encounter diseases, which originate from the indigenous flora but to which exotics have not co-evolved defence mechanisms. This chapter shows that is it possible to phase out WHO Hazard Class I products without creating new problems. Low toxicity alternatives to pest management in cocoa production in West Africa are being developed. Smallholder cocoa farmers will need to be able to access information and knowledge to become better informed managers of their farms, whereas other stakeholders in the IPM network will need to re-focus their current strategies. A long-term process of re-education and the re-organization of farmer support systems should be considered to promote more sustainable cocoa production.

In Chapter 13, Carol Shennan, Tara Pisani Gareau and Robert Sirrine discuss an agroecological approach to pest management in the US. This involves the application of ecological knowledge to the design and management of production systems so that ecological processes are optimized to reduce or eliminate the need for external inputs. There are many potential approaches to deal with different pests in different types of cropping systems. Any single ecological approach does not provide a 'silver bullet' to eliminate a pest problem. Successful management requires a suite of approaches that together create an agroecosystem where pest populations are maintained within acceptable levels. Ecological pest management (EPM) seeks to weaken pest populations while at the same time strengthening the crop system, thus creating production systems that are resistant and/or resilient to pest outbreaks. Despite the evolution of US agriculture toward intensive, large-scale monocultures maintained by high-cost, off-farm inputs, farmers do have an increasing variety of cultural and biological management tools available that can maintain low levels of pest damage with little use of external inputs. The chapter illustrates the different methods and approaches that are being used in farming systems across the US. At the same time, it is clear that there is still a long way to go. Knowledge gaps still exist, and these are important constraints on the widespread use of EPM.

In Chapter 14, Stephanie Williamson and David Buffin discuss the transition to safe pest management in a variety of industrialized agricultural systems. Over the last decade or so, integrated pest and crop management has become increasingly common in North America, Europe and Australasia. However, there are many interpretations of what constitutes IPM, ICM, Integrated Production and Integrated Farming, which makes it harder to assess progress. Some reasons for the limited uptake of integrated approaches may include low levels of understanding among farmers or a lack of incentives to change established practice. However, some retailers, such as the Co-operative Group and Marks and Spencer, have prohibited the use of many pesticides on crops grown for them. Five case studies of IPM are discussed in detail: apples and pears in Belgium, pesticide-free arable in Canada, healthy-grown potatoes in Wisconsin, vining peas grown for Unilever in the UK, and arable crops and field vegetables cultivated for the Co-operative Group.

In Chapter 15, Harry van der Wulp and Jules Pretty review policy and market trends that are converging towards more sustainable production systems that will be less dependent on pesticides. The chapter describes how national policies, international conventions and aid programmes are shaping pest and pesticide management. An emerging new agenda for crop protection in the next decade indicates that there can be further reductions in reliance on pesticides. These processes encourage the phasing out of hazardous products, whilst phasing in alternative approaches and less hazardous products. The many examples described in this book demonstrate that there is an enormous potential for reductions in the use of pesticides. With the necessary political will, backed up by consumer awareness and appropriate market responses, it should now be possible to detox agriculture.

Jules Pretty
University of Essex, Colchester
November 2004

Acknowledgements

The editor wishes to thank all those who contributed to this book. Particular thanks go to Kevin Gallagher and Harry van der Wulp for their continuing support and advice. Part of the background research for this book was completed by the Centre for Environment and Society at the University of Essex, with financial support provided by the Global IPM Facility within the framework of its role to stimulate international discussion on pest and pesticide management. The views expressed in this book, however, do not necessarily reflect those of the Global IPM Facility.

All authors of chapters are experts in their own disciplinary or regional areas, and so take responsibility only for their own work. The content of this book does not necessarily reflect the views of any of the organizations associated with the authors of chapters, nor of any organizations funding research that was used for the book. Authors of some of the individual chapters wish to make the following specific acknowledgements.

Chapter 2: Some of the information in this chapter appeared in an article by Kishi and LaDou in the *International Journal of Occupational and Environmental Health*, 7 (2001), 259–265. Much of the chapter is based on a research project for the Global IPM Facility. Other components in the chapter appeared in a 2002 report, *Initial Summary of the Main Factors Contributing to Incidents of Acute Pesticide Poisoning*, for the Intergovernmental Forum for Chemical Safety Working Group on Acutely Toxic Pesticides. The author thanks the following for their help and advice: Peter Kenmore, Kevin Gallagher and Manuela Allara at the Global IPM Facility, and Russ Dilts, Andrew Bartlett, Helen Murphy, Norbert Hirschhorn, Richard Clapp, Carolyn Hessler-Radelet, Carol Rougvie, Ingrid Martonova, James McNeil, Junko Yasuoka, Andrew Watterson, Joe LaDou, Catharina Wesseling, Berna van Wendel de Joode, Leslie London, Douglas Murray, Peter Kunstadter, Alan Hruska, Monica Moore, Barbara Dinham, Cathleen Barns, Judy Stober, Nida Besbellin and Donald Halstead.

Chapter 3: The IPM case studies used for the analysis of low-cost pest management options were drawn from a dataset developed as part of the SAFE 1 and 2 projects at the University of Essex, funded by the UK Department for International Development (DFID).

Chapter 4: The author would particularly like to thank her African partners PAN Africa and OBEPAB for providing details of cases of pesticide poisonings in the field and of the conditions that farmers work under, which make incidents of poisoning virtually inevitable. Special thanks go to Simplice Davo Vodouhê who took great pains to send documentation ensuring the accuracy of the information used. Monica Moore drew the author's attention to important information, and Jules Pretty and Harry van der Wulp helped with editorial comments and improvements.

Chapter 6: This chapter is based on a literature review by the authors as part of a collaborative research project (CPP/R7500) by CABI and NRI funded by the UK Department for International Development (DFID). The authors are grateful to John Mumford for valuable comments on the chapter and to Tony Little for help in preparing the original literature review, from which this chapter has been adapted. The views expressed are not necessarily those of DFID nor of project collaborators other than the authors. We also acknowledge the invaluable contribution of Tony Little, Md Arif Ali, Martin Kimani and Leonard Oruko to the fieldwork on farmer decision-making in India and Kenya, which formed another output of the project reported separately.

Chapter 7: This chapter is based on an earlier paper entitled 'Pressure facing failure of knowledge based action' in the proceedings of an OECD Cooperative Research Program Workshop on 'Agricultural Production and Ecosystems Management',12–16 November, Ballina, Australia, OECD, Paris. An earlier version of the chapter appeared in 2003 in the *International Journal of Agricultural Sustainability (IJAS)*, 1(1), 73–88. The author would like to acknowledge the constructive editorial comments and substantive suggestions of three anonymous IJAS reviewers.

Chapter 8: The authors would like to thank K. L. Heong of the International Rice Research Institute for assistance in providing papers and previews of the Rice IPM CD-ROM and express their appreciation to Ricardo Labrada (FAO), N. H. Huan (Plant Protection Department of Vietnam), Henk van den Berg (Community IPM/FAO), Clive Elliott (FAO) and Chan Paloeun (Cambodian Agricultural Research and Development Institute) for valuable comments and inputs on the rice IPM case study.

Chapter 9: The research projects mentioned in the case studies were funded by the Directorate General for International Cooperation (DGIS), the Netherlands, the Gatsby Charitable Foundation, the German Federal Ministry for Economic Cooperation and Development (BMZ) and the GTZ-IPM Horticulture Project.

Chapter 10: The studies reported here were originally financed by the Rockefeller Foundation. Subsequent investigations were conducted with the support of the USAID Soil Management and IPM Collaborative Research Support Programs as well as the FAO, the Ecosystem Health Program of the Canadian International

Development Research Council, and the Dutch Fund for Eco-Regional Research. The authors would like to acknowledge the contribution of the INIAP and EcoSalud team in Carchi, and in particular the efforts of Lilián Basantes, Mariana Pérez, Myriam Paredes, Jovanny Suquillo, Luis Escudero and Fernando Chamorro. The authors would also like to thank the Carchi Consortium and the individuals and communities that participated in different aspects of this effort.

Chapter 11: This chapter is based on research carried out by Pesticide Action Network UK in collaboration with PAN Africa (Senegal), Organisation Beninoise pour la Promotion de l'Agriculture Biologique (OBEPAB), Safe Environment Group (Ethiopia) and Community Action Programme for Sustainable Agriculture and Rural Development (Ghana). The author would also like to thank research collaborator Seth Gogoe and advisor Anthony Youdeowei in Ghana, as well as the project steering group. This work was kindly funded by the Development Directorate of the European Union, the UK Department for International Development (DFID) and the Rowan Trust. The views expressed are those of the author alone.

Chapter 12: This chapter is based on work funded by the Swiss Development Cooperation (SDC). The authors wish to thank various CABI colleagues (Julie Flood, Mark Holderness and Keith Holmes) and Global IPM Facility partners (Kevin Gallagher and Harry van der Wulp) for useful comments on earlier manuscripts.

Chapter 14: The authors thank the following colleagues for providing additional information and comments for the case studies: Sarah Lynch of WWF-US, Liz Chadd of Unilever, Allison Schoofs of the University of Manitoba, Kevin Barker of the Co-operative Group, David Gardner of farmcare, and Jacques Denis of GAWI.

List of Terms, Acronyms and Abbreviations

ai	active ingredient
agrobiological	A biological product used for pest, disease or weed management (see www.agrobiologicals.com)
ARDS	adult respiratory distress syndrome
BC	biological control
bn	billion (a thousand million, or 10^9)
BPH	brown planthopper
Bt	Bacillus thuriengiensis
CGIAR	Consultative Group on International Agricultural Research
CIP	International Potato Center (US)
CPB	cocoa pod borer
DBM	diamondback moth
DDT	dichlorodiphenyltrichloroethane
DNOC	dinitro-o-cresol
EA	Environment Agency (UK)
EC	European Commission
EPA	Environmental Protection Agency (US)
EPM	ecological pest management
EU	European Union
FAO	UN Food and Agriculture Organization (Rome)
FFS	farmer field school
GM	genetically modified
GMHT	genetically modified herbicide tolerant (crop)
GMO	genetically modified organism
ha	hectare
IARCs	international agricultural research centres
ICIPE	International Centre of Insect Physiology and Ecology
ICM	integrated crop management
IP	integrated production
IPM	integrated pest management
IPPM	integrated production and pest management
IRM	insecticide resistance management
IRRI	International Rice Research Institute
kg	kilogramme

M	million (as in Mha = million hectares, Mkg = million kilogrammes)
mg	milligramme
MRL	maximum residue levels
N	nitrogen
NED	new encounter disease
OC	organochlorine
OECD	Organisation for Economic Co-operation and Development
OP	organophosphate
OPM	organic pest management
PAN	Pesticide Action Network
PAR	Participatory Action Research
PAHO	Pan American Health Organization
PCB	polychlorinated biphenyl
pest	This term is commonly used in the book to include insects, diseases, weeds, snails, birds and rats
pesticide	This term is commonly used in the book to include insecticides, fungicides, herbicides, acaricides, miticides and nematicides
PFP	pesticide-free production
PIC	prior informed consent
POPs	persistent organic pollutants
SSA	sub-Saharan Africa
UNEP	United Nations Environment Programme
USAID	United States Agency for International Development
USDA	United States Department of Agriculture
WHO	UN World Health Organization

Chapter 1

Pesticide Use and the Environment

Jules Pretty and Rachel Hine

A BRIEF HISTORY OF PESTICIDES

Pesticides are intended to kill unwanted organisms. Most act by interfering with a variety of biochemical and physiological processes that are common to a wide range of organisms. Besides target pests, weeds and fungi, they also affect wildlife and human health. Some can be lethal, and many can cause illness at sublethal levels. But the risks differ greatly from pesticide to pesticide. Some are acutely toxic but produce no long-term effects, whilst others are of long-term health or environmental concern. Much of the information on these side-effects of pesticides remains contested, and so there is no agreement about how much harm they cause.

Pesticides are not modern inventions, as they have long been used to control pests and diseases in agriculture (Carson, 1963; Conway and Pretty, 1991; Cremlyn, 1991; Dinham, 1993; van Emden and Peakall, 1996). In 2500 BC the Sumerians used sulphur compounds for insect control. Later, seeds were treated by Chinese farmers with various natural organic substances to protect against insects, mice and birds, whilst inorganic mercury and arsenic compounds were used to control body lice. The Greek and Roman writers Aristotle, Homer and Cato describe a variety of fumigants, oil sprays and sulphur ointments used by farmers, and Pliny recommends the use of arsenic as an insecticide. However, natural pesticides did not come into common use until the agricultural revolution of 17th–18th century Europe. Nicotine was used in the 1600s, and was followed by the discovery of the wood preservative properties of mercuric oxide in the early 1700s, and of the fungicidal properties of copper sulphate in the early 1800s.

By the mid-19th century, rotenone from the roots of derris and pyrethrum from chrysanthemum flowers had been discovered, and these were accompanied by a rapid growth in the use of inorganic products, particularly of arsenic. Paris Green (copper arsenite) was first used in 1867, coming into such common use by the early 20th century in the US that it led to the world's first legislation to control pesticides. Bordeaux mixture (copper sulphate and lime) was discovered to be effective against powdery mildew in 1882. The local custom in France was to treat roadside vines with the mixture to prevent theft, and it was noticed that these vines also escaped infestation with the disease.

The early part of the 20th century saw the increased use of many dangerous products derived from arsenic, cyanide and mercury. Most were broad-acting in their effect on pests and diseases. Some, such as iron sulphate, were found to have selective herbicidal properties against weeds. Calcium arsenite came to replace Paris Green, and by the 1920s arsenic insecticides were in widespread use. This provoked considerable public anxiety about residues of these products on fruit and vegetables.

Against this disturbingly toxic background, the 1930s saw the beginning of the era of synthetic organic products. This decade saw the introduction of alkyl thiocyanate insecticides, the first organic fungicide, salicylanilide, dithiocarbamate fungicides, and later chloranil, before Paul Muller made the remarkable discovery in 1939 of the insecticidal properties of DDT (dichlorodiphenyltrichloroethane). It was first manufactured in 1943, and was initially valuable for delousing people to prevent the spread of typhus, and for the control of malarial mosquitoes. DDT was soon followed by the manufacture of several chlorinated hydrocarbon compounds, including aldrin, endrin, heptachlor, and the recognition of the herbicidal activity of phenoxyacetic acids, such as MCPA and 2,4-D. At that time, all of these synthesized products were valued for their persistence in the environment.

Organophosphates (OPs) emerged from wartime research on nerve gases. The first product that came into commercial use was parathion, an effective insecticide that was soon also found to be highly toxic to mammals. Malathion then came into wider use after 1950, as it had very low mammalian toxicity. OPs block cholinesterase, the chemical that transfers nerve impulses across synapses, and so their effect is primarily on the nervous system. The advantage of the OPs is that they are rapidly degraded in the environment to non-toxic secondary compounds – unlike the organochlorines (OCs). In a very short time, both OCs and OPs were being used in most countries of the world and on almost every crop. The immediate benefits were obvious, but it gradually became apparent that many of these new products also had severe drawbacks. They were affecting wildlife and people in ways that had not been anticipated. Later generations of pesticide products included the carbamates and synthetic pyrethroids. With some exceptions, these products were generally relatively less toxic to humans compared to the previous generation of OPs and OCs.

Over time, pesticide products have tended to become less broad-ranging in their effects and more targeted towards pests, weeds or diseases. However, such specificity does come at a cost. Broad-effect pesticides are both cheaper to manufacture and can be sold to more farmers. Specific products inevitably have smaller markets. The role of commercial pressures in pesticide development and use are discussed further in Chapter 4. However, a large number of new pest management technologies have become available in recent years, many using the term agrobiologicals (Chapter 5). These products are mostly available only in OECD countries and a few developing countries, such as China and India.

How Much Pesticide is Used?

In the past 50 years the use of pesticides in agriculture has increased dramatically and now amounts to some 2.56 billion kg per year. The highest growth rates for the world market, some 12 per cent per year, occurred in the 1960s. These later fell back to 2 per cent during the 1980s, and reached only 0.6 per cent per year during the 1990s. In the early 21st century, the annual value of the global market was US$25 billion, down from a high of more than $30 billion in the late 1990s. Some US$3 billion of sales are in developing countries (CropLife, 2002). Herbicides account for 49 per cent of sales, insecticides 25 per cent, fungicides 22 per cent, and others about 3 per cent (Table 1.1).

A third of the world market by value is in the US, which represents 22 per cent of active ingredient use. In the US, however, large amounts of pesticide are used in the home/garden (17 per cent by value) and in industrial, commercial and government settings (13 per cent by value). By active ingredient, US agriculture uses 324 million kg per year (which is 75 per cent of all reported pesticide use, as this does not include sulphur and petroleum products). Use in agriculture has increased from 166 million kilogrammes (Mkg) in the 1960s, peaked at 376 Mkg in 1981, and has since fallen back. However, expenditure has grown. Farmers spent some US$8 billion on pesticides in the US in 1998–1999, about 4 per cent of total farm expenditures. This had increased from $3.6 billion in 1980.

Table 1.1 *World and US use of pesticide active ingredients (average for 1998–1999)*

Pesticide use	World pesticide use		US pesticide use	
	(Million kg ai)	%	(Million kg ai)	%
Herbicides	948	37	246	44
Insecticides	643	25	52	9
Fungicides	251	10	37	7
Other[1]	721	28	219[2]	40
Total	2563	100	554	100

Note:
1 Other includes nematicides, fumigants, rodenticides, molluscicides, aquatic and fish/bird pesticides, and other chemicals used as pesticides (e.g. sulphur or petroleum products)
2 Other in the US includes 150 Mkg of chemicals used as pesticides (sulphur or petroleum)
Source: OECD (2001a); EPA (2001)

Industrialized countries accounted for 70 per cent of the total market in the late 1990s, but sales are now growing in developing countries (Figure 1.1). Japan is the most intensive user per area of cultivated land. The global use of all pesticide products is highly concentrated on a few major crops, with some 85 per cent by sales applied to fruit and vegetables (25 per cent), rice (11 per cent), maize (11 per cent), wheat and barley (11 per cent), cotton (10 per cent) and soybean (8 per cent) (UK Crop Protection Association, 2001).

There is also considerable variation from country to country in the kinds of pesticide used. Herbicides dominate the North American and European domestic markets, but insecticides are more commonly used elsewhere in the world. In the US in the late 1990s, 14 of the top 25 pesticides used are herbicides (by kg ai), with the most commonly used products being atrazine (33–36 Mkg), glyphosate (30–33 Mkg), metam sodium (a fumigant, 27–29 Mkg), acetochlor (14–16 Mkg), methyl bromide (13–15 Mkg), 2,4-D (13–15 Mkg), malathion (13–15 Mkg), metolachlor (12–14 Mkg), and trifluran (8–10 Mkg). Glyphosate and 2,4-D were the most common products used in domestic and industrial settings (Environmental Protection Agency (EPA), 2001). In Asia, 40 per cent of pesticides are used on rice, and in India and Pakistan, some 60 per cent are used on cotton. India and China are the largest pesticide consumers in Asia. Pesticide consumption in Africa is low on a per hectare basis.

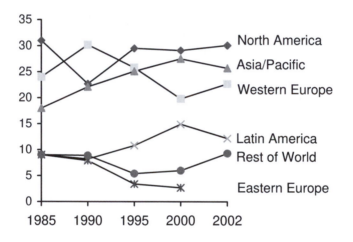

Figure 1.1 *Geographical distribution of agrochemical use, 1985–2002 (%)*

Source: Crop Protection Association UK, annual reports 1986–2001 (from company analyst Wood MacKenzie); Agrow, 2003a (from Allan Woodburn Associates)

PESTICIDE OVERUSE AND OTHER MISUSE

An important issue centres on the relationships between the use, overuse and potential misuse of pesticides and any potential harm that might occur as a side-effect. Pesticides and their formulations are licensed by governments for use subject to strict conditions, and it is generally assumed that, if they are used in accordance with instructions, harm to the environment and human health should not occur – notwithstanding the emergence of later evidence that might show a previously unknown effect. Thus, do pesticide problems only occur when they are overused or otherwise misused?

Table 1.2 *Pesticide use by OECD country (1999)*

Country	Annual pesticide use (Mkg)
US	324
Italy	167
Australia	120
Japan	65
Mexico	36
UK	35
Germany	35
Spain	34
Turkey	33
Canada	29
Russia	17
Portugal	12
Poland	10

Source: OECD (2001a)

Overuse can occur when farmers are advised to spray on a routine calendar basis, rather than when pest problems exceed an economic threshold where the cost of treatment is not greater than the pest losses incurred. Routine applications minimize the time and cost of decision-making: 'spray at weekly intervals' is, for example, a powerful extension message. Pesticides are also often seen as a simple insurance premium against crop failure. Direct subsidies to reduce the retail cost of pesticides tend to encourage overuse, although this is less of a problem at the present time, as most countries have eliminated direct pesticide subsidies.

Further problems often occur when safety instructions are either missing or in an inappropriate language. In a remote Quechua-speaking village in Peru, some 42 children were poisoned with 18 fatalities in 1999 when methyl parathion was supplied by an international company with Spanish instructions for its use on vegetables (Peruvian Congressional Committee, 2001). Such labelling problems, whether deliberate or accidental, are common in many developing countries.

Another problem comes from pressures from consumers, or more directly from supermarkets and retailers, for cosmetically perfect produce. This has become particularly important in industrialized countries, where the appearance to the consumer or processor greatly influences the price farmers receive. Such cosmetic control puts a high premium on blemish-free produce, which means that an extra pesticide application may be justified even when the risk of downgrading a food product is small.

In the US, cosmetic control has been especially prevalent on citrus produce. Blemishes on the skins of fruits reduce the returns to farmers, even though they may not reduce yields nor affect nutrient content, storage or flavour. The citrus rust mite, for example, causes russetting or bronzing on oranges and in the 1970s most of Florida's orange groves were being sprayed for rust mites, at an annual cost of some US$40–50 million. Oranges from treated orchards even sold at a premium, even though yields were the same as in untreated orchards.

In some contexts, however, pesticides are deliberately misused, particularly for fishing, resulting in substantial environmental and health problems. The Mexican Environmental Enforcement Agency has documented the illegal use of deltamethrin and coumaphos for marine fishing in the Pacific state of Michoacan (PROFEPA, 2001). Deltamethrin is a pyrethroid known to be highly toxic to aquatic animals, and coumaphos is an organophosphate rated as highly toxic to humans. Both are used to fish for langostino, a lobster-like crustacean served in expensive restaurants. Inland pesticide fishing has also been documented in Mexico, with reports of ill-health in adults and children linked to the consumption of contaminated fish and shrimp.

NATIONAL REGULATIONS AND INTERNATIONAL CODES

During the course of the 20th century, most countries developed systems to regulate the use and misuse of pesticides. Although the first legislation for pesticides was the United States Insecticide Act of 1910, which prohibited the manufacture, sale or transport of mislabelled or adulterated chemical substances, the first provision for compulsory registration was not made until the Federal Insecticide, Fungicide and Rodenticide Act of 1947 (FIFRA). Registration of pesticides was administered by the United States Department of Agriculture (USDA) and manufacturers and distributors were required to seek approval for all their products. Authorities were also able to restrict the uses of a particular product, and require that specific products only be applied by certified operators.

At about the same time, the UK government acted in response to the deaths of seven operators using dinitro-o-cresol (DNOC) in 1947 with the introduction of the Agriculture (Poisonous Substances) Act of 1952. This sought to promote safe working practices and protect workers from acutely toxic pesticides. But the two countries subsequently took quite different courses of action until the mid-1980s. In the US, further legislation made registration procedures more comprehensive, while in the UK controls were based on a voluntary approach. In 1972, the US FIFRA was amended to establish a more comprehensive registration procedure. From this point, before they could be marketed, new products were subjected to stricter standards than existing products, with which they would be competing. In effect, this meant that new pesticides were generally less hazardous. At the same time, already approved products had to pass through special re-registration, because of the inadequacy of the older methods of safety assessment. Inevitably, the new review process has entailed lengthy delays. This means that some new products – less hazardous than the existing products they might replace – have been denied registration, in some circumstances leaving the more hazardous products on the market.

In the UK, registration has for many years been through a non-statutory Pesticides Safety Precautions Scheme, in which manufacturers, distributors and importers undertook not to introduce new pesticides or new uses of pesticides until safety clearance had been granted by government. In return, industry was guaranteed complete confidentiality regarding all the safety data it submitted.

The scheme had the advantage of being flexible and adaptable, permitting rapid responses to events and new knowledge and providing an opportunity for policy to be developed in an anticipatory way. However, the scheme was replaced in the mid-1980s by a statutory approach under the Food and Environment Protection Act (FEPA), which now provides powers to control many aspects of sales, supply, use, distribution and marketing pesticides, to set residue limits, and to provide information to the public.

In the European Union (EU), many pesticides are now being phased out under the Pesticides Directive 91/414. As a result of this directive, manufacturers have to defend their products, and prove they meet today's higher environmental and health standards. In mid 2003, 320 pesticide active ingredients were taken off the market, and a further 49 were given temporary derogation. Roughly a further 150 appear likely to be voluntarily withdrawn by industry. Adding to the 19 already banned, this indicates that some 60 per cent of all products (some 500 in number) that were on the market in the early 1990s will no longer be available by 2008 (EC, 2002a).

Most developing countries only began to seriously consider the need for control of pesticides in the late 1960s and early 1970s. Today, there are still only a minority with strong legislation, appropriate resources and the means of enforcement. Few countries have the primary health care and occupational health systems necessary to detect and treat pesticide poisoning (although this is also true of most industrialized countries), nor the agricultural training and extension services that can ensure high standards of proper pesticide application. Residues in food and the environment can be high, but the lack of pollution monitoring and published data reduces the sense of urgency.

The last three decades of the 20th century also saw increased efforts to incorporate some risk-reducing measures, standards and regulations into international agreements. Following a decade of growing concerns over persistent organochlorines, especially DDT, high-profile court cases and important scientific breakthroughs on causality, particularly on birds of prey, the Stockholm conference on the human environment in 1972 was the first international meeting to address concerns about pesticides.

This was followed by the OECD's recommendations on the assessment of the potential environmental effects of chemicals in 1974. These built on the earlier establishment of an enduring principle in environmental pollution, namely the 1972 Polluter Pays Principle, which was defined by the OECD in this way:

The principle to be used for allocating costs of pollution prevention and control measures to encourage rational use of scarce environmental resources and to avoid distortions in international trade and investment is the so-called Polluter Pays Principle. The Principle means that the polluter should bear the expenses of carrying out the above mentioned measures decided by public authorities to ensure that the environment is in an acceptable state. In other words, the costs of these measures should be reflected in the cost of goods and services which cause pollution in production and/or consumption.

The first international code regarding pesticides, however, was not agreed until 1985, when the UN Food and Agriculture Organization (FAO) and United Nations Environment Programme (UNEP) were able to broker amongst national governments the first International Code of Conduct on the Distribution and Use of Pesticides. This provided guidance for pesticide management for all public and private entities engaged in pesticide distribution and use. It was revised and updated in the 1990s, and adopted at the 123rd session of the FAO Council in 2002.

The first Code was followed in 1992 by the United Nations Rio Conference on Environment and Development (the Earth Summit), where the signing of Agenda 21 by all countries present seemed to put sustainable development clearly on the international agenda. Chapter 14 of Agenda 21 addressed Sustainable Agriculture and Rural Development, and Chapter 19 dealt with Chemicals. The principles for sustainable forms of agriculture that encouraged minimizing harm to the environment and human health were agreed. However, progress has not been good, as Agenda 21 is not a binding treaty on national governments, and all are free to choose whether they adopt or ignore such principles. Nonetheless, an important outcome of Rio was the establishment of the UN Global IPM Facility in 1995, the aims of which are to provide international guidance and technical assistance for integrated pest management across the world.

However, the Rio Summit did recommend that the voluntary prior informed consent (PIC) clause in the FAO code of conduct become an international convention. This accepted that industrialized countries had greater resources to test and assess pesticide hazards than developing countries, and so it should be incumbent on exporting countries to notify importing authorities of data on known hazards. The Rotterdam Convention on Prior Informed Consent procedure was adopted in 1998 for certain internationally traded chemicals and pesticides.

Meanwhile, concerns over the persistence of certain organic products had been growing, particularly following their discovery in remote regions of the world and in the tissues of humans and animals not directly exposed to pesticides. The Stockholm Convention on Persistent Organic Pollutants (POPs) was signed in 2001. This is a treaty to protect human health and the environment from POPs. The UNEP plans to expand this agreement to a phasing out of 12 POPs that are particularly persistent and prone to bioaccumulate. These are aldrin, chlordane, dieldrin, dioxins, DDT, endrin, furans, heptachlor, hexachlorobenzene, mirex, polychlorinated biphenyls (PCBs) and toxaphene. Nine of these are agricultural pesticides.

A growing number of countries are now reporting reductions in pesticide use as a result of the adoption of agricultural sustainability principles. These have occurred as a result of two types of very different approaches:

1   policy-led and primarily top-down pesticide reduction programmes in industrialized countries, such as in Sweden, Denmark, the Netherlands and some provinces of Canada;
2   farmer-field school led and policy-enabled community IPM in rice programmes, beginning in South East Asia, subsequently spreading throughout Asia and then to other continents.

REDUCING RISKS

In the past 30 years, a simple classification system for pesticides by acute hazard has come into regular use. This was first developed by the World Health Organization (WHO), and approved by its 28th World Health Assembly in 1975. Guidelines were first issued in 1978, and these have been revised and reissued at two-yearly intervals since then. The classification system is very simple, and based on acute risk to human health – the risk of single or multiple exposures over a short period of time (Table 1.3). The main measures are the acute oral and dermal toxicity of products to the rat, since this is a standard procedure in toxicology. These are measured by the LD_{50} value – a statistical estimate of the number of mg of toxicant per kg of bodyweight required to kill 50 per cent of a large population of the test animals.

The WHO classifies pesticides into four classes of risk, Class Ia (extremely toxic), Class Ib (highly toxic), Class II (moderately toxic), and Class III (slightly hazardous), plus 'active ingredients unlikely to present acute hazard in normal use' (IPCS, 2002). The active ingredients of cholinesterase-inhibiting pesticides often belong to WHO Class Ia, Ib or II. In industrialized countries, most WHO Class Ia and Ib pesticides are now either banned or restricted.

Table 1.3 *Classes of pesticides in WHO classification scheme*

| | LD_{50} for rat (mg/kg body weight) | | | |
| | Oral | | Dermal | |
Class of pesticide	Solids[1]	Liquids	Solids	Liquids
Ia – extremely hazardous	<5	<20	<10	<40
Ib – highly hazardous	5–50	20–200	10–100	40–400
II – moderately hazardous	50–500	200–2000	100–1000	400–4000
III – slightly hazardous	>500	>2000	>1000	>4000

1 This refers to the physical state of the active ingredient.
Source: IPCS, 2002

The WHO does not include specific symbols to help farmers or other pesticide users to identify these classes in its recommendations. However, it does state that the Class Ia and Ib products should bear a symbol indicating high hazard (usually a skull and cross-bones), and a key word, such as poison or toxic. Examples of pesticides in each of the four classes are shown in Table 1.4. The WHO lists nearly 1000 pesticide active ingredients in its guidelines, some 260 of which are now officially designated as obsolete or are discontinued. The revised and updated UN International Code of Conduct now supports IPM strategies that encourage natural pest control mechanisms, urges avoidance of Class Ia and Ib products, and preferably Class II too, and sets out stronger product steward-ship strategies and collection systems for empty containers.

Table 1.4 *WHO classification system for pesticides and examples of products in each category (out of a total of 845 products listed)*

Class Ia	Class Ib	Class II	Class III	Unclassified	Obsolete or discontinued
29 listed in WHO guidelines	61 listed in WHO guidelines	122 listed in WHO guidelines	122 listed in WHO guidelines	247 listed in WHO guidelines	264 listed in WHO guidelines
Aldicarb	Azinphos-ethyl	Bendiocarb	Alachlor	Amitrole	Aldrin
Difenacoum	Carbofuran	Carbaryl	Dicamba	Atrazine	Camphechlor (or
Fonofos	Chlorfenvinphos	Carbosulfan	Dichloro-benzene	Axoxystrobin	toxaphene)
Hexachloro-benzene	Dimeton-S-methyl	Copper sulphate	Fenthiocarb	Benomyl	Chloranil
Mevinphos	Monocrotophos	2,4-D	Glufosinate	Carbendazin	Chlordimeform
Parathion	Nicotine	DDT	Ipobenfos	Diethofencarb	Dimeton-S
Phenylmercury	Paris Green	Deltamethrin	Isoproturon	Diuron	Dinoseb
acetate	Strychnine	HCH	Malathion	Iprodione	Endrin
Phosphamidon	Triazaphos	Hepatchlor	Mecoprop	Gibberelic acid	Dieldrin
	Warfarin	Paraquat	Metalaxyl	Mancozeb	Leptophos
		Permethrin	MCPA	Pentanochlor	Mirex
		Pirimicarb	Nitrapyrin	Simazine	Promecarb
		Rotenone	Pimaricon	Sulphur	2,4,5-T
		Thiodicarb	Resmethrin	Temephos	
			Thiram	Trifluorsulfuron-methyl	

Although the organic sector has established many very successful agricultural systems across the world that do not rely on synthetic pesticides (Scialabba and Hattam, 2002), the great majority of farmers still rely on the use of some pesticides. Many of these are still Class Ia, Ib and II products, and represent a significant risk to human and environmental health. The key challenge, therefore, is to find ways to transform agricultural systems towards sustainability in which adverse effects are not caused.

The revised International Code of Conduct on the Distribution and Use of Pesticides now defines IPM in this way:

> *IPM means the careful consideration of all available pest control techniques and subsequent integration of appropriate measures that discourage the development of pest populations and keep pesticides and other interventions to levels that are economically justified and reduce or minimize risks to human health and the environment. IPM emphasizes the growth of a healthy crop with the least possible disruption to agro-ecosystems and encourages natural pest control mechanisms* (FAO, 2002).

Despite apparently good national and international intentions and commitments, and even though there is often adequate availability of less hazardous alternatives, large quantities of undesirable and harmful pesticides continue to be used. These include the following:

1   Products with acute toxicity hazards – pesticides classified as WHO Hazard Class Ia, Ib and II continue to be marketed and used under conditions that pose high risk to mixers, applicators and farm families who generally are not able to protect themselves adequately because they do not have access to affordable protective clothing or because climatic conditions discourage its use.
2   Products with chronic health hazards – these include probable carcinogens and potential hormonal analogues, and products that may cause birth defects or suppress the auto-immune system.
3   Persistent pesticides – persistent pesticides and those with persistent breakdown products that continue to cause contamination problems after application as they spread through the ecosystem and food chain.
4   Products that disrupt ecosystems – broad-spectrum pesticides that affect beneficial organisms and wildlife, and products highly toxic to pollinators, fish or birds.

The concept of phase-out targets for pesticides is to remove from use the most hazardous products first, given that there are cost effective alternatives already on the market. The priority target criterion is acute mammalian toxicity in view of the high incidence of farmer poisoning, especially where protective clothing is not available or too costly to use. Other target criteria include chronic health hazards and hazards to ecosystems. The idea is that the phasing out of undesirable products and the phasing in of new products should be accompanied by policy measures that support this process. Such policy changes may include removal of subsidies on products scheduled for phasing out, taxation of products with high social costs, financial incentives to encourage local development and production of new products, incentives to encourage partnerships between local producers in developing countries and producers of new products in OECD countries, a review of lists of registered pesticides, the establishment, monitoring and enforcement of maximum residue limits, and investment in farmer training through farmer field schools.

But is there really a need for such an international initiative designed to reduce the harm arising from pesticide use and/or misuse? To some, the answer to this question is still a resounding no (Avery, 1995; Schmitz, 2001). Arguments in support of the current use of pesticides usually centre on the costs that would occur if farmers could no longer use them. Cut down or remove pesticides, and crop yields and farmer profits would fall, food supply would be disrupted and national food systems undermined. The emergence of so-called high-yield agriculture is taken to mean that wildlife has been saved, as low-yield systems would require more hectares to feed people.

It is further argued that current levels of pesticides are the only way to meet future food needs given population growth towards 9 billion by 2050, and that the produce quality demanded by consumers can only be met with industrialized methods of farming. Any opposition to such sensible and progressive approaches is taken to come only from alarmists, scare-mongerers and luddites seeking to prevent scientific progress. Another approach is to argue that any alternatives, whether organic or more sustainable in other ways, are low-yielding, require their own forms of harmful pest control, and produce foods with high levels of their own toxic products. Finally, it is commonly stated that companies are already acting responsibly with regard to pesticides. They use product steward-ship as a way to meet legal requirements, and are involved in 'safe-use' schemes to reduce the harm to farmers and the environment (Vorley and Keeney, 1998; Murray and Taylor, 2000).

As discussed in Chapter 3, some of these arguments have played an import-ant role in the framing of an economic assessment of pesticides. If there is a likelihood of adverse effects, then all we should do is assess the benefits and make a rational choice for their use provided they exceed the costs. A key problem centres on what is and what is not included in these equations. Bromley (1994) use the term 'the language of loss' to describe the narratives relating to the costs, impacts and losses that industry would suffer were it to be regulated to reduce environmental pollution or harm to health: *'discussing pesticides in this manner frames the debate in a way that distorts the choice problem. Some might even suggest that the framing is not accidental. An alternative frame would seek to identify least-cost production alternatives with reduced pesticide use.'* Many of these argu-ments of the language of loss are addressed both implicitly and explicitly throughout this book.

PESTICIDES IN THE ENVIRONMENT

When a pesticide is applied to crops, most of the product is either taken up by plants and animals or is eventually degraded by microbial and other chemical pathways. But some is dispersed to the environment: some is vaporized to be eventually deposited in rainfall, some remains in the soil, while some reaches surface and ground water by run-off or leaching. In this way, some persistent products such as organochlorines have been discovered in most environments of the world.

Pesticides have long been detected in rainfall, and can travel long distances. Lindane found in a remote Japanese lake with no inflows of surface or ground water appeared to have travelled 1500 km from China or Korea (Anderson, 1986). Concentrations can be very high: organophosphates were found at concentrations of 10–50 µg l^{-1} in rainfall in the US, well above today's maximum acceptable levels of 0.1 µg l^{-1} for drinking water (Glotfelty et al, 1987). But even at very low concentrations, the total cumulative loading on natural environments can be huge: 0.005 µg l^{-1} of DDT in rainfall over Canada in the 1970s put an annual loading on Lake Ontario of some 80 kg from rainfall alone (Conway and Pretty, 1991).

Pesticides in ground water, surface waters and drinking water have become a serious and increasingly costly environmental side-effect of pesticide use. Pesticides reach water by leaching, run-off, transport on soil particles, and rapid flow though cracked soils and field drains. Most pesticides found in the environment come from surface run-off or leaching. The proportion lost is usually of the order of 0.5 per cent of the amount applied, but can sometimes rise to 5 per cent (Conway and Pretty, 1991; Vorley and Keeney, 1998). The early generations of pesticides, the organochlorines, arsenicals and paraquat, are strongly adsorbed to soil particles, and tend only to be lost when the soil itself is eroded. This can then be a significant source of pollution, such as when aldrin and dieldrin, formerly used on bulb fields and long since banned, reappeared in water-courses in south-west England in the 1990s (RCEP, 1996).

In the EU, numerous ground water supplies now exceed the maximum admissible concentration of 0.1 µg l^{-1} for any individual product, or 0.5 µg l^{-1} for total pesticides. In the mid-1990s, groundwater samples with residues above 0.1 µg l^{-1} ranged from about 5 per cent in Denmark to 50 per cent in Italy, Spain and the Netherlands (Agrow, 1996). In the US, some 9900 wells out of 68,800 tested between 1971 and 1991 had residues exceeding EPA standards for drinking water. Some products have been found long after their supposed cessation of agricultural use. In the UK, the greatest contamination by pesticides is under farmland on chalk, although it is important to note that farming is not the only source. Industry of various types has been implicated in point-source pollution of very high concentrations at several locations.

In the US, many pesticides continue to be found in groundwater, even though regulations have increased alongside better knowledge of their side-effects (Kolpin and Martin, 2003). The National Water-Quality Assessment (NAWQA)

Program of the US Geological Survey, for example, analysed 500 sites in 19 hydrologic basins in the 1990s, and found a common presence of the organochlorine products, DDT, total chlordane, dieldrin, and total PCBs. Organochlorine pesticide concentrations were higher in agricultural regions with histories of high use (Wong et al, 2000). Residues of DDE were detected in sediments at 39 per cent of sites and in fish at 79 per cent of sites. More recently, surveys of pesticides in more than 1400 wells up to 2001 found that a great deal of groundwater contains compounds at higher concentrations than 0.1 µg l⁻¹. Between 0.1 and 1.0 per cent of wells contained alachlor, carbofuran, cyanazine, 2,4-D, dicamba oxamyl, and tebuthiuron at above 0.1 µg l⁻¹; 1–3 per cent contained bromacil, diuron, metolachlor, norflurazon, prometon and simazine; and 13.6 per cent contained atrazine at above this limit (Kolpin et al, 2002; Kolpin and Martin, 2003). Except for carbofuran and oxamyl, all of these are herbicides.

PEST AND WEED RESISTANCE

Another cost of pesticide overuse is induced resistance in pests, weeds and diseases. Resistance can develop in a pest population if some individuals possess genes that give them a behavioural, biochemical or physiological resistance mechanism to one or more pesticide products. These individuals survive applications of the pesticide, passing these genes to their offspring so that with repeated applications the whole surviving population soon comes to be resistant. Unfortunately, natural enemies evolve resistance to pesticides more slowly than herbivores, mainly because of the smaller size of natural enemy populations relative to pests, and their different evolutionary history. The coevolution of many herbivores with host plants that contain toxic secondary compounds thus means they have metabolic pathways easily adjusted to produce resistance.

The first case of resistance was detected in 1914, but the main growth started to occur in the 1950s. By the late 1990s, some 2645 cases of resistance (species × products) in insects and spiders had been recorded, involving more than 310 pesticide products and 540 different species (MSU, 2000; Bills et al, 2003). During the 1990s, there was a 38 per cent increase in products to which one or more arthropod species is now resistant, and a 7 per cent increase in arthropod species that are resistant to one or more pesticides.

Resistance has also developed in weeds and pathogens. Before 1970, few weeds were resistant to herbicides but, by the late 1990s, at least 180 could withstand one or more products. Some 150 fungi and bacteria are also known to be resistant to one or more products (Georghiou, 1986). New problems continue to emerge, particularly resistant weeds. Blackgrass resistant to one or more herbicides has now been found on 750 farms in 30 counties of the UK, about 3.7 per cent of the country's 20,000 arable farms (Pretty, 1998). In Canada, resistant wild oats infects 1.2 million hectares of Manitoba cropland, and in Australia more than 3000 large wheat farms covering one million hectares have weed biotypes resistant to most herbicides (Vorley and Keeney, 1998).

PESTICIDES AND WILDLIFE: THE DIFFICULTIES IN ESTABLISHING CAUSALITY

Environmental contamination by pesticides is now widespread, with some serious implications for wildlife and humans. Adverse effects can arise in many different ways. Pesticides may come into direct contact with wildlife causing death or injury. They may contaminate sources of food, or alternatively eliminate other sources so indirectly threatening certain individuals and occasionally whole species. They may disrupt internal hormonal regulation, causing physiological and behavioural changes. The published literature on these effects is now substantial, and testament both to the wide ranging effects of pesticide products and to the continuing uncertainty over precise mechanisms and causalities. Although all pesticide products are tested for their toxicity before consent is granted for their commercial use, a full understanding of their effects in the field has often taken many years to unravel.

It is not our intention here to review all the recorded effects of pesticides on wildlife and the environment. The data are simply too numerous to summarize accurately in one place. All products are widely tested before registration, and so their class of acute toxicity in the WHO classification is known. Restrictions on the use of specific products are intended to protect vulnerable plants and animals, as well as limit exposures of humans. However, problems may arise when pesticides are used outside the limits of the restrictions, or when the restrictions do not anticipate an adverse effect or causal pathway.

Pesticide direct effects have been observed in the field on most classes of animals, including bees and other beneficial insects, birds, fish, amphibians and reptiles, and mammals. Even though individuals of these groups may be affected, it is not always clear whether there may be effects on whole populations and thus on whole ecosystems. Equally, however, it would be wrong to assume that all negative effects on individuals do not translate into population effects. The problem for ecological studies is that it is very difficult to disentangle the specific effects of pesticides from a variable background of fundamental changes to habitats and ecosystems brought about by the wider effects of modern farming, or other threats to the environment, such as industrial pollution or climate change.

One example is the decline in abundance of striped bass and Chinook salmon in California, which was recently thought to be a consequence of the escape of pesticides in drainage water from the 200,000 hectares of cultivated rice in the Sacramento valley. Products used by rice farmers are known to be toxic to these fish, but in recent years pesticide concentrations have declined to below toxic levels, and yet populations of the fish have not recovered (Byard, 1999). Other factors may be playing a key role, such as industrial pollutants, water habitat changes through dams and diversions, the introduction of exotic competitive fish species, changes in food sources, and housing developments. Nonetheless, California is one of the most intensive users of pesticides, at some 85–110 million kg of active ingredient (ai) per year, and use has increased by 10 per cent during the 1990s, twice as fast as elsewhere in the US (EPA, 2001).

In general, much more is known about the effects of individual products on individual target organisms, and much less on indirect or cumulative effects on whole ecosystems. This is why the effects of pesticides on wildlife continue to cause occasional surprise, despite extensive regulatory testing. One of the best known examples of the long-term effects of pesticides is of predatory birds and persistent organochlorines. It illustrates clearly how difficult it has been to establish causality, and how long it takes to do so.

Not long after the persistent organochlorines came into widespread use in the 1940s, populations of several species of predatory birds began to decline dramatically in both North America and Europe. A clear pattern was shared by sparrowhawks, ospreys, bald eagles, barn owls and brown pelicans, although the most notorious fall, and perhaps the best documented, was that of the peregrine falcon. There were 5000–9000 nesting pairs of peregrines in the 1930s in the US. The start of the decline was later shown to have been in the late 1940s, and numbers continued to fall catastrophically to the 1970s, when only 32 nesting pairs could be confirmed in the whole continent (Ratcliffe, 1980). In the UK, the decline did not begin until the 1950s, with peregrines disappearing in southern England by 1961, falling to 20 per cent of former levels in the north, and 70–90 per cent in Scotland, where agriculture was much less intensive.

The first signs of reasons for these declines occurred in the 1950s, when ornithologists began to notice that peregrine breeding was becoming less successful. A report on the Hudson River Valley population, formerly one of the most abundant, indicated they produced no young in any year of the early 1950s. But Kiff (1998) noted that a paper given at the American Ornithological Union in 1953 had 'elicited not a single question or comment from the assembled ornithologists'. Yet by the early 1960s, anecdotal evidence had it that not a single peregrine had fledged in the north-east US, and Hickey's survey of 14 states in 1964 found not one of 133 eyries to be inhabited (Hickey, 1988).

The first indications of clear causality occurred when Ratcliffe (1958) reported the discovery of broken peregrine eggs during 1951–56. By the early 1960s, there was speculation that this was caused in some unknown fashion by organochlorine pesticides. The eggshell thickness of herring gulls was then found to be directly correlated with DDE content, the stable breakdown product of DDT, and similar relationships were later found for peregrines, and other falcons and hawks (Cade et al, 1971; Peakall et al, 1976; Hickey, 1988). Ratcliffe (1970) then surveyed British egg collections dating back to 1900, and demonstrated clearly that eggshells after 1947 had become significantly thinner. This remarkable visual proof of a connection to the increased use of organochlorines was later repeated for British sparrowhawks for 1880–1975 by Newton (1979).

It then became clear that the population crashes were a combination of reproductive failure resulting from eggshell thinning and adult deaths caused by the bioaccumulation of cyclodien pesticides, such as dieldrin and aldrin, which were used for cereal seed treatment. The seeds, when consumed by seed-eating birds, such as pigeons, did not contain enough of a dose to kill these birds, but did accumulate to fatal levels when they were in turn predated by peregrines and other birds of prey. Although there still remains some unresolved controversy over the relative roles of DDE and cyclodienes in the declines (Newton, 1986), it was clear by the 1970s and 1980s that a group of causes and effects had indeed been identified.

The final proof would come with recovery after the products ceased to be used. Predatory birds in the UK and North America have substantially recovered in numbers, although not to 1930s levels. DDT was banned in the US in 1972, and dieldrin was gradually phased out from the 1960s to 1974. Nonetheless, DDT and DDE residues in the environment are very persistent, and have not entirely disappeared. Remarkably, it took nearly half a century of scientific and policy effort to establish some degree of causality and to make important policy decisions to remove these products from use.

As indicated earlier, it is impossible to generalize about all pesticides and their direct, indirect, acute and chronic effects on wildlife. Some are known to be highly toxic to fish, such as pyrethroids, but are relatively harmless to mammals. Others affect pollinators, such as bees, but not other insects. Some remove certain organisms from ecosystems, and so have indirect effects on the success of others. Some cause changes in whole habitats. Again, unravelling such consequences can take many years. Another example comes from the use of herbicides in arable temperate systems, their effects on arable weeds and their seed production, and consequent declines in farmland birds in western Europe.

There is now clear evidence that the abundance and diversity of farmland birds has fallen in recent decades in the intensively farmed landscapes of Europe, particularly the UK (Campbell and Cooke, 1995; Suárez et al, 1997; Defra, 2003). The declines are not associated with the direct effects of pesticides, but do coincide with continuing intensification of farming practices. It is, however, the indirect effects that appear to be significant, with herbicides removing weeds from arable fields and their margins, with consequent adverse effects on birds relying on seeds or insect herbivores as food, or with insecticides directly removing insect sources of food. The first confirmed link of this type was for grey partridge, the declines of which were clearly shown to be caused by herbicide use (Potts, 1996). The declines of a further 20 or so species, including tree sparrow, song thrush, skylark, linnet, bullfinch and blackbird, are now associated with these changes in the abundance of foods in intensively farmed landscapes, and such has been the cultural significance in the UK that farmland birds have been adopted by the government as one of its headline indicators for sustainable development.

The importance of herbicides and their effects on arable weeds and their seed production has been dramatically illustrated by the UK Farm Scale Evaluations of three GM crops: maize, beet and oil seed rape. Conducted over three years on more than 60 farms across England and Scotland, these powerful ecological experiments were able to show precisely how different forms of crop management affect the amount of food available to desirable farmland wildlife (Champion et al, 2003; Firbank et al, 2003).

Similar changes in the abundance and diversity of farmland birds have been recorded in Canada, where species richness and abundance is greater in heterogeneous farmed landscapes compared with those dominated by just wheat, maize and soybean (Jobin et al, 1996). The mixed landscapes contained more habitat types, such as wetlands, woodland, hedgerows, old fields, pastures and hayfields, and more mixed farm enterprises. It is now clear that direct effects on wildlife are likely to be less common in industrialized counties, now that pesticide products are generally safer. However, their indirect ecological effects appear to remain significant.

ENDOCRINE DISRUPTION: A NEW REASON TO BE CAUTIOUS

The direct effects of pesticides are much easier to establish than indirect effects. If wildlife is harmed or killed, and this can be observed, then it may be possible to establish both an association and causality. Indirect effects tangled up with the natural variability in numbers of animals and plants in their habitats are much harder to identify, as shown by the cases of predatory birds and farmland songbirds above. The effect of DDE on eggshell thinning was actually the first recognized case of endocrine disruption in wild populations. This is now recognized as a growing problem, and a further reason to be cautious about some pesticides (OSTP, 1996; NAS, 1999; EPA, 2000; OECD, 2001b; Defra, 2003). The endocrine system is the communication system of glands, hormones and cellular receptors that guide the development, growth, reproduction and behaviour of organisms. Endocrine glands include the pituitary, thyroid and adrenal glands, the female ovaries and male testes. Thus an endocrine disrupter exerts its effect by mimicking or interfering with the actions of hormones.

Many important hormones, such as oestrogen, progesterone, testosterone and thyroxin, are associated with high-affinity receptor proteins in target cells, and when these hormones come into contact with these receptors, they provoke a series of effects. This high affinity is important, as exogenous chemicals can also bind to these sites, either minimizing the effects of the natural hormones or blocking the sites, so preventing proper cell signalling. The chemicals that mimic or block sex steroid hormones are commonly called environmental oestrogens or anti-oestrogens, and these are the most studied of all disrupters.

Several expert working groups of scientists from Europe and North America now conclude that there is increasing evidence that biologically-active concentrations of endocrine disrupting chemicals are having adverse effects on wildlife reproductive health, and possibly on humans too (Colborn et al, 1993; Crisp et al, 1998; NAS, 1999; EPA, 2000). Many products have been reported to possess endocrine disrupting capacity, including some natural products (e.g. coumestrol from clover), pesticides (e.g. dieldrin, DDT, endosulfan), medical drugs (e.g. tamoxifen), and commercial and industrial chemicals (e.g. alkylphenols, phthalates, PCBs and some metals). Endocrine disrupters can mimic or block natural female sex hormones (and so are termed oestrogens or anti-oestrogens), mimic or block male sex hormones (androgens or anti-androgens), interfere with sex steroid systems, or disrupt pituitary, thyroid and interregnal hormone systems.

Like other problems with pesticides, there is stronger causal evidence from laboratory studies, but no more than evidence of associations between the presence of certain chemicals and observed adverse effects in mammals, birds, reptiles, fish and molluscs in the natural environment. The main abnormalities include sex differentiation with feminized or masculinized sex organs, changed sexual behaviour, and altered immune function. But, as Vos et al (2000) put it, *'for most reported effects, the evidence for a causal link with endocrine disruption is weak or non-existent'*. Nonetheless, the laboratory studies using realistic exposure levels seem to suggest that such causality will eventually be established. Human health risks that may be associated with exposure are still unknown and therefore controversial.

Not all pesticides are endocrine disrupters, and many industrial pollutants are more widespread in the environment and thus more of a threat, such as PCBs. Table 1.5 contains a list of some products known to have endocrine disrupting effects. At an international scientific meeting at the Swiss Federal Institute of Science and Technology in 1999, a number of common products were further identified as being potential endocrine disrupters (ENDS, 1999). These included the confirmed disruptors metiram, procymidon and vinclozolin, and the potential disrupters benomyl, carbofuran, deltamethrin, glyphosate and penconazole.

Table 1.5 *Chemicals with widespread distribution in the environment reported to have reproductive and endocrine disrupting effects*

Herbicides	Fungicides	Insecticides	Nematicides	Industrial chemicals
2,4-D	Benomyl	ß-HCH	Aldicarb	Dioxin
2,4,5-T	Hexachlorbenzene	Carbaryl	DBCP	PBBs
Alachlor	Mancozeb	Dieldrin		PCBs
Atrazine	Maneb	DDT,		Pentachlorophenol
Nitrofen	Tributyl tin (TBT)	metabolites		(PCP)
		Endosulfan		Phthalates
		Lindane		Styrenes
		Parathion		

Source: Colborn et al, 1993

Although laboratory-based studies have shown that certain chemicals cause endocrine disruption, only a small number of field studies have found the effects of disruption in individuals, and only a very limited number have observed effects on populations and communities. This is not unique for endocrine disrupter research, but *'is rather a situation characteristic for eco-epidemiology in general'* (Vos et al, 2000). Two of the most significant continuing problems relate to seals in the Baltic Sea and frogs in California.

The widespread and worrying disease syndrome in Baltic grey and ringed seals that has caused population decline has been clearly linked to high body concentrations of PCBs, DDT and their metabolites (Jensen and Jansson, 1976;

Bergman and Olsson, 1985; Bergman, 1999). Autopsies found uterine lesions and tumours causing sterility in female seals (30 per cent of grey, 70 per cent of ringed), together with a range of non-reproductive symptoms, including damage to brains, bones and guts, and decreased skin thickness. Over the past two decades, as body burdens of the endocrine disrupters have declined, so there has occurred a recorded improvement in the reproductive performance of both seal species in the Baltic (Bergman, 1999). Despite these clear links between ill-health, population levels and concentrations of chemicals, the underlying mode of action of both PCB and DDT compounds is still not fully understood (Vos et al, 2000). Again, this shows how difficult it is to establish beyond doubt the threats, mechanisms and effects (on individuals and populations), even in a well-researched case.

As notorious as the seals is the case of frogs in the US. Many amphibian species experienced substantial declines in abundance and distribution over the period of agricultural intensification in the 20th century (Fisher and Shaffer, 1996; Hayes et al, 2002). Many factors are involved, perhaps most importantly the draining of wetlands and loss of habitats. Amongst a range of other hypotheses, including increasing UV-B radiation and climate change, is the possibility that agricultural pesticides have altered the growth and survival of frogs. Once again, laboratory studies have found a range of subtle effects occurring in whole plant and amphibian communities exposed to common herbicides such as atrazine (Diana et al, 2000), and in frogs exposed to levels of atrazine now found in the environment (Hayes, 2000; Hayes et al, 2002). Regional geographic studies have also found pronounced associations between declines, regions of high pesticide use, and the presence of pesticide residues in frog species (Datta et al, 1998; Davidson et al, 2001).

The most significant recent research by Hayes and colleagues (2002) found that male frogs exposed to very low levels of atrazine, the most commonly used herbicide in the US (27 million kg are applied annually), developed symptoms of hermaphroditism and demasculinity, together with suffering significantly reduced levels of the hormone testosterone. The exposure levels causing these effects were of the order of 0.1 to 1.0 μg kg^{-1} (or ppb), whilst the allowable level in drinking water is 3 ppb, and short-term exposures of 200 ppb are not considered a risk under current regulations. Concentrations in surface waters in intensively cultivated agricultural regions regularly exceed 200 ppb, and can reach 2300 ppb.

Reproductive effects have also been reported in other aquatic reptiles and amphibians. All crocodiles, many turtles and some lizards lack sex chromosomes (sex is organized after fertilization), and so eggs exposed to oestrogenic compounds, such as some PCBs, produce significantly more females (Crain and Guilette, 1997). A noted case is that of alligators in Lake Apopka in Florida. A major spill of a pesticide mixture containing dicofol and DDT in 1980, combined with years of agricultural and municipal run-off, led to elevated levels of endocrine disrupting chemicals in the tissue of alligators, the occurrence of various developmental abnormalities in both males and females, and a 50 per cent decline in juvenile numbers (Vos et al, 2000).

Fish have also been observed to have suffered endocrine disruption in some environments. A notable example in the UK is male flounders exposed to sewage effluents, which have developed the protein vitellogenin that is normally only required in egg yolks (Harries et al, 1997; Allen et al, 1999). Other fish studies indicate a significant inverse relationship between salmon catches in Canadian watersheds with pesticide applications, although here the chemical of concern is nonylphenol, a solvent used with pesticides rather than the active ingredient itself (Fairchild et al, 1999).

Added to this evidence of effects on wildlife, there have also been growing concerns about the potential effects of endocrine disrupters on human health and reproductive performance. Concerns about declining sperm counts in males in industrialized countries have drawn attention to the potential role for endocrine disrupters. There is some evidence that certain products (e.g. DDT and HCH) could play a role in breast cancer, or others (e.g. the fungicides vinclozolin and procymidone) could be anti-androgenic (Steinmetz et al, 1996; Vos et al, 2000). However, the WHO and UNEP concluded in 2002 that evidence for hormone disruption in humans is no better than weak, and does not match the more confirmed cases of birds and thinned eggs, alligators in Florida, and sexual differentiation in frogs (ENDS, 2002).

It is now clear that many compounds are endocrine disrupters, but not all are agricultural. There are notable problems from effluents from sewage treatment works and the paper industry. Moreover, some agricultural pesticides that are endocrine disrupters are no longer in use, although they still persist in the environment. There are many confirmed laboratory studies showing causality, although most field studies as yet only show no more than associations.

CONCLUDING COMMENTS

Pesticides are now widely used in food production systems across the world, and increasingly, in some countries, in the home and garden. Some 2.5 billion kg of active ingredients are applied each year, amounting to an annual market value of some US$30 billion. Just over a fifth of all pesticides are used in the US. However, most pesticide markets in industrialized countries are no longer growing, and companies are looking to developing countries for increased sales. More than 800 products are in regular use worldwide.

Pesticides have become ubiquitous in environments worldwide, some reaching hazardous levels for humans. Pest resistance has become increasingly common, with 2645 cases of resistance in insects and spiders recorded in the late 1990s. The problem for regulators is that causality is often difficult to establish. This is graphically shown by the amount of scientific effort required to understand the effects of pesticides on wild bird populations. A further reason to be cautious now comes from concerns about the endocrine disrupting properties of some pesticide products.

The recently revised UN International Code of Conduct on the Distribution and Use of Pesticides has provided new guidance on integrated pest management for agricultural systems, and is complemented by the Rotterdam Convention on Prior Informed Consent and the Stockholm Convention on Persistent Organic Pollutants. International agencies and national governments are increasingly targeting WHO Class Ia (extremely hazardous) and Ib (highly hazardous) products as the first priority for replacement. The second priority is Class II (moderately hazardous) pesticides. Many effective alternatives do exist.

Chapter 2

The Health Impacts of Pesticides: What Do We Now Know?

Misa Kishi

INTRODUCTION

With the exception of antipersonnel chemicals such as war gases, pesticides are the only toxic chemicals that we deliberately release into the environment, which, by definition, are intended to cause harm to some living thing. (Keifer, 1997)

We know that pesticides cause many public health problems, but their true extent remains unknown. There are several reasons for this. Some health outcomes from pesticide poisoning are not easily recognized, especially when there is a time lag between exposure and outcomes. Scientific methods for studying pesticide effects are more suitable for dealing with the effects of a single agent in a temperate climate, although many developing countries are in the tropics where multiple pesticides are routinely mixed and used in cocktails. Existing data are often from studies on healthy, young male subjects, even though the majority of people in developing countries do not fall into this category. All these conditions make it very difficult for us to know the true extent of the adverse health impacts of pesticides.

This chapter assesses the existing data on the human health impacts of pesticides. Its primary focus is on the problems facing people in developing countries. While citizens in industrialized countries are mainly concerned with low-level exposures to the general public, farmers and agricultural workers in developing countries are exposed to many dangerous products that are almost impossible to use safely under field conditions. As a result, they are more likely to develop symptoms of pesticide poisoning. This chapter also includes studies of non-agricultural workers (such as pesticide factory workers, vector-control workers) and children, as well as selected data from industrialized countries and environmental exposures.

THE REAL EXTENT OF ACUTE PESTICIDE POISONING

Acute pesticide poisoning is a serious public health problem in developing countries, where many farmers still use highly toxic products the use of which is neither banned nor restricted. Hazardous pesticides can be manufactured in industrialized countries and then exported to developing countries (Smith, 2001), or the active ingredient can be exported and then manufactured into the end product.

Cholinesterase-inhibiting pesticides, namely organophosphates (OPs) and carbamates, are the most common causes of severe acute pesticide poisonings (Jeyaratnam et al, 1987; Lum et al, 1993; McConnell and Hruska, 1993; Wesseling et al, 1993; Keifer et al, 1996; Wesseling et al, 2000; IFCS, 2003). According to the World Health Organization (WHO) classifications of pesticides by hazards (IPCS, 2002), the cholinesterase-inhibiting pesticides identified as causal agents in these poisonings often belong to Classes Ia (extremely hazardous) and Ib (highly hazardous). (The difference between OPs and carbamates is that the duration of carbamates' inhibition of cholinesterase in the nervous system is shorter and regeneration of enzyme activity occurs within a few hours (Box 2.1). Compared with OP intoxication, carbamate poisonings are usually less severe.

Box 2.1 *How cholinesterase-inhibiting pesticides work on the nervous system*

Cholinesterase is an enzyme that hydrolyzes neurotransmitter acetylcholine into inactive fragments of choline and acetic acid at the completion of neurochemical transmission. Acetylcholine is essential for nerve transmission in the central nervous system and in the somatic and parasympathetic nervous systems. Organophosphates and carbamates inhibit cholinesterase by phosphorylating the active site of the enzyme. As exposure levels to organophosphate and carbamate pesticides increase, cholinesterase activity decreases. This leads to accumulation of acetylcholine at synapses, which in turn causes overstimulation and disruption of transmission. Symptoms of mild to moderate poisoning are headache, dizziness, blurred vision, weakness, uncoordination, muscle fasciculation, tremor, diarrhoea, abdominal cramping, and occasionally chest tightness, wheezing and productive cough. Symptoms of severe intoxication include incontinence, convulsions and unconsciousness.

Source: Tafuri and Roberts, 1987; Costa, 1997

Paraquat (WHO Class II, moderately hazardous), a widely used nonselective contact bipyridyl herbicide, is also known as an important cause of acute pesticide poisonings (Wesseling et al, 1997a). The pesticide industry claims that paraquat is most unlikely to cause serious health problems under correct conditions of use. Some studies, however, found that occupational poisonings among agricultural workers are common (Wesseling et al, 2001; Murray et al, 2002) and can cause serious poisonings, including deaths (Wesseling et al, 1997b).

In Costa Rica, fatal occupational paraquat poisonings were documented after oral contact, dermal absorption and possible inhalation, including adults as well as children, presenting either renal or liver impairment, followed by adult respiratory distress syndrome (ARDS) or pulmonary oedema (Wesseling et al, 1997b).

Endosulfan (WHO Class II, moderately hazardous), an organochlorine insecticide, has been identified as the cause of occupational poisonings both in developing countries and the developed world (Brandt et al, 2001; Murray et al, 2002) (see Chapter 11). It is important to note that the use of mixtures is common in agricultural practices. Some studies, both in developing countries and the US, showed that pesticide poisonings are often caused by mixtures of pesticides (Blondell, 1997; Cole et al, 2000; Keifer et al, 1996).

How many people are poisoned by pesticides each year? Regrettably, there is no easy answer to this important question. We know that almost all deaths due to acute pesticide poisoning occur in developing countries, even though these countries consume only a tenth of the world value of pesticides (see Chapter 1). The WHO has a long-standing estimate that three million cases of severe pesticide poisoning occur each year (comprising two million suicides, 700,000 occupational poisonings, and 300,000 accidental poisonings), resulting in 220,000 deaths (WHO, 1990). Suicides remain a significant cause of death (Eddleston, 2000), and, although not caused by agricultural use, they cannot be entirely separated from the easy access rural workers and their families have to these products (Wesseling et al, 1997a).

These WHO estimates for worldwide pesticide poisonings are likely to capture only the tip of the iceberg, as they are solely based on confirmed hospital registries. The estimates are based on a calculation of a recorded versus unrecorded incident ratio of 1:6 (WHO, 1990). Thus they probably overestimate the proportion of suicides and underestimate the actual number of pesticide poisonings (Wesseling et al, 1997a; London and Bailie, 2001). Based on surveys in four Asian countries, Jeyaratnam estimated that if all levels of severity are included, 3 per cent of agricultural workers in developing countries, or 25 million people, suffer from pesticide poisoning each year (Jeyaratnam, 1990).

However, specific country studies show higher rates of poisoning. Nine per cent of the Indonesian farmers participating in a prospective study recalled at least one pesticide poisoning during the previous year (Kishi et al, 1995). In Costa Rica, the estimate of the annual incidence of symptomatic occupational poisoning among agriculture workers was 4.5 per cent (Wesseling et al., 1993). Yet it is also important to note that directly observed poisoning rates are much higher than self-reported rates. In the Indonesian study, 21 per cent of the spraying operations resulted in three or more neurobehavioural, respiratory or intestinal signs or symptoms, which was taken as a functional definition of poisoning. However, only 9 per cent of the farmers in the study had reported pesticide poisoning over the past year. One reason for this discrepancy is that farmers are likely to ignore the symptoms of pesticide poisoning or to not take them seriously, because they accept that becoming sick is simply an unavoidable part of their work (Kishi et al, 1995).

Another factor that leads to the underreporting of pesticide poisoning is that the focus of epidemiological studies in developing countries has mainly been on male farmers who apply pesticides, although women are equally at risk (London et al, 2002). Similarly, there is little available information about occupational pesticide poisonings among children. According to a review by the Pan American Health Organization (PAHO), some 10–20 per cent of all poisonings involve children under 18 years of age (Henao et al, 1993). In Nicaragua, nearly one in five work-related poisonings involved children under 16 years of age (McConnell and Hruska, 1993).

At the household level, women and children are at considerable risk. Women are often engaged in different types of agricultural labour, such as planting, weeding and harvesting. They spend long hours in the fields where pesticides are being sprayed, or work in the fields immediately after pesticides have been applied. In some areas, women can be in charge of pesticide application (Kimani and Mwanthi, 1995; Murphy et al, 1999), and even in areas where men tradition-ally do the spraying, the migration of men to the cities, as well as injuries, sickness and the deaths of male family members due to war and diseases such as AIDS, can leave women to do the spraying (London et al, 2002). It is also common for children to help on the farm, including applying pesticides (Harari et al, 1997). A further risk arises from contaminated clothes washed by women, often mixed with other laundry (Kishi et al, 1995). Pesticides are commonly stored in the home within the reach of children. In Indonesia, 84 per cent of respondents store pesticides in their homes, 75 per cent in living or kitchen areas, and 82 per cent within reach of children (Kishi et al, 1995). In Ghana, 31 per cent of respondents store pesticides in the bedroom for security reasons, as pesticides are expensive and not always readily available, with 18 per cent storing them elsewhere in the house (Clarke et al, 1997).

WHO is currently developing new estimates for acute pesticide poisoning through its Project on the Epidemiology of Pesticide Poisoning in India, Indo-nesia, Myanmar, Nepal, Thailand and the Philippines (Nida Besbellin, pers. comm.). In the trial implementation phase, data were collected using a new harmonized pesticide exposure record format, and medical staff were instructed on the collection of information, on the diagnosis and treatment of cases of pesticide exposure and on the use of the poisoning severity score (PSS). This trial phase has confirmed that pesticide poisoning is a serious public health problem. The data demonstrated the magnitude of the problem due to intentional poison-ing, but did not appear to reflect the situation concerning occupational and accidental exposures. Population-based studies are now required in order to collect information about cases that are not in hospital records. The second stage of the study will include such studies and a surveillance protocol is being developed for community-based studies.

WHY ARE SO MANY PEOPLE POISONED?

Hazardous products are supposed to be used by trained personnel wearing protective equipment, as the skin is the major route for occupational exposure to pesticides, followed by inhalation and oral intake. However, studies in Indonesia (Kishi et al, 1995) and Ecuador (Cole et al, 2000) demonstrate that it is common for agricultural workers in developing countries to apply pesticides barefoot with little or no protective equipment. Other studies from Tanzania, Kenya, Indonesia and Costa Rica have produced similar results (van Wendel de Joode et al, 1996; Murphy et al, 1999; Ohayo-Mitoko et al, 1999).

There are a number of further ways in which agricultural workers can be exposed to pesticides. Kitchen spoons, bottle caps and bottles are commonly used to mix the concentrated chemicals, and these implements can easily contaminate workers' bare skin. Furthermore, if farmers spray pesticides into the wind, or if the target is tall, farmers may find themselves walking into pesticide mists before the droplets have fully settled on the crops, thus wetting their skin and clothes. In addition, pesticide backpacks are often ill-maintained, and therefore leak, resulting in the skin and clothing being soaked with pesticides. Long-sleeved shirts and trousers, and handkerchiefs worn as masks, may be used as protective measures, but are largely ineffective. When these become wet with pesticides, they may enhance absorption through the skin and mouth.

Contrary to the assumption that 'farmers handle pesticides in a risky manner because of the lack of knowledge about dangers of pesticides', most farmers do generally know that pesticides are toxic (Eisemon and Nyamete, 1990; McDougall et al, 1993; Clarke et al, 1997; Murphy et al, 1999; Aragon et al, 2001; Kunstadter et al, 2001; Kishi, 2002). However, this knowledge does not necessarily translate into behaviour that mitigates effects, as there are often structural barriers or other reasons that override farmers' concerns about safety when applying pesticides. Indeed, researchers have found in Central America *'a vast array of structures which create a context in which unsafe practice may be the sensible, if not the only possible line of action. . . .The inappropriate use of pesticides is driven by many complex factors'* (Murray and Taylor, 2000).

This situation is similar in other developing countries. A study in Ghana showed that:

> *[a]lthough farmers claim knowledge of health risks from pesticides, they do not generally use personal protective measures, the predominant reasons given being that the protective equipment is out of their financial reach and uncomfortable to use under the prevailing hot and humid climatic conditions* (Clarke et al, 1997).

A further problem is that washing facilities are rarely located close to agricultural fields, so agricultural workers cannot wash themselves properly until they get home. They therefore spend long hours in agricultural fields wearing contaminated clothing, and eat, drink and smoke with pesticide-soaked hands.

Environmental exposure in agricultural communities is another area of concern. There are numerous, commonly observed practices that demonstrate the high risks of pesticide exposure among rural communities in developing countries, even though there has been too little research that has focused on this area. For instance, people who live in agricultural communities commonly use irrigation canals and streams for daily activities such as washing and bathing, yet empty pesticide containers dumped in fields and irrigation canals are a common occurrence. PAHO reviewed several Latin American countries and concluded that '*it is common to find residues of organochlorine and organophosphorus compounds in drainage, well, and river water*' (Henao et al, 1993), while a later study in South Africa found widespread pollution of farm area surface and ground water with low levels of endosulfan, with a variety of other pesticides (chlorpyrifos, azinphos-methyl, fenarimol, iprodione, deltamethrin, penconazole and prothiofos) found to exceed drinking water standards (London et al, 2000).

In addition, people who live close to agricultural fields can be exposed to pesticides when fields are sprayed from the air. In Nicaragua, the aerial drift of pesticides provokes lower cholinesterase levels and increased numbers of symptoms from pesticide poisoning in people living nearby (Keifer et al, 1996). In Israel, an association was found between exposure and symptoms among children and adults living in kibbutzim affected by drift exposure (Richter et al, 1992). Another study in Nicaragua found cholinesterase depression among children in the communities near an airport where organophosphates were loaded and unloaded from airplanes (McConnell et al, 1999b). In El Salvador, an association was found between the two-week prevalence of acute symptoms in children (including the detection of urinary metabolites of organophosphates), with an adult in the same household who had recently applied methyl parathion (Azaroff, 1999; Azaroff and Neas, 1999).

For such an important public health problem, it is surprising that there still remains so little research on pesticide exposure and adverse health effects in developing countries. As WHO estimates indicate, the problems may be much more severe than previously supposed. With pesticide use set to continue to rise in many agricultural systems, it is clearly important to find ways to reduce exposure to hazardous products.

THE CHRONIC EFFECTS OF PESTICIDE EXPOSURE

The estimates of numbers of people poisoned by pesticides refer only to acute pesticide poisoning. They do not address the chronic effects of exposure, which include cancer, neurological and reproductive effects, respiratory and skin disorders, and impaired immune functions (WHO, 1990; Keifer, 1997; Krieger, 2001).

Chronic effects can occur through either low-dose, long-term exposures or high-dose, short-term exposures and both conditions are likely to occur in developing countries. In industrialized countries, by contrast, the focus of concern has generally shifted from occupational exposures to low-level exposures among the general public (Fleming and Herzstein, 1997). However, there are growing concerns about pesticide effects on family members of agricultural workers in some countries. In the US, for example, several studies are investigating the health effects on both farmers and their spouses and children. These include a large prospective Agricultural Health Study, carried out by the National Cancer Institute, and the National Institute of Environmental Health Sciences, which investigates health effects on farmers, spouses, and children in North Carolina and Iowa (Alavanja et al, 1996; Galden et al, 1998; see also www. aghealth.org).

Other examples include a research programme on children's pesticide exposure in the farm environment by the University of Washington (Fenske et al, 2000) and a study of pesticide exposures and their effects on pregnant women and their children by the University of California at Berkeley (Eskenazi et al, 1999). In addition, the Natural Resources Defense Council has reported on the effects of pesticides on the children of farmers and farm workers (Solomon and Motts, 1998). Some health professionals and researchers have begun to study the effects of pesticides on inner-city children (Landrigan et al, 1999), as well as the possible adverse effects of environmental pesticide exposures on children's learning and behavioural development (Schettler et al, 2000; Schettler, 2001).

PESTICIDES AND CANCER

The findings of a number of occupational studies on farmers have been conducted worldwide, although mainly in industrialized countries, with consistent findings. A review of the carcinogenicity of pesticides shows that farmers experience higher than expected rates of cancers of the lymphatic and blood system, lip, stomach, prostate, brain, testes, melanoma, other skin cancers and soft tissue sarcoma (Zahm et al, 1997). While studies of female farmers and female farm-family members have not been conducted as extensively as those of male farmers, it appears that they too have excesses of cancers of the lymphatic and blood system, lip and stomach, as well as ovarian cancer and possibly cervical cancer. Another review (Solomon et al, 2000) shows similar results, but the authors interpret that the excess of skin cancers and cancer of the lip are more likely to be caused be exposure to ultraviolet light than pesticides. Zahm and colleagues (1997) concluded that in spite of the limited data, '*there is strong evidence that selected phenoxyacetic acid herbicides, triazine herbicides, arsenical insecticides, organochlorine insecticides, and organophosphate insecticides play a role in certain human cancers.*' The US Agricultural Health Study recently found a small but significant increase in prostate cancer risk for pesticide applicators and farmers than the general population of North Carolina and Iowa (Alavanja et al, 2003).

In developing countries, studies of pesticide exposure and cancer are rare (London et al, 2002). In Costa Rica, a retrospective cohort study found raised cancer risk among banana plantation workers who were exposed to the nematicide, dibromochloropropane (DBCP). Male workers showed an increased risk of melanoma and penile cancer, while female workers showed an increased risk of cervical cancer and leukaemia (Wesseling et al, 1996). Two studies in Colombia and Mexico (Olaya-Contreras et al, 1998; Romieu et al, 2000) found exposure to organochlorine insecticides was a risk factor for female breast cancer, but two others in Mexico and Brazil did not reach the same conclusion (Lopez-Carrillo et al, 1997; Mendonca et al, 1999).

Though the aetiology of childhood cancer is not well understood, associations have been found between parental and infant exposures to pesticides and childhood brain tumours, leukemia, non-Hodgkin's lymphoma, sarcoma and Wilms' tumour (Daniels et al, 1997; Zahm and Ward, 1998; Gouveia-Vigeant and Tickner, 2003). Studies of childhood cancer in relation to pesticide exposures are rarely conducted in developing countries. In Brazil, however, a case-control study examined associations between parental exposures to pesticides and the risk of Wilms' Tumor, and found associations between paternal and maternal exposures to farm work with frequent pesticide use and Wilms' tumour in their children (Sharpe et al, 1995). Owing to the limited geographic extent of these studies, it remains difficult to draw wider conclusions.

NEUROLOGICAL EFFECTS

Cholinesterase-inhibiting pesticides are known to cause persistent neurological and neurobehavioural damage following acute exposures or long-term exposure to low doses. Persistent damage to the central and peripheral nervous systems has been found following episodes of poisoning in developing countries by cholinesterase-inhibiting pesticides (Rosenstock et al, 1991; McConnell and Magnotti, 1994; McConnell et al, 1999a; Miranda et al, 2002a, 2002b; Wesseling et al, 2002). Studies in Ecuador found adverse effects on the peripheral and central nervous system in agricultural workers who apply pesticides as well as in other farm members who are likely to be indirectly exposed to pesticides (Cole et al, 1997b, 1998a) (see Chapter 10).

Other pesticides are also causing neurological problems. In Brazil, an association was found between occupational exposure to maneb and chronic neurological impairment (Ferraz et al, 1988), and in Costa Rica, neurotoxic effects occurred in DDT-exposed vector-control sprayers (van Wendel de Joode et al, 2001). In Mexico, preschool children in a farming community where pesticides were heavily used were compared with children in another farm community that used little or no pesticides. Compared with the less exposed children, the children in the community with high pesticide use showed decreased stamina, short-term memory impairment, difficulties in drawing, and had problems with hand–eye coordination (Guillette et al, 1998).

REPRODUCTIVE AND OTHER EFFECTS

A number of pesticide products are known or suspected to be reproductive toxicants, and those who are occupationally exposed to pesticides or who live in agricultural communities again appear to be at greater risk. A number of studies conducted in developing countries that document the adverse impacts of pesticides on reproductive health are summarized in Table 2.1.

There are many other health effects from exposure to pesticides. Impaired immune functions from pesticide exposure have been widely reported in

Table 2.1 *Reproductive health studies on pesticides in some developing countries*

Country	Main findings
Chile	An association between maternal pesticide exposure and congenital malformations.
China	An increased risk of small-for-gestational-age and threatened abortion among those exposed to pesticides occupationally.
	A higher than expected number of congenital anomalies in the central nervous system among women exposed to pesticides during the first trimester of pregnancy.
Colombia	An increase in the prevalence of abortion, prematurity and congenital malformations amongst female workers and the wives of male workers in floriculture.
	An increased risk of birthmarks, especially haemangioma.
Costa Rica	High rates of male infertility among banana workers exposed to dibromochloropropane (DBCP) in the 1970s.
India	A high frequency of abortions and stillbirths among workers in a grape garden exposed to organochlorine and organophosphate pesticides.
	Abortions, stillbirths, neonatal deaths and congenital defects among cotton field workers.
Sudan	A higher incidence of stillbirth in farm families exposed to pesticides.

Source: Whorton et al, 1977; Rita et al, 1987; Restrepo et al, 1990a, 1990b; Rupa et al, 1991; Thrupp, 1991; Zhang et al, 1992; Taha and Gray, 1993; Levy et al, 1999; Rojas et al, 2000

developing countries (Repetto and Baliga, 1997a; 1997b). This has enormous implications for the life expectancies of millions of farmers, particularly women, who are affected by HIV (Page, 2001). Respiratory problems are also common. An association before respiratory impairment with long-term occupational exposure to a variety of organochlorine (OC) and OP pesticides was found among sprayers in mango plantations in India, and a reduction of pulmonary function and frequent complaints of respiratory symptoms occurred among farm workers exposed to various OPs in Ethiopia (Rastogi et al, 1989; Mekonnen and Agonafir, 2002). In addition, associations between paraquat and long-term respiratory health effects were found in Nicaragua and South Africa (Castro-Gutierrez et al, 1998; Dalvie et al, 1999). Chlorothalonil is a risk factor for dermatitis among banana plantation workers in Panama (Penagos et al, 1996; Penagos, 2002), as is the fungicide maneb in Ecuador (Cole et al, 1997a).

OBSTACLES TO CHANGE: ECONOMIC AND POLITICAL FACTORS

There are many obstacles to the reduction of harm arising from the misuse of pesticides in developing countries. Most of the barriers are economic and political, including a number of factors that make hazardous pesticides readily available and relatively inexpensive. For many years, governments subsidized pesticides to reduce their cost to farmers, although this has become uncommon today. Multinational companies have the resources and influence to promote the use of their products, and often play a key role in framing national pesticide policies. Some development assistance programmes have also played a role in promoting pesticide use (Tobin, 1994, 1996). Local sellers and distributors directly advise farmers on how much pesticide they should use, even though their incentives are financial – the more they sell, the more they earn. All these factors influence farmers' perceptions of pesticides as agricultural 'medicine', as they are referred to in a number of languages, and the common belief that pesticides are necessary for healthy and plentiful crops.

Underpinning these problems is a more fundamental concern. National health ministries and international health organizations tend to promote a 'health paradigm' for dealing with pesticides. Common 'solutions' include improvement of diagnosis and treatment of pesticide poisoning, health education and the dissemination of information about the dangers of pesticides, and promotion of personal protective equipment. These are similar 'solutions' to those actively promoted by the pesticide industry (McConnell and Hruska, 1993) in their 'safe use' programmes, with the underlying assumption that *a linear relationship exists between the transfer of knowledge and changes in behaviour* (Murray and Taylor, 2000).

The problem is that people may know something is correct, but may not be able to act. Although millions of dollars have been spent, *there is no evidence that widespread "safe use" programs have greatly affected pesticide exposure and morbidity* (Wesseling et al, 1997a). A seven-year research program by Novartis on the adoption of safe and effective practices found that *despite the increase in the number of farmers adopting improved practices, a large number still did not do so even though they were aware of the health risks* (Atkin and Leisinger, 2000).

As indicated earlier in the chapter, farmers generally do know the dangers of the pesticides, but this knowledge alone is not sufficient to change their behaviour. Their first priority is usually economic survival, which generally overrides concerns for health (Aragon et al, 2001; Kishi, 2002). A survey of sugarcane farmers in Fiji found that 26 per cent were very concerned about the health risks of pesticide use, but indicated that the perceived benefits outweighed the risks (Szmedra, 1999). In Sri Lanka, the perception among farmers is that heavy pesticide use is essential for good crops, and so farmers do their best to minimize or deny the health problems from their 'necessary' exposure to pesticides (Sivayoganathan et al, 1995).

Even if farmers were able to obtain and use protective equipment, the problems of pesticide exposure to family members, to people living in agricultural communities, and to the general public, would not be solved. The environmental and agroecosystem problems will remain too – including the adverse impacts on fish, animals and natural predators, as well as pest resistance to pesticides.

NEED FOR SURVEILLANCE SYSTEMS

Surveillance is one of the most important tools for understanding the extent of public health problems and for controlling occupational hazards. It is defined as:

> *the ongoing systematic collection, analysis, and interpretation of health data essential to the planning, implementation, and evaluation of public health practices, closely integrated with the timely dissemination of these data to those who need to know* (NIOSH, 2001).

However, for a variety of reasons, including underreporting and lack of surveillance systems, the true extent of pesticide poisoning in developing countries remains unknown (Murray, 1994; Murray et al, 2002).

There are a number of reasons for underreporting. One is that many of the agricultural workers who develop pesticide-related symptoms do not visit health facilities. A possible explanation for this is that they are likely to ignore their pesticide-related illness and not take it seriously (Kishi et al, 1995). A study in Kenya found a similar explanation among female workers. The majority of acute pesticide poisoning cases referred to the health facilities were male, and a possible reason was that:

> *women either ignored the symptoms, or did not feel that the heath conditions were more serious to warrant health care than their daily activities. Apparently, many of the women either ignored the symptoms or relied on self medication. Indeed, many women reported that the symptoms were neither acute nor were they life threatening. In fact, complaints related to pesticides exposures were considered "minor" health problems by the women* (Kimani and Mwanthi, 1995).

Moreover:

> *the low literacy rate among the women and other limiting factors such as distance, and low social economic status may have been the reason why many of them did not seek health care attention after acute exposure to the pesticides.*

Other causes for underreporting include ignorance on the part of medical personnel, objections to performing extra paperwork, and the absence of supervisory feedback. Misdiagnosis by medical personnel is another problem: the symptoms of mild and intermediate levels of pesticide poisonings (e.g. dizziness, nausea, headache) are non-specific and are easily misdiagnosed as flu. Misdiagnosis may even occur among farmers who suffer from severe intoxication that leads to death. Loevinsohn (1987) reported that in a rice-growing area with intensive pesticide use in the Philippines, increases in the mortality rate caused by strokes were correlated with pesticide exposure.

As indicated earlier, currently available data overestimates the proportion of suicides by pesticides in developing countries. In an intensive surveillance project, London and Bailie (2001) point out that hospital and health authority sources in South Africa overestimated the proportion of cases resulting from suicides and greatly underestimated the proportion of occupational poisonings. London and Bailie also argue that these patterns of pesticide poisonings may seriously misinform policymakers about priorities for regulating pesticide use.

The problem of underreporting in surveillance can be improved by training medical personnel at different levels. After implementing active surveillance in Nicaragua, reported cases increased from seven to 396 (Cole et al, 1988). Even so, it was found that at least a third of poisoning cases in Nicaragua did not seek primary health care (Keifer, 1997). In Central America, efforts at designing and implementing surveillance systems became part of the activities in the PLAGSALUD project (Keifer et al, 1997; Murray et al, 2002). PLAGSALUD and the ministries of health created a list of 12 pesticides that caused acute pesticide poisonings: aldicarb, aluminium phosphide, carbofuran, chlorpyriphos, endosulfan, etoprophos, methamidophos, methomyl, methylparathion, monocrotophos, paraquat and terbufos. The PLAGSALUD project then conducted a region-wide under-reporting study in 2000, which found that the majority (76 per cent) of acute pesticide poisonings were work-related, followed by accidental poisonings and by suicides. It also indicated that rates of underreporting of pesticide poisonings were at least 98 per cent (Murray et al, 2002).

LIMITATIONS IN TOXICOLOGY AND EPIDEMIOLOGY

Traditional risk assessment tends to deal with healthy males and one-time exposures to a single pesticide (Carpenter et al, 2000). The reality in developing countries, however, is that people are exposed to many products that may interact, and many of those exposed are not healthy, young males. Children and foetuses are recognized as the groups most sensitive to pesticides (Fleming and Herzstein, 1997), and the adverse health effects of pesticides can be further aggravated by poor nutrition, dehydration and infectious diseases (WHO, 1990; Repetto and Baliga, 1997a). Also, some pesticides are endocrine disrupters with the potential to cause adverse effects in both wildlife and humans, even at small doses (see Chapter 1).

Pesticides are often applied in combinations or mixtures, and there is some evidence that exposure to mixtures are associated with higher rates of case fatality and morbidity (Jeyaratnam, 1982; Kishi et al, 1995; Cole et al, 2000). The interactions of pesticides can be inhibitory, additive or synergistic. Recent in vitro and animal studies demonstrate synergistic effects among some pesticides and their possible roles in the aetiology of certain diseases (Thiruchelvam et al, 2000; Payne et al, 2001).

Much risk assessment also assumes that climatic conditions under which pesticides will be used are similar to those where most assessments were developed. Furthermore, some researchers suggest that a greater risk of adverse health effects exists in poorer communities (Tinoco-Ojanguren and Halperin, 1998), while others suggest that the risks for pesticide poisoning for women are underestimated (Kimani and Mwanthi, 1995; London and Bailie, 2001; London et al, 2002).

It is the high-risk groups in every country who disproportionately bear the burden of pesticide exposure (WHO, 1990; London and Rother, 2000). Even in industrialized countries, migrant or seasonal farm workers, who are often minorities and from low-income groups, are the most highly exposed populations. This higher-risk group is rarely included in research on acute and chronic health effects related to pesticides (Moses et al, 1993; Zahm and Blair, 1993).

Exposure assessment is important for epidemiology, and measurement of plasma or red blood cell cholinesterase levels is the most common and potentially least expensive biomarker for detecting exposure to OP and carbamate exposure (Wilson et al, 1997). This method is used widely in developing countries. Such measurement of cholinesterase is meant to be a tool to prevent acute organophosphate intoxication. It has been reported that symptoms usually do not appear until there is a 50 per cent decrease in cholinesterase activity compared with preexposure baseline levels (Tafuri and Roberts, 1987). Therefore, by monitoring their cholinesterase levels, workers can be removed from exposure before symptoms develop.

However, the normal range of cholinesterase levels is quite broad, and individuals with a high normal value could lose half of their cholinesterase activity while remaining within the normal range. Due to this high interindividual variability in baseline cholinesterase activity, some have questioned whether the levels of cholinesterase enzymes accurately measure the exposure or effects relating to OPs and carbamates (Fleming and Herzstein, 1997; Wesseling et al, 1997a).

In developing countries, the scarcity of resources and appropriate tools to carry out epidemiological studies thus presents a major challenge. This can limit study design, time period and sample size, all of which affect results. For example, if the number of study participants is small, the study could fail to detect an association even when there is one. In other words, when an epidemiological study on the effects of pesticides shows no effects, this does not necessarily mean that the pesticides in question do not cause adverse effects.

Furthermore, even when there is an increased number of studies that show associations between exposure to pesticides and outcomes, not all studies will demonstrate the associations. In this situation, different scientists draw different conclusions, even when they are looking at the same set of study results. Some may find a significant association, whereas others may declare there is limited evidence or no clear evidence (Osburn, 2000). Some may even go on to conclude that absence of evidence means an absence of adverse effects. But as Watterson described (1988), this is not the case:

> Sometimes medical and civil service staff ... argue that the absence of evidence of pesticide poisoning is evidence of the absence of pesticide poisoning... This argument is simply illogical. If people do not know how to identify cases of pesticide poisoning; if symptoms of such poisoning can be easily confused with influenza, cold and other common ailments; if we do not know how to

measure non-acute exposures to various pesticides; if we do not know what effect long-term low-level exposure to pesticides can have on people, we are simply not in a position to state that "there is evidence of absence of pesticide poisoning." We have an absence of evidence about pesticide poisoning; we do not have evidence of absence about sub-chronic, chronic, neurological, genetic and reproductive effects of agro-chemicals.

WHAT CAN BE DONE?

In order to avoid the adverse effects of pesticide use, other ways of reducing exposure need to be explored more widely, including the reduction of pesticide use without accompanying crop losses, and more effective regulation of the distribution and use of hazardous products. A key issue centres on whether pesticides are really needed to grow all crops. There is growing evidence that pesticide use can be reduced while maintaining stable or even increased agricultural production through implementation of integrated pest management (IPM) programmes, and there are many organizations that have been successfully promoting IPM throughout the world (Pretty, 2002). Since the concept of IPM was developed in the 1950s, its theory and practice have evolved and, as a result, the term IPM has come to have different meanings. Antle and Wagenet (1995) warn that, *'despite the public perception that integrated pest management techniques reduce or eliminate pesticide use, many IPM techniques are based on economic thresholds for pesticide application that do not explicitly consider either environmental or human health impacts.'* Thus IPM based on economic thresholds can establish pesticides as predominant elements and is indistinguishable from 'intelligent pesticide management'.

In contrast, there are other types of IPM programmes that put farmers at the centre. Over the past two decades in Asia, UN Food and Agriculture Organization (FAO) has taken a leading role in developing and supporting farmer-centred, ecologically-based IPM programmes, in collaboration with various governmental and non-governmental organizations. As a result, over two million farmers in 12 Asian countries have completed IPM training at season-long farmer field schools. Through participatory IPM training, farmers learn and develop the skills to critically analyse and manage their agroecosystem in order to grow healthy crops (Dilts, 1998; Matteson, 2000; Settle et al, 1996; Useem et al, 1992). These IPM farmer field schools are now being pilot-tested in Latin America and Africa under the coordination of the FAO-based Global IPM Facility.

The methods and skills learned can be applied not only to solving agricultural problems, but also to other types of problems. For example, in Cambodia a pilot programme named 'Farmer Life Schools' organized through the network of the farmers who had completed IPM farmer field schools, was conducted in 2000 to help communities address other critical social issues, including HIV/AIDS (Sokuthea, 2002). Another important spin-off has led to programmes for community-based pesticide surveillance conducted by farmers who have completed the IPM Farmer Field Schools (Murphy, 2001; Murphy et al, 2002).

The farmer self-surveillance studies have been piloted in several south-east Asian countries, including Cambodia, Thailand and Vietnam. They take a different approach from conventional public health research. Rather than being the subjects of research conducted by outside experts, agricultural workers are placed at the centre. Workers who are exposed to risks examine their own working environment and practices and make their own decisions about actions that can be taken to reduce risks. Thus, data on the incidence of pesticide poisoning is analysed and used at a local level. In most public health research, on the other hand, outside experts come to the agricultural communities, collect data from the field, analyse it, draw conclusions and make recommendations which may or may not be practical or acceptable to agricultural workers. Giving feedback to the people who are experiencing occupational risks and assisting them to prevent pesticide-related illness is important and ethical, but it does not always happen.

While the importance of cross-disciplinary health and agricultural approaches has been recognized, the majority of public health researchers still do not look beyond the health sector. It is important to note, however, that a small but growing number of public health researchers have begun emphasizing the importance of pesticide use reduction through implementation of pest control methods and recommend it as a solution to the wider problems of pesticide poisoning (WHO, 1990; McConnell and Hruska, 1993; London and Myers, 1995; Kishi et al, 1995; Clarke et al, 1997; Ohayo-Mitoko et al, 1997; Richter and Safi, 1997; Wesseling et al, 1997a; van der Hoek et al, 1998; Lowry and Frank, 1999; McConnell et al, 1999b; Murphy et al, 1999; Ohayo-Mitoko et al, 1999; Cole et al, 2000; Kishi and LaDou, 2001; Ngowi et al, 2001; Wesseling et al, 2001).

There have been two studies to evaluate the value of IPM training in reducing pesticide use and its impacts on farmers' health in Nicaragua (Corriols and Hruska, 1992; Hruska and Corriols, 2002) and in Indonesia (Kishi, 2002). The Nicaraguan study found that IPM training is effective in reducing pesticide use and is associated with a lower incidence of acute pesticide poisoning and less cholinesterase inhibition, compared with the farmers without IPM training. In Indonesia, knowledge of the health risks of pesticide use was not sufficient to change farmers' behaviour, as their main concern is crop damage that can lead to economic ruin. However, IPM Farmer Field School training did offer farmers a viable alternative, by demonstrating the economic, agricultural, health and environmental advantages of eliminating unnecessary pesticide use. A strong message to public health professionals is that for the interventions to be effective, they must use appropriate methods, meet the community's priorities and values, and offer feasible alternatives.

CONCLUDING COMMENTS

This chapter confirms that pesticides do harm human health, although their effects are not widely recognized and their true extent remains unknown. This is true in both industrialized and developing countries, and for both their acute and long-term effects. However, the extent of the problem is far greater in developing countries. In industrialized countries, the focus has shifted from occupational exposure to the effects of long-term, low-level exposures to the general population. While the problems of acute effects in developing countries have been recognized to a certain extent, the perception promoted by the pesticide industry is that the number of acute pesticide poisonings due to suicides is greater than occupational poisonings. However, this is not supported by the data. For a variety of reasons, including underutilization of health facilities by agricultural workers, inability of health personnel to diagnose pesticide poisoning, and lack of understanding of the importance of reporting, underreporting of occupational poisoning is very common. This in turn misleads policy-makers. Furthermore, even when no data exist on the adverse effects of pesticides, it cannot necessarily be assumed that there are no problems. Several recent studies showed that the incidence of pesticide poisoning is much higher than previously thought.

What are the best solutions to these pesticide problems? The conventional approach is to reduce exposure by promoting protective equipment and health education to agricultural workers. This approach has been actively promoted by industry, often working together with government officials, in 'safe use' projects. However, these approaches have not eliminated pesticide poisoning. Not only has the availability of protective equipment been questioned, along with its impracticality in hot, humid tropical climates, but an increasing number of studies show that knowledge of the dangers of pesticides is not sufficient condition to change agricultural workers' behaviour.

The solution is regulatory control to eliminate the use of pesticides that fall in WHO Hazard Class I and restricted access to the more problematic products in Class II. This should be combined with efforts to reduce reliance on pesticides through ecologically based, farmer-based IPM and other forms of sustainable farming, including organic farming. Despite the widely held notion that pesticides are needed for high yields, experience has shown that stable or even increased yields can be achieved through implementation of IPM and organic farming. Further research is required to understand the factors that determine successful implementation and sustainability of IPM Farmer Field Schools, together with better analysis of the health and environmental costs of pesticide use.

Chapter 3

Paying the Price: The Full Cost of Pesticides

Jules Pretty and Hermann Waibel

INTRODUCTION

Since modern pesticides came to be successfully promoted for use in agriculture from the mid 20th century, the economics of their benefits and costs has been under scrutiny. Without them, some argue, there would have been widespread famine. With them, others contend, there has been unacceptable harm to environments and human health. The debates remain polarized today. In this chapter, we explore the considerable challenges surrounding the measurement of the costs of the side-effects of pesticides, detail the costs in selected countries, and analyse more than 60 integrated pest management (IPM) projects in 26 countries to illustrate the benefits that can arise from the adoption of a variety of integrated pest management and organic systems.

Though the use of pesticides is known to have many short- and long-term consequences, there are surprisingly few data on the environmental and health costs imposed on other sectors and interests. Agriculture can negatively affect the environment through overuse of natural resources as inputs or through their use as a sink for pollution. Such effects are called negative externalities because they are usually non-market effects and therefore their costs are not part of market prices. Negative externalities are one of the classic causes of market failure, whereby the polluter does not pay the full costs of his/her action, and therefore these costs are known as external costs (Baumol and Oates, 1988; Pretty et al, 2000, 2003a).

Externalities in the agricultural sector have at least four features: (i) their costs are often neglected; (ii) they often occur with a time lag; (iii) they often damage groups whose interests are not well represented in political or decision-making processes; and (iv) the identity of the source of the externality is not always known. For example, farmers generally have few incentives to prevent pesticides escaping to water-bodies, the atmosphere and to nearby nature as they transfer the full cost of cleaning up the environmental consequences to society at large. In the same way, pesticide manufacturers do not pay the full cost of all their products, as they do not suffer from any adverse side-effects that may occur.

Partly as a result of lack of information, there is little agreement on the economic costs of externalities in agriculture. Another reason for externalities is methodological. Some authors suggest that the current system of economic calculations grossly underestimates the current and future value of natural capital (Abramovitz, 1997; Costanza et al, 1997, 1999; Daily, 1997; Pretty et al, 2000, 2001). Such valuation of ecosystem services remains controversial because of methodological and measurement problems (Georghiou et al, 1998; Hanley et al, 1998; Farrow et al, 2000; Carson, 2000; Pretty et al, 2003a) and because of its role in influencing public opinions and policy decisions.

In this chapter, we explore both the full costs of current pesticide use and some of the benefits arising from the use of alternative practices. We also review results of cost–benefit studies on pesticide use and the debate that has often followed from such studies.

ECONOMIC STUDIES ON PESTICIDE BENEFITS

Most economic studies that have tried to assess the benefits of pesticides are based on a comparison of two extreme scenarios, usually current use compared with drastic or complete reductions in use. They then ask how much such reductions would cost farmers and the agricultural industry (Knutson et al, 1990; Frandsen and Jacobsen, 2001; Kaergård, 2001; Schmitz, 2001).

Not surprisingly, these studies conclude that substantial losses would arise from yield reductions, the need for additional food imports, and greater consumer risks through exposure to increased mycotoxins on foods. For example, costs arising from a complete ban were put at US$18 billion in the US (Knutson et al, 1990), and from a 75 per cent ban at 1 per cent of GDP in Germany (Schmitz, 2001). Schmitz (2001) stated: '*without these chemicals, even more people in the world would go hungry and the food security of several poor countries could no longer be guaranteed*'. Frandsen and Jacobsen (2001) also concluded from a study of Danish agriculture that '*the economic analyses clearly illustrate that pesticides are a crucial input factor in the crop sectors and a complete ban on the use of pesticides would probably imply drastic changes in agricultural production.*' Although inappropriate for making any policy recommendations, such studies implicitly promote the belief that pesticide reductions are detrimental to the economy.

A major problem with macro studies on pesticide use reduction scenarios is the lack of sector level data which can adequately describe the complex relationships between the pest ecosystem on the one hand and the economic system on the other. For example, time series data on pesticide use by cropping system are often unavailable or inaccessible to public organizations. Also, little information on the geographic distribution of pest, pathogenic and beneficial organisms exists, nor are there sufficient data on crop loss representing the situation in farmers' fields (Waibel et al, 1999a, 1999b). Consequently, changes in yield as a result of pesticide reduction become exogenous rather than predicted values in the economic models used (Knutson et al, 1990).

Several critical methodological objections have been raised about studies of this kind (Pearce and Tinch, 1998). First, the benefit estimates rely on 'expert' opinion about the loss of crop yield that would occur if pesticide use were entirely banned. Thus, the choice of experts is crucial to the findings. The now notorious Knutson et al (1990) study asked 140 crop science experts to estimate crop losses in the US should no pesticides be permitted, yet did not consult organic experts nor those who believed pesticide use reduction was viable for at least some farmers (Ayer and Conklin, 1990; Buttel, 1993). The same shortfall appears in the study by Schmitz (2001) for Germany.

Another problem is that these studies tend to be based on unrealistic assumptions about immediate and enforced reductions in pesticide use. Frandsen and Jacobsen (2001) use a complete ban scenario to illustrate how marked the effects on Danish agriculture would be. Yet few believe a total ban of all pesticides at any one time would be sensible nor feasible. Transitions are much more likely to occur more steadily, with new technologies gradually replacing old more harmful and costly ones. Therefore, Pearce and Tinch (1998) rightly conclude that benefit–cost ratios derived from 100 per cent chemical reduction scenarios are of no use for drawing policy conclusions.

In addition, pesticide benefit studies also ignore the empirical evidence that shows both IPM and organic programmes in the field can produce transitions to low or even zero pesticide use without crop yield losses. The usefulness of economic models to estimate the effects of pesticide reduction scenarios is highly dependent on realistic estimates of the substitution elasticities for various technologies. Since such data can rarely be taken from past observed decisions and the behaviour of the economic agents, they heavily depend on independent expert knowledge of the alternatives to pesticides. Often such estimates are '*based on agronomic expert knowledge*' (Knutson et al, 1990), the conclusions depend crucially on the extent to which experts think alternative technologies are effective and efficient. Agroecological approaches are now known to offer significant opportunities to reduce aggregate pesticide use (Pretty, 1995, 2002; Uphoff, 2002; McNeely and Scherr, 2003), but if such evidence is ignored, then a circular argument is constructed in which the belief that agricultural yields can only be increased with pesticides is strengthened, together with a further belief that their removal inevitably results in yield reductions.

Since the economic models are generally unsuitable to factor in the negative externalities for environment and human health, the benefit–cost ratio of pesticides is always overestimated. Taking these costs into account, the benefit–cost ratios of even the extreme scenarios fall from high estimates of 40 (Pimentel et al, 1992, 1993) to –2.1 in the US (Steiner et al, 1995) and to only 1.5 in Germany (Waibel et al, 1999a, 1999b). Hence, the claims that external costs are irrelevant in the light of excessively high benefits are questionable.

Bromley (1994) gave such benefit analyses the moniker of the 'language of loss' to illustrate how an important policy and practice debate is framed in a particular way to sustain only certain interests. Such language matters enormously. Waibel et al's study of externalities in Germany (Waibel et al, 1999a, 1999b; Waibel and Fleischer, 2001) was originally commissioned by the country's then

Ministry of Agriculture, but when the results were presented in 1997, the Ministry refused permission for publication of the research. It grudgingly changed its mind in 1998 only after pressure from the media. Waibel and Fleischer (2001) summarize the responses of various stakeholder groups to their results, showing how some key groups sought to undermine the credibility of the independent research (Table 3.1).

Table 3.1 *Reactions of various stakeholder groups in Germany to the publication of findings on the externalities of domestic agriculture*

Stakeholder groups	Reactions from each stakeholder group
Chemical industry	Benefits of pesticides are underestimated
Federal Farmers' Association	Study is 'harmful' to farmers
Regulatory Agencies	External costs are overestimated
Ministry of Agriculture	Study is unscientific; later amended to 'scientifically new'
Plant protection experts	Insufficient data to conduct such a study
Green NGOs	External costs tend to be underestimated

Source: Waibel and Fleischer (2001)

In conclusion, the assumed, perceived and claimed benefits of pesticides as suggested by some of the economic studies may have well contributed to the lack of efforts to undertake studies to assess quantitiative evidence of their external costs. Policy-makers, especially in developing countries, often consider these costs a minor issue relative to the huge benefits they attribute to pesticides. Keeping the debate in a polarized mode is a strategy to resist change. It is promoted by those groups whose interests would suffer from serious efforts to reduce pesticides toward their social optimum.

VALUING PESTICIDE EXTERNALITIES

In a perfect world, and given the right incentive structure, technologies and practices should be in use that result in minimal external costs. A variety of factors prevent this from becoming a reality, including lack of incentives to innovate, technological lock-in, adverse incentives from companies, and misguided advice from pesticide producers and retailers (Wilson and Tisdell, 2001). Hence, although difficult to quantify, pesticide costs tend to include a substantial proportion of costs that must be borne by society. To promote policy change towards the socially optimal level of pesticide use, it is necessary to make these costs explicit.

Despite the fact that it has been common knowledge for more than half a century that many pesticides cause harm to the environment and to human health, it is remarkable that there is an almost complete absence of a full costing of a single product, let alone for the current level of pesticide use on a worldwide basis. The problem is that we do not know the marginal benefits and costs but, at best, only the total values for the current situation. Hence, the only conclusion we could therefore draw from such analysis is that, if costs exceed benefits, we should stop using pesticides. However, this is a rather theoretical scenario. What is necessary instead is to assess the effects of changes in pesticide use, that is consider to what extent pesticide use can be reduced so that the cost of that change does not exceed the benefits.

Why is it not possible to put this theoretical concept into practice? The first problem is that there are so many pesticides and risks vary enormously from product to product, and generalizations cannot be made about their fate in the environment, their persistence, and their acute and chronic toxicity (EPA, 2001; OECD, 2001). A comprehensive scientific approach requires costs and risks to be assessed for each product in each of the agricultural systems in which they might be used. Putting a cost on adverse effects would also have to account for the values of resources in each of these contexts, so that the externalities could be internalized. We would also need to know the technological alternatives for pest management, in particular the implications of non-pesticide control strategies for natural resources and the environment.

In spite of these methodological difficulties and shortage of data, there have been attempts to conduct aggregate estimates of pesticide externalities at country level. Here we summarize recent studies of agriculture in China, Germany, the UK and the USA (Pimentel et al, 1992; Steiner et al, 1995; Ribaudo et al, 1999; Waibel et al, 1999a, 1999b; Pretty et al, 2000, 2001; Norse et al, 2001; Waibel and Fleischer, 2001; EA, 2002; Tegtmeier and Duffy, 2004). Following the framework used in these studies, we assign the environmental and health costs of pesticide use to five categories:

1   drinking water treatment costs (including increased costs of water substitution and monitoring costs);
2   pollution incidents in water courses, fish deaths, and revenue losses in aquaculture and fishing industries;
3   health costs to humans (farmers and farm workers; rural residents; food consumers);
4   adverse effects on on- and off-farm biodiversity (pest resistance, loss of beneficial insects, fish, wildlife, pollinators, domestic pets);
5   adverse effects on climate from energy use during the manufacture of pesticides.

1 Some pesticides (together with nutrients, soil, farm wastes and micro-organisms) escape from farms to pollute ground and surface water. Costs are incurred by water delivery companies (and passed on to their customers) when they treat water to comply with drinking water standards set out in legislation (standards in the EU for pesticides are 0.1 µg litre^{-1} for a single product and 0.5 µg l^{-1} for all pesticides). These costs would be much greater if the policy goal were complete removal of all residues. Consumers also incur avoidance costs when they pay to switch to bottled water.

2 A variety of agricultural wastes can disrupt surface water systems: cattle and pig slurry, silage effluent and dairy wastes cause eutrophication, and toxic pesticides (particularly sheep dips) can kill aquatic life. In the late 1990s, UK agriculture accounted for an average 2600 incidents per year, of which about 50 caused extensive fish kills (EA, 2002). Pesticides in water courses can result in a reduction in the economic value of a fishery (Mason, 2002), although the greater problems arise from eutrophication and fish deaths caused by algal blooms (Ribaudo et al, 1999; Pretty et al, 2003a). Thus the livelihoods of those involved in commercial fishing can be adversely affected by agricultural pollution. Monitoring costs are further incurred by governments and the private sector for assessing pesticide residues in food and the environment, and for the administration of schemes and grants to reduce pollution and for advisory services. It could, however, be argued that any form of agriculture would still incur such monitoring costs.

3 Pesticides can affect workers engaged in their manufacture, transport and disposal, operators who apply them in the field and their families, and the wider public. Estimates for the external health costs of pesticides are almost certainly considerable underestimates, owing to differing risks per product, poor understanding of chronic effects (see Chapter 2), weak monitoring systems and misdiagnoses by doctors (Repetto and Baliga, 1996; Pearce and Tinch, 1998). It is very difficult to say exactly how many people are affected by pesticide poisoning each year. According to voluntary reporting in the UK, some 100–200 incidents occur each year, of which very few are fully substantiated. However, recent government research indicates significant under-reporting (HSE (Health and Safety Executive), 1998a, 1998b). One survey of 2000 pesticide users found that 5 per cent reported at least one symptom in the previous year about which they had consulted a doctor. A further 10 per cent had been affected (mostly by headaches), but had not consulted a doctor. Fatalities from pesticides at work in Europe and North America are rare – one a decade in the UK, and eight a decade in the US. In California, where there is the most comprehensive system of reporting in the world, official records show that annually 1200–2000 farmers, farm workers and the general public were poisoned as a result of pesticide application during the 1980s and 1990s (CDFA, passim; Pretty, 1998).

4 Modern farming is known to have had a severe impact on wildlife: in the UK, 170 native species became extinct during the 20th century, including 7 per cent of dragonflies, 5 per cent of butterflies and 2 per cent of fish and mammals. In addition, 95 per cent of wildflower-rich meadows were lost, 30–50 per cent of ancient woods, 50 per cent of heathland, 50 per cent of lowland fens and wetlands and 40 per cent of hedgerows (Pretty, 2002). Species diversity is also declining in the farmed habitat itself. Draining and fertilizers have replaced floristically-rich meadows with grass monocultures, overgrazing of uplands has reduced species diversity, and herbicides have cut diversity in arable fields. Farmland birds have particularly suffered: the populations of nine species fell by more than a half between 1970 and 2000. Some of these problems are caused by the wider struc-

tural changes to agriculture; others are caused by the use of pesticides – particularly where herbicides remove arable weeds that are important sources of food for insects and birds (Firbank et al, 2003).

5 As an economic sector, agriculture also emits carbon to the atmosphere by the direct and indirect consumption of fossil fuel (Leach, 1976; OECD/IEA, 1992; Pretty, 1995; Cormack and Metcalfe, 2000; Pretty et al, 2003b). With the increased use of nitrogen fertilizers and pesticides, pumped irrigation and mechanical power in industrialized agricultural systems, agriculture has become progressively less energy efficient over time. These sources account for 90 per cent or more of the total direct and indirect energy inputs to most farming systems. Thus low-input or organic rice systems in Bangladesh, China and Latin America are 15–25 times more energy efficient than irrigated rice produced in the US. For each tonne of cereal or vegetable from high-input systems, 3–10 GJ of energy are consumed in its production. But for each tonne of cereal or vegetable from low-input systems, only 0.5–1.0 GJ are consumed (Pretty, 1995).

We use standard data on the energy used for pesticides to calculate the avoided carbon emissions by reducing or changing practices: 238 MJ of direct and indirect energy are used to manufacture 1 kg of herbicide, 92 MJ kg^{-1} for fungicides, and 199 MJ kg^{-1} for insecticides. The amount of carbon emitted per unit of energy depends on the use made of non-renewable and renewable resources in the domestic energy sector in question. These vary from 24 kg C GJ^{-1} for coal, 19 kg C GJ^{-1} for oil, and 14 kg C GJ^{-1} for natural gas (DTI, 2001). Thus herbicides indirectly emit 4.64 kg C per kg active ingredient, fungicides 1.80 kg C kg ai^{-1}, and insecticides 3.74 kg C kg ai^{-1} (Pretty et al, 2003b). Using OECD (1999) data on the relative use of these different classes of products (22 per cent of all pesticides used are insecticides, 33 per cent are fungicides, 45 per cent are herbicides), this gives a weighted average of 3.31 kg C emitted kg ai^{-1}. The external costs of a tonne of carbon emitted to the atmosphere has been calculated to be US$44.70 or £29.80 (Pearce et al, 1996; Eyre et al, 1997; Holland et al, 1999), and so the use of 1 kg of pesticide active ingredient imposes external climate change costs of $0.148.

It is clear that the categories mentioned are not all-inclusive: For example, insect outbreaks arising from pesticide overuse can affect other farmers, including those using low levels of pesticides or none at all. Therefore, the summary of published empirical evidence as shown in Table 3.2 contains a diverse picture of cost estimates for pesticide externalities in China, Germany, the UK and the US. In the late 1990s in China, pesticide use was 1000 Mkg ai (of which some 30 per cent was used on rice), 27 Mkg in Germany, 23 Mkg in the UK, and 425 Mkg in the US. This sample of four countries comprises more than half of all the 2.56 billion kg of pesticides used globally each year.

The data for China are derived from a detailed analysis of rice production in two provinces, Hunan and Hubei, by Norse et al (2001), who calculated that China's aggregate rice production of 198 M tonnes imposed 13–49 billion Yuan (US$1.6–5.9 billion) of environmental and health costs. This is equivalent to

66–247 Yuan (US$8.1–30.4) per tonne of rice or about 8–20 per cent of the variable costs of rice production. Pesticides account for over a third of these external costs while the cost of pesticides is well below 10 per cent of the material costs of rice production. Overall, based on a number of studies, it is known today that, especially in developing countries, pesticides cause significant human health risks (Crissman et al, 1998). Valuing health risks, for example in the Philippines, shows that in rice human health and pesticides costs are at a ratio of 1:1 (Rola and Pingali, 1993).

Table 3.2 *Cost category framework for assessing full costs of pesticide use (million US$ per year)*

Damage costs	China[1]	Germany	UK	US
1 Drinking water treatment costs	nd[2]	88	182	897
2 Health costs to humans (farmers, farm workers, rural residents, food consumers)	900[3]	14[4]	2[4]	132
3 Pollution incidents in water courses, fish deaths, monitoring costs and revenue losses in aquaculture and fishing industries	nd	51	6	128
4 Adverse effects on on- and off-farm biodiversity (fish, beneficial insects, wildlife, bees, domestic pets)	350[5]	9	64	280
5 Adverse effects on climate from energy costs of manufacture of pesticides	148	4	3	55
TOTALS	**1398**	**166**	**257**	**1492**

1 China costs are just for rice cultivation
2 nd = no data
3 Range of health costs in China given in original study as US$500–1300 million
4 Does not include any costs of chronic health problems
5 Range for adverse effects on biodiversity given as US$200–500 million

Sources: adapted from Pimentel et al, 1992; Steiner et al, 1995; Ribaudo et al, 1999; Waibel et al, 1999a, 1999b; Pretty et al, 2000, 2001; EA, 2002; Norse et al, 2001; Waibel and Fleischer, 2001; Tegtmeier and Duffy, 2004

The study by Pimentel et al (1992) was the first assessment of pesticide externalities in the US. It was followed by two wider analyses of agricultural externalities (Steiner et al, 1995; Ribaudo et al, 1999), and recently updated by a comprehensive assessment by Tegtmeier and Duffy (2004). We have adjusted some data to remove private costs borne by farmers together with some obvious over-overestimates (see Pretty et al, 2001). The data for Germany are based on a study (Waibel et al, 1999a, 1999b; Waibel and Fleischer, 2001) that put the annual pesticide externalities at 252 M DM for the late 1990s, not including the former East Germany. The UK data are derived from a study of the total externalities from all UK agriculture (Pretty et al, 2000, 2001), later adjusted downwards to £1.54 billion for the year 2000 (EA, 2002; Pretty et al, 2003a). All data are converted to US$ for this chapter.

In both the German and UK studies, deliberately conservative estimates of costs were used, particularly for chronic pesticide poisoning of humans, for which there is no data, even though some effects are known to exist. If, for example, the assumptions made in the US studies about the cancer risk of pesticides were utilized in this calculation, the external costs in Germany would almost double. Also, most of the calculated costs are actual damage costs and do not include returning the environment or human health to pristine conditions. In addition, costs may be underestimates of people's willingness to pay to avoid these externalities. Usually pesticide issues rank high in public fears about environmental and health hazards. Also, the costs of pesticide resistance are excluded from the four country cases. For Germany, for example, Fleischer (2000) calculated the present value of atrazine resistance at DM 5.51 per ha using a discount rate of 1 per cent.

As shown in Table 3.3, in Chinese rice production the external costs from pesticides exceeds their market value (ratio 1.86). This may be a good reflection of the situation in developing countries. In the three industrialized country studies, the ratio is less than unity (Germany = 0.36; UK = 0.31; US = 0.19). The costs per unit of arable and permanent cropland vary from US$8.80 ha^{-1} to $47.2 ha^{-1}. However, even with the conservative estimates from the case studies, it is clear that the external costs per unit of cropland can no longer be ignored. However, these aggregate data hide large variations among crops, with conventional cotton and some vegetables likely to be imposing the greatest costs.

Table 3.3 *Pesticide externalities in selected countries*

Country	Value of pesticides (M US$ yr^{-1})	Externalities (M US$ $^{-1}$)	Crop area (M ha)	Externalities per hectare (US$ ha^{-1} yr^{-1})
China	750	1398	30 (rice only)	47.2
Germany	450	166	12	13.8
UK	825	257	11	23.4
US	8000	1492	169	8.8

1 Arable and permanent crop data from FAO (2003). The total arable and crop area in China is 135.6 M ha. The figure for the US does not count the 15.3 M ha idled each year.
2 Pesticide use in China (rice only) is 300 Mkg yr^{-1}, in Germany 27 Mkg yr^{-1}, in the UK 23 Mkg yr^{-1}, and in the US 425 Mkg yr^{-1}.

These data could be used to calculate the external costs per kg of pesticide active ingredient applied. However, as shown in Chapters 1 and 2, there are many hundred different formulations of pesticides, with widely varying risks. The external costs of Class Ia and Ib products are likely to be much greater than those in Class III. At present, these differences are unknown. However, if we did assume parity in the adverse effects of pesticides, then the external costs would be US\$4.66 kg^{-1} in China, \$6.14 kg^{-1} in Germany, \$11.17 kg^{-1} in the UK, and \$3.51 kg^{-1} in the US, with an average of \$4.28 kg^{-1} active ingredient.

With some 2.5 billion kg applied worldwide, this would suggest annual costs in the order of US\$10.7 billion, if the conditions in these four countries are representative of those in other developing and industrialized countries. The total value of the world market for pesticides was some US\$25 billion in 2002. In OECD countries, where use of pesticides amounts to some 955,000 kg per year, the annual external costs amount to US\$3.84 billion (using the data from Germany, the UK and the US). These rough data show that external costs are a significant part of the total costs of pesticides. Hence, programmes that can reduce pesticide use are likely to generate public benefits. One such possibility is the introduction of integrated pest management (IPM). In the next section we therefore explore some of the available evidence on the impact of such programmes on a global scale.

THE COSTS AND BENEFITS OF IPM

Recent IPM programmes, particularly in developing countries, are beginning to show how pesticide use can be reduced and pest management practices can be modified without yield penalties. In principle, there are four possible trajectories of impact if IPM is introduced (Figure 3.1):

1 both pesticide use and yields increase (A);
2 pesticide use increases but yields decline (B);
3 both pesticide use and yields decline (C);
4 pesticide use declines, but yields increase (D).

The assumption of conventional agriculture is that pesticide use and yields are positively correlated. For IPM, the trajectory moving into sector A is therefore unlikely but not impossible, for example in low-input systems. What is expected is a move into sector C. While a change into sector B would be against economic rationale, farmers are unlikely to adopt IPM if their profits would be lowered. A shift into sector D would indicate that current pesticide use has negative yield effects or that the amount saved from pesticides is reallocated to other yield increasing inputs. This could be possible with excessive use of herbicides or when pesticides cause outbreaks of secondary pests, such as observed with the brown plant hopper in rice (Kenmore et al, 1984).

In Figure 3.2, we present data from 62 IPM initiatives in 26 developing and industrialized countries (Australia, Bangladesh, China, Cuba, Ecuador, Egypt, Germany, Honduras, India, Indonesia, Japan, Kenya, Laos, Nepal, the Netherlands, Pakistan, the Philippines, Senegal, Sri Lanka, Switzerland, Tanzania, Thailand, the UK, the US, Vietnam and Zimbabwe). We used an existing dataset that audits progress being made on yields and input use with agricultural sustainability approaches (for research methodology, see Pretty and Hine, 2001;

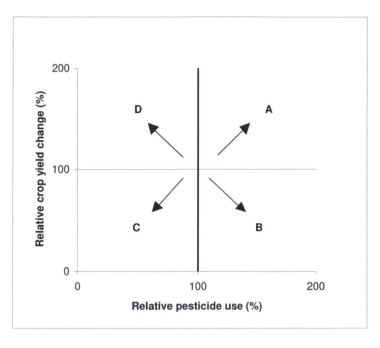

Figure 3.1 *Possible relationships between pesticide use and crop yields*

Pretty et al, 2003a). The research audited progress in developing countries, and assessed the extent to which farmers were increasing food production by using low-cost and locally available technologies and inputs.

The 62 IPM initiatives to cover some 25.3 M ha, that is less than 1 per cent of the world crop area, and directly involve some 5.4 million farm households. The evidence on pesticide use is derived from data on both the number of sprays per hectare and the amount of active ingredient used per hectare. In this analysis, we do not include recent evidence on the effect of genetically modified crops, some of which have resulted in reductions in the use of pesticides, such as of herbicides in the UK (Champion et al, 2003) and China (Nuffield Council on Bioethics, 2004), and some of which have led to increases, such as in the US (Benbrook, 2003).

There is only one sector B case reported in recent literature (Feder et al, 2004). Such a case has recently been reported from Java for farmers who received training under the popular FAO Farmer Field School model. However, the paper does not offer any plausible explanation for this result but does point out that there were administrative problems in implementing the project that was funded by the World Bank. The cases in sector C, where yields fall slightly while pesticide use falls dramatically, are mainly cereal farming systems in Europe, where yields typically fall up to some 80 per cent of current levels when pesticide use is reduced to 10–90 per cent of current levels (Pretty, 1998; Röling and Wagemakers, 1998).

Sector A contains ten projects where total pesticide use has indeed increased in the course of IPM introduction. These are mainly in zero-tillage and conserva-

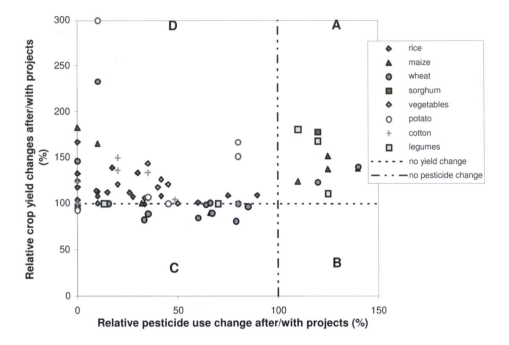

Figure 3.2 *Association between pesticide use and crop yields (data from 81 studies of crops, 62 projects, 26 countries)*

tion agriculture systems, where reduced tillage creates substantial benefits for soil health and reduced off-site pollution and flooding costs. These systems usually require increased use of herbicides for weed control (de Freitas, 2000), although there are some examples of organic zero-tillage systems in Latin America (Peterson et al, 2000). Over 60 per cent of the projects fall into category D where pesticide use declines and yields increase. While pesticide reduction is to be expected, as farmers substitute pesticides by information, yield increases induced by IPM is a more complex issue. It is likely, for example, that farmers who receive good quality field training will not only improve their pest management skills but also become more efficient in other agronomic practices such as water, soil and nutrient management. They can also invest some of the cash saved from pesticides in other inputs such as higher quality seeds and inorganic fertilizers.

When drawing conclusions from this database, some factors call for a cautious interpretation while others suggest that the benefits from IPM could be even greater than those suggested on yield and pesticide changes alone. On the negative side, one problem is that in most cases no statistical analysis was possible to test whether the observed differences in yield and pesticide use were significant. On the other hand, reduced pesticide use equates into a reduction of external costs whose magnitude depends on the baseline situation.

The baseline against which change is measured has been current practices, which as we have shown are known to cause some harm to environments and human health. A change that specifically reduces the use of those pesticides causing harm (e.g. WHO Class I and II products) can create large benefits for farmers and other groups in society. In the light of these results, the question must now be: why, if IPM shows clear benefits to farmers, are reduced pesticide use approaches and technologies not more widely in use? Or, as Wilson and Tisdell (2001) have put it: why are farmers still using so many pesticides? While, as cited above, these authors have provided some possible explanation, other questions remain in the light of successful country cases where pesticide reductions have actually worked. In the next section we therefore summarize these cases.

CURRENT EVIDENCE OF PESTICIDE REDUCTIONS AT COUNTRY LEVEL

A growing number of countries are now reporting reductions in pesticide use as a result of the adoption of agricultural sustainability principles. These have occurred as a result of two types of very different approaches:

1 policy-led and primarily top-down pesticide-reduction programmes in industrialized countries, such as in Sweden, Denmark, the Netherlands and some provinces in Canada;
2 farmer-field school led and policy-supported community IPM in rice programmes, beginning in south-east Asia, and then spreading throughout Asia and on to other continents (see Chapters 8–11).

Several OECD countries set ambitious national targets in the mid-1990s to reduce the use of inputs. Sweden's aim was to reduce input consumption by 20 per cent by the year 2000. The Netherlands also sought a cut in pesticide use by 50 per cent by the year 2000 as part of its Multi-Year Plan for Crop Protection. The cost of this reduction programme was estimated at US$1.3 billion, most of which was raised by levies on sales. Denmark aimed for a 50 per cent cut in its pesticide use by 1997, a plan which relied mostly on advice, research and training. Canada set itself a target for a 50 per cent reduction in pesticide use by 2000 in Quebec and by 2002 in Ontario. In the US, the administration announced in 1993 a programme to reduce pesticide use whilst promoting sustainable agriculture. The aim was to see some form of IPM on 75 per cent of the total area of farmland by the year 2000.

Supplemented by other policy measures, such as new regulations, training programmes, provision of alternative control measures and reduced price support, there have been some considerable reductions in input use. In Sweden, pesticide consumption fell by 61 per cent between 1981–1985 and 1996–2000 (from 23 to 9.3 million kg ai); in Denmark, by 40 per cent between 1985 and 1995 (from 7 to 4.3 million kg ai); in the Netherlands, by 41 per cent between 1985 and

1995 (from 21.3 to 12.6 million kg ai), and in Ontario by 40 per cent (from 8.8 to 5.2 million kg). However, the full significance of these apparent sharp falls in use is disputed. In Sweden, half the decline was attributed to the introduction of new lower dose products, such as the use of sulphonurea products applied only at 0.004–0.006 kg ai ha^{-1} instead of phenoxyherbicides applied at 1–2 kg ai ha^{-1}. In Denmark, the reduction was not accompanied by a cut in the frequency of application, which remains at the 1981 level of 2.5 doses ha^{-1} yr^{-1}. Success has been achieved without a diminished dependence on pesticides, which should really embody the basic concept for pesticide reduction.

Another analysis of Swedish pesticide consumption used LD$_{50}$ values as an indicator of acute toxicity, and this showed that the changes had resulted in a fall from 38,000 acute toxicity equivalents to 11,000 and then to 8700 by the end of the 1990s (Ekström and Bergkvist, 2001). However, another measure, the hectare–dose method (the quantity of active ingredient applied per ha) showed a substantial increase – from 1.6 million in 1981–1985 to 4.3 million in 1996–2000.

Scientific studies in industrialized countries underline the findings from pesticide reduction programmes that pesticide reduction can be beneficial to society's goals. For example, Mullen et al (1994) found both private and public benefits generated from IPM adoption for early leaf spot on peanuts in Virginia. The total savings in external costs were estimated to be US$844,000 per year for 59,000 households, on top of which farmers benefited from a small but important reduction in inputs costs. Brethour and Weerskink (2001) calculated that the 40 per cent reduction in pesticide use between 1983 and 1998 in Ontario produced benefits of CAN $305 per household. Aggregated across all 3.78 M households in the province, the value of the environmental risk reduction was $1.18 billion.

While IPM has had mixed success in the industrialized world, it has received much more attention in the developing world. Here, the discovery by Kenmore et al (1984) that pest attack on rice was proportional to the amount of pesticides used had a significant impact. They found that pesticides were killing the natural enemies (spiders, beetles, parasitoids) of insect pests, and when these are eliminated from agroecosystems, pests are able to rapidly expand in numbers. This led in 1986 to the banning by the Indonesian government of 57 types of pesticides for use on rice, combined with the launching of a national system of farmer field schools to help farmers learn about the benefits of biodiversity in fields. The outcomes in terms of human and social development have been remarkable, and farmer field schools are now being deployed in many parts of the world.

In Bangladesh, for example, a combined aquaculture and IPM programme is being implemented by CARE with the support of the UK government and the European Union. Six thousand farmer field schools have been completed, with 150,000 farmers adopting more sustainable rice production on about 50,000 hectares (Barzman and Desilles, 2002). The programmes also emphasize fish cultivation in paddy fields, and vegetable cultivation on rice field dykes. Rice yields have improved by about 5–7 per cent, and costs of production have fallen owing to reduced pesticide use. Each hectare of paddy, however, yields up to 750 kg of fish, a substantial increase in total system productivity for poor farmers with very few resources. Similar effects are seen with rice aquaculture in China (Li, 2001).

Such substantial changes in pesticide use are bringing countries economic benefits in the form of avoided costs. One of the first studies to quantify the social costs of pesticide use was conducted at the International Rice Research Institute (IRRI) in the Philippines. Researchers investigated the health status of Filipino rice farmers exposed to pesticides, and found statistically significant increased eye, skin, lung and neurological disorders. Two-thirds of farmers suffered from severe irritation of the conjunctivae, and about half had eczema, nail pitting and various respiratory problems (Rola and Pingali, 1993; Pingali and Roger, 1995). In addition, the authors showed that in a normal year no insecticide application was better than researcher recommended economic thresholds or even farmers' routine practices.

A so-called 'complete protection' strategy, with nine pesticide sprays per season, was not economical in any case. When health costs were factored in, insecticide use in rice became completely uneconomical. As Rola and Pingali (1993) and put it:

> the value of crops lost to pests is invariably lower than the cost of treating pesticide-related illness and the associated loss in farmer productivity. When health costs are factored in, the natural control option is the most profitable pest management strategy.

Any expected positive production benefits of applying pesticides are clearly overwhelmed by the health costs.

Other economic studies have calculated the economic benefits to farmers and wider society of IPM and pesticide reduction programmes. Cuyno et al (2001) showed that IPM in onion production in the Philippines reduces pesticide use by 25–65 per cent without reducing yields. Farmers benefited through increased incomes, and it was estimated that some US$150,000 worth of benefits were created for the 4600 residents of the five villages within the programme area.

Through various multi-lateral agreements, most countries in the world have indicated that they are in favour of the idea of agricultural sustainability. Clearly, there are now opportunities to extend policy-led programmes including farmer field schools for pesticide reduction and to increase farmers' knowledge of alternative pest management options across diverse agricultural systems.

CONCLUDING COMMENTS

The external environmental and health costs of pesticides are rarely addressed when discussing pesticide use in agriculture. Studies that have tried to assess the benefit of pesticides have often been flawed in that they have tended to use unrealistic scenarios about use reduction, and have not based estimates of yield changes on recent evidence of IPM from the field. Data from four countries are incorporated into a new framework for pesticide externalities, and this shows that total annual externalities are US$166 million in Germany, $257 million in the UK, $1398 million in China (for rice only) and $1492 million in the US. These externalities amount to between US$8.8 and $47.2 per hectare of arable and permanent crops in the four countries, or an average of $4.28 per kg of active ingredient applied. These suggest that the 2.5 billion kg of pesticides used annually worldwide (some 400 g ai per capita per year or 1.3 g per capita per day) currently impose substantial environmental and human health costs, and that any agricultural programmes that successfully reduce the use of pesticides that cause adverse effects create a public benefit by avoiding costs.

We analysed 62 IPM initiatives from 26 countries to illustrate what trajectories yields and pesticide use have taken. There is promising evidence that pesticide use can be reduced without yield penalties, with 54 crop combinations seeing a 35 per cent increase in yields while pesticide use fell by 72 per cent. A further 16 crop combinations saw small reductions in yield (7 per cent) with 59 per cent reductions in pesticide use, and 10 saw an average 45 per cent increase in yields accompanied with a 24 per cent increase in herbicide use. A number of countries have adopted policies to reduce pesticide use, either policy-led target pro-grammes or farmer training (field school) led community IPM programmes. Promising progress indicates that greater efforts should be made to extend such policy support across all industrialized and developing countries. Also, govern-ment commitment to sustainable agriculture with a minimum use of pesticides needs to be steady and long term. An important goal in this regard is that pest control options be maintained and if possible expanded. There are lessons that can be learned from the 'pesticide story' that are equally valid for upcoming or already seemingly shining stars on the crop protection heavens. That is, to avoid dependence on any perceived 'silver bullets'.

Chapter 4

Corporations and Pesticides

Barbara Dinham

INTRODUCTION

Pesticides are now developed, manufactured and sold mainly by multinational corporations, which play a major role in influencing farmers' decisions on pest management. These corporations will be critical actors in any efforts to develop safer pest management systems. However, changes are taking place in the agrochemical industry at an accelerated pace, with the main companies merging to consolidate profits and research costs. The same companies are investing in genetically modified (GM) crops and these products now account for an increasing level of sales and profit. The strategies of agrochemical companies, together with policies promoting pesticide use, have a significant influence on agricultural development. This chapter assesses recent developments in the pesticide market, with a particular focus on what these mean for developing countries.

COMPANIES AND THEIR MARKETS

Pesticides were a highly profitable business during the 1960s and 1970s, with an average growth rate of 10 per cent per year. However, sales gradually levelled off, and by the 1990s growth was averaging only 0.6 per cent per year (Bryant, 1999). By 2002, the market had fallen back from its peak of US$30 billion to US$27.8 billion (Agrow, 2003a). Pressure on sales comes from the high research costs, tighter regulation and limited room for market expansion in the US, European Union (EU) and Japanese markets. These pressures have led to consolidation and many mergers, so that just six companies now control some 75–80 per cent of the world's agrochemical market (Tables 4.1 and 4.2). Some 20 per cent of the pesticide market is made up of several Japanese companies and, increasingly, generic producers who manufacture out-of-patent products, the biggest of which is the Israeli company Makhteshim-Agan.

The 'big six' companies, Syngenta, Bayer, Monsanto, BASF, Dow and DuPont, have also developed GM seeds to ensure they maintain a strong position in the agricultural input market. Towards the end of the 1990s, several corporations sought to create synergies in their research in both agriculture and pharma-

ceuticals by creating 'lifescience' companies. The strategy did not bring significant benefits, and many later segregated or spun off the less profitable agricultural chemical side, further driving consolidation in the industry. In 2001, the
agrochemical industry association changed its name to CropLife International,
and broadened its remit to include both pesticides and agricultural biotechnology in order, according to its Director, to *'ensure that the scientific and technological innovation of the plant science industry can continue to benefit all those who need
it, all over the world, while at the same time rewarding the pioneering work of our
companies'* (Verschueren, 2001). In the early 2000s, sales from the seeds and the
genetic modification sector grew rapidly, while agrochemical sales were static
or fell.

In 2002, these companies collectively earned over US$4.5 billion from sales
of GM seeds, led by DuPont, Monsanto and Syngenta. The seed sector is much
less concentrated, with ten firms controlling 30 per cent of the worldwide
US$24.4 billion commercial seed market (ETC, 2001). Over time, seed marketing
has shifted towards the promotion of combinations of seeds and pesticides
(Goulston, 2002).

Table 4.1 *Agrochemical and GM seed sales of the leading companies, 2000–2002
(US$ million)*

| Company | Agrochemical sales | | | GM seed sales |
	2000	2001	2002	2002
Syngenta[1] (Swiss/UK)	5,888	5,385	5,260	938
Bayer[2] (German)	2,252	2,418	3,775	85
Monsanto (US)	3,885	3,755	3,088	1,585
Aventis (German/French)	3,701	3,842	(see Bayer)	(see Bayer)
BASF (German)	2,228	3,105	2,787	0
Dow (US)	2,271	2,612	2,717	na
DuPont (US)	2,009	1,814	1,793	1,920
Sales of top companies	22,234	23,034	19,420	4,528
Total market	29,200	27,104	27,790	

1 Figure for 2000 is estimated from combined sales of Novartis and AstraZeneca.
2 Following a takeover of Aventis, the 2002 figure represents combined sales.
Source: Agrow, 27 July 2001, No 381; Agrow, March 2003 28 No 421; Jarvis and Smith, 2003.

Strategies for increasing pesticide sales include developing and promoting new
chemical products and selling services, while extending the life of those older
products with significant markets. As the research costs of older products have
already been recouped, these sales are very profitable. While outside the scope
of this chapter, many of the products have non-agricultural applications. The
market was valued at US$10.5 billion in 1998 and was growing at around 4–5
per cent a year. It includes sales in forestry, as well as home and garden use, golf
courses, municipal and railway use, and industrial control uses for vermin and
insects (Bryant, 1999).

Table 4.2 *Halving of major agrochemical companies, 1994–2003*

Beginning 1994	By 1997	By 1999	2000–2003
DowElanco (US)	DowElanco	Dow Agro Sciences	**Dow** AgroSciences has 9.4% of Dow Chemical sales
DuPont (US)	DuPont	DuPont	**DuPont** Crop Protection has 15.5% of DuPont sales
Monsanto (US)	Monsanto	Monsanto	**Monsanto**. Pharmacia bought 80% in 2000; became independent agribusiness company in 2002
Bayer (EU-G) Hoechst (EU-G) Schering (EU-G) Rhône-Poulenc (EU-Fr)	Bayer AgrEvo Rhône-Poulenc	Bayer Aventis	**Bayer** CropScience (G) represents 12% of Bayer sales. Took over Aventis in 2002
Ciba Geigy (Swiss) Sandoz (Swiss)	Novartis (Swiss, merger 1996) (acquired Merck)	Novartis	**Syngenta** AG (Swiss) formed in 2000
Zeneca (ex-ICI)(EU-UK)	Zeneca	AstraZeneca (UK-Swedish merger) (1999)	
BASF (EU-G) Cyanamid (US) [purchased Shell Agriculture (UK/Neth.) in 1993]	BASF Cyanamid bought by AHP, 1994 (US)	BASF Cyanamid (US)	**BASF** Corporation Agricultural Products (US) has 18% of BASF(G) sales

Source: Agrow's Top 20: 2003 Edition, Richmond, UK, 2003

A number of developing countries are producers and exporters of pesticides, with both national companies and subsidiaries of the major producers producing out-of-patent active ingredients. India and China are the largest producers of generic products, closely followed by Argentina. In 1996, India had approximately 125 companies manufacturing more than 60 technical grade pesticides. India is the world's largest organophosphate producer, and companies make many hazardous products, including some banned elsewhere. These include pesticides classified by the World Health Organization (WHO, 2001) as extremely and highly hazardous (Class Ia/Ib) dichlorvos, monocrotophos, parathion methyl, phorate, terbufos, zinc phosphide; moderately hazardous (Class II) active ingredients chlorpyriphos, cypermethrin, alpha cypermethrin, DDT, deltamethrin, dimethoate, ethion, fenvalerate, lambdacyhalothrin, paraquat, profenophos and quinalphos, plus the deadly fumigant aluminium phosphide (Agrow, 1996).

China is the world's second largest agrochemical producer by volume, some 450,000 tonnes in 2000, of which 35 per cent was exported. More than 2000 companies are involved in pesticide production and packaging, and much production is geared to highly toxic products, such as methamidophos for cotton and rice crops (Tyagarajan, 2002). The huge US$1.4 billion market is attracting foreign companies to invest directly or in joint ventures (Agrow, 2003b), and industry analysts suggest these may push out the smaller local companies (Tyagarajan, 2002). Syngenta opened a paraquat plant in 2001 as a joint venture with a Chinese company, and within two years China became the biggest market for the product outside the US (Syngenta, 2002a).

CORPORATE RESEARCH BUDGETS AND SPHERES OF INFLUENCE

A driving force behind consolidation in the industry is the cost of research and development, which typically amounts to around 10 per cent of a company's sales (Table 4.3). Both biotechnology and the discovery of new active ingredients requires long-term, costly research. Without this, the big six could not keep their lead on the market, maintain the range of agrochemical products, and meet modern regulatory standards. Despite the rapid growth in GM technology since the first commercial cultivations in 1996, all companies still predict that pesticides will remain central to their industry for the foreseeable future. The average rate of introduction of new active ingredients since 1980 has been 12.7 products per year. In 1997 and 1998, 19 and 13 new active ingredients were introduced (Phillips and McDougall, 1999). Companies have sought to increase the speed of discovery for new active ingredients by using genomics and rapid screening processes. While development costs increased by 21 per cent between 1995 and 2000 to US$184 million, and the lead-time from identification to sale grew from 8.3 to 9.1 years, the cost of registering a new chemical was reduced by 15 per cent over this period (Phillips McDougall, 2003).

The six research-based agrochemical companies between them operated budgets of some US$3.2 billion to take products from identification to market in 2001–2002. By contrast, the budget available for agricultural research in developing countries is roughly an order of magnitude less. In 2003, the budget of the 16 research centres that are part of the Consultative Group on International Agricultural Research amounted to US$330 million (CGIAR, 2003).

Advertising plays a key role in persuading farmers to buy products. A recent analysis of the trends in pesticide advertising has demonstrated how product promotion targets prevalent social and cultural values (Kroma and Flora, 2003). Advertisements in the 1940s to the 1960s tended to stress science, and typical product names were simazine, isotox and lindane. Almost all (95 per cent) of the pictorial metaphors reflected the positive chemical attributes of the pesticide. In the 1970s and 1980s, the emphasis shifted to the domination of nature. Brand names like Prowl, Marksman, RoundUp, Lasso, Bullet, Warrior and Pounce proliferated. The herbicide Prowl was portrayed by a snarling feline, ready to spring and destroy obstacles in its path. In another advert, symbolized by a wolf, the narrative says 'there can only be one leader of the pack'. By the 1990s,

Table 4.3 *Research budgets of agrochemical corporations (average 2001–2002)*

Company/year (US$ million)	Budget (as a % of sales)	Research
Aventis (taken over by Bayer in 2002)	433	12.1
BASF	356	12
Bayer CropScience	527	13.4
Dow AgroSciences	282	8.5
DuPont (1996 latest available figure)	200	8
Monsanto	560	14
Syngenta	710	11.3
Total research expenditure	**3200**	

Source: Agrow's Top 25, 2001 and 2003 Editions; Agrow's Ag-Biotech Top 20: 2003 edition.

criticisms about the high ecological and social costs had begun to have an effect, and brand names and pictorial images came to emphasize harmony and working with nature, with names like Beacon, Permit, Fusion, Resolve, Resource, Harmony and Accord. Harmonious ecological images were common, such as one silhouetting farm animals with a woman and child playfully using a water pump. The narrative of another suggests 'Best against grass. Best for the land'. In developing countries, products were often advertised with scantily clad women, or with extravagant claims for improved yields. Both practices were stopped, eventually, by the advertising guidelines in the FAO International Code of Conduct of Conduct on the Distribution and Use of Pesticides.

GROWING MARKETS IN DEVELOPING COUNTRIES

With declining prospects for growth of pesticide sales in rich markets, developing country markets are increasingly a target, particularly those in the newly-industrializing countries in Asia and Latin America. Detailed usage figures are not readily available, but insecticides, often the most acutely toxic of products, have formed the main share of sales in these regions (Bryant, 1999). Fruit and vegetables, cereals (wheat, barley), rice, maize, cotton and soybeans account for approximately 85 per cent of sales (Figure 4.1). Except for cereals, these crops are widely grown by smallholders in developing countries. Rice is predominantly grown in Asia, with some 90 per cent grown by small-scale farmers. Cotton accounts for 10 per cent of total pesticide use, including approximately 25 per cent of insecticide use. Fruit and vegetables are grown worldwide, and the expansion of the fresh produce for the export industry in developing countries has accelerated pesticide use in these crops (Dinham, 2003b).

The growing need to protect human health and the environment led to the adoption of the International Code of Conduct on the Distribution and Use of Pesticides (FAO Code of Conduct) by governments in 1995 (FAO, 2002) (see Chapter 1). Since then, it has been amended twice, in 1999 and 2002, to improve

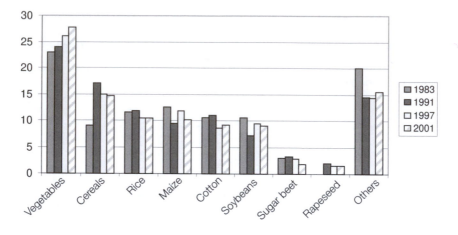

Figure 4.1 *Crops and pesticide use 1983–2001 (%)*

Source: Crop Protection Association UK. Annual reports 1983–2001 (supplied by company analyst Wood MacKenzie)

the guidance to governments and industry. The pesticide industry association, CropLife International, whose membership includes all the major agrochemical corporations, has made implementation of the FAO Code of Conduct a condition of membership. While pesticide laws and regulations have improved in developing countries as a result of this Code, the human and financial resources to implement its provisions are often lacking.

In industrialized countries, companies are required to register each formulation of a pesticide active ingredient for use on each crop. Developing countries generally operate simpler registration systems, although even these can be difficult to enforce. The *Pesticide Manual*, the most comprehensive record of active ingredients, lists 812 active ingredients available on the market and 598 superseded actives (Tomlin, 2000). Regulatory authorities in industrialized countries have tended to raise standards over time, with the European Union, for example, removing 320 registrations in July 2003. In some cases, the risks of adverse effects to health or the environment result in a ban or severe restriction, in other cases companies have not supported the registration because of low sales or the products have been superseded and it is not economically worthwhile to invest in research to fill the data gaps. Nonetheless, if products are manufactured in the European Union, they can still be exported, and products may continue to be marketed in developing countries.

Thus stricter regulation and phasing out (sometimes banning) of older products in industrialized countries encourages companies to extend their life elsewhere, and farmers in developing countries generally want cheaper products. Highly toxic organophosphate insecticides like parathion methyl (WHO Class Ia) marketed since the 1950s, and monocrotophos (WHO Class Ib), first registered in 1965, are still widely used in developing countries.

The Asian region has been a particular target for agrochemical sales, and the large rice and cotton markets are significant users of pesticides. Of the US$2.4 billion global rice agrochemical market in 2001, Japan accounted for 45 per cent of sales (spending $635 ha^{-1}). Markets in other Asian countries, where poorer farmers are applying the products, have recently increased. For example, sales in Vietnam rose from US$46 million in 1996 to $68 million in 2001 (Agrow, 2003c). In China per hectare pesticide costs in rice cultivation increased steadily from 1980 to reach US$13 ha^{-1} in 1998. In spite of vast differences in expenditure, average rice yields in China (6.3 t ha^{-1}) are comparable with those of Japanese farmers (6.7 t ha^{-1}) (Watkins, 2003).

Yet much of this pesticide market growth may be unnecessary and could be curtailed with better access to information and training for farmers. There have been demonstrated successes with integrated pest management (IPM) strategies for rice and vegetables that have reduced pesticide dependence (see Chapter 8), which suggest that many farmers elsewhere are using unnecessarily large amounts of pesticides.

The demand for cheaper products leads to the paradox that while it appears that the major markets are industrialized countries, price is not a guide for comparing levels of pesticide use between industrialized and developing countries. Pesticide costs relative to farm incomes are far lower in industrialized countries. On the one hand, this means application rates are higher in rich countries: on rice, application rates average 11.8 kg aiha^{-1} in Korea and 19.4 kg ai ha^{-1} in Japan, but only around 1.0 kg ai ha^{-1} in Vietnam (FFTC, 1998). But poor farmers in developing countries spend a higher percentage of household income on pesticides, and in certain crops like rice, vegetables and cotton, excessive and unnecessary spraying is frequently recorded. The International Rice Research Institute (IRRI) has ascribed the rapid increase in pesticide, mainly insecticide, use on rice in the Philippines over the last 30 years to aggressive marketing by pesticide producers, coupled with the increasing use of high yielding varieties (IRRI, 2003). Some products are predominantly sold in developing countries: 75 per cent of the market for the herbicide paraquat, first marketed in 1969, is in Asia, Central and South America (Dinham, 2003a), even though the product is associated with significant adverse health effects (Wesseling et al, 1997a, 2001).

In the 1970s and 1980s in Asia, many governments played a central role in pesticide production and distribution. Pesticides were often subsidized, helping the market to grow (Farah, 1994). During the 1990s, there was a change in policy as the International Monetary Fund and World Bank promoted trade liberalization, and most countries now leave production and marketing to the private sector. Farmers can buy a greater range of products, but their information comes largely from the company and retailer. Pesticides are often sold in small kiosks, markets, shops selling food and other household goods, and by travelling salesmen. Village pesticide dealers are rarely well informed about the best pesticide to use for a particular pest problem, or how to use it without unnecessary risk to health. For example, in Cambodia 95 per cent of 1000 farmers surveyed learned to use pesticides from neighbours or traders who themselves were unaware of pesticide risks, and were often unable to read labels. However, the traders were perceived to have a status 'similar to a doctor' (EJF, 2001).

A recent study of four African countries found a proliferation of informal trading and lack of appropriate advice to farmers from the private sector suppliers (Chapter 11). Another in Asia found that the shift from government role in distribution of pesticides in the 1970s and 1980s to private sector distribution in the 1990s meant that most information for farmers about pesticides came from the company making the product, via the retailer. It observed that: *'Most village pesticide dealers are not well informed about what pesticide to use for a particular pest problem, or how to use it safely. Since they are interested in selling as much of their products as possible, they have no interest in minimizing pesticide use'* (FFTC, 1998).

Nonetheless, the agrochemical industry is aware of its poor public image. In 1992, the European chemical industry association CEFIC carried out a survey with over 8000 men and women in eight European countries, and followed this every second year up to 1998: interviews with over 7000 men and women showed a slow but continuous decline in their perception that industry sets high safety standards (Dewar, 2000). What, then, can the industry do?

THE FAO CODE OF CONDUCT AND COMPANY PRODUCT STEWARDSHIP

The FAO Code of Conduct sets standards and guidelines for pesticide use in developing countries, calling on governments, the agrochemical industry, food industries, and public interest groups to assist in its implementation. But a 1995 survey by the FAO for the tenth anniversary of the Code of Conduct found no improvements to health, and a deterioration of environmental standards. Nevertheless, almost all governments do now have pesticide laws and regulatory systems, although implementation and enforcement remains problematic. A revised Code of Conduct was adopted in 2002, with many improvements. It recommends in Article 7.5 that *'Prohibition of the importation, sale and purchase of highly toxic and hazardous products, such as those included in WHO classes Ia and Ib (35), may be desirable'*, and in Article 3.5 that *'Pesticides whose handling and application require the use of personal protective equipment that is uncomfortable, expensive or not readily available should be avoided'* – a precaution that would rule out use of most pesticides in WHO Class II. It has new standards on advertising, and it calls for the licensing of pesticide retailers to ensure they are sufficiently well informed to assist farmers in purchasing their products. The terminology referring to 'safe use' was removed to avoid incorrect perceptions about the risks of using hazardous products.

Earlier definitions of IPM in the FAO Code of Conduct had emphasized strategies that included pesticide use, but the 2002 revision adopted a new definition:

> *IPM means the careful consideration of all available pest control techniques and subsequent integration of appropriate measures that discourage the development of pest populations and keep pesticides and other interventions*

to levels that are economically justified and reduce or minimize risks to human health and the environment. IPM emphasizes the growth of a healthy crop with the least possible disruption to agroecosystems and encourages natural pest control mechanisms.

The FAO Code of Conduct is voluntary but influential. To be effective, its guidelines need to be incorporated into national regulations, and resources need to be allocated for implementation. The industry association, CropLife International, expects its members to comply with the FAO Code of Conduct. In 2003, its members adopted the revised Code, which promotes responsibility to account for product impacts throughout their lifecycle. To implement the Code, companies developed product stewardship guidelines, which state that products must be registered for use on each crop, lowest toxicity formulations should be encouraged, expert medical advice must be available on a 24 hour basis, and protective clothing must be available and convenient. However, most pesticide use in developing countries at field level still does not comply with these company requirements, as product stewardship policies stop company obligations at distributor level (Dinham, 1993). Companies will need to upgrade their recommendations in line with the provisions of the revised FAO Code of Conduct.

Some companies have made pledges to address pesticide poisoning concerns in developing countries. Bayer indicated that it would phase out WHO Class I products, but, in 1991, it commissioned a new plant at Hammatnager in India to produce a dust formulation of methyl parathion – a Class Ia pesticide responsible for many deaths in developing countries. Syngenta has also pledged to phase out older products, including the organophosphate monocrotophos, but is still a producer, and is targeting sales of paraquat (Class II) in Latin America and Asia. Dow's organophosphate chlorpyrifos, now severely restricted in the US, and virtually eliminated from home and garden use after the most extensive scientific review of a product ever conducted by the Environmental Protection Agency (EPA), is widely used in developing countries.

Companies have largely avoided liability for the adverse effects of products. However, there is a growing recognition of the need to address the difficulty of redress. Principle 13 of the Rio Declaration states that:

> *States shall develop national law regarding liability and compensation for the victims of pollution and other environmental damage. States shall also cooperate in an expeditious and more determined manner to develop further international law regarding liability and compensation for adverse effects of environmental damage caused by activities within their jurisdiction or control to areas beyond their jurisdiction.*

The European Commission (EC) has initiated an amendment to the Product Liability Directive (85/374/EEC) to include chemicals, including agrochemicals, under legislation for liability for 'defective' products. Sellers could be liable for negligence if found to fail in their duty not to create unreasonable risks, and for pesticides, this would mean producers could be liable if any commercialized product turned out to be unsafe, even if it had been approved by the authorities. A voluntary system of guidelines has been established under the International Standards Organisation (ISO) 14000 for Environmental Management Systems Standards, which seeks to reduce harmful effects of business activities on the environment. The guidelines cover obligations for auditing, evaluation of performance, labelling and lifecycle assessments that governments can draw on to implement as national standards: companies can elect to be 'ISO certified'.

Victims of pesticide poisoning have, to date, had limited success holding manufacturers accountable under legal action, but greater recognition of rights and responsibilities of corporations, corporate social responsibility and issues of liability, could become a driver of product stewardship. In Florida, an Appeal court in 2003 upheld a ruling in favour of a woman who had been exposed to Benlate (the fungicide benomyl) when seven weeks pregnant. Her son was born with microphthalmia, a rare birth defect involving severely underdeveloped eyes. The jury returned a verdict that held DuPont strictly liable, and that both DuPont and the farm spraying benomyl were negligent. The total award was US$4 million, with 99.5 per cent to be paid by DuPont (Castillo *v*. DuPont, 2003).

Legal action is very difficult to undertake in developing countries. In Peru, a lawsuit was initiated against Bayer when 24 children died after exposure to methyl parathion: the case was one of misuse (mixed with milk drunk by the children), but the victims families held that such a toxic product should not have been sold where the prevailing socio-economic conditions prevented safe use. The case was not upheld, but the families are considering further action (Rosenthal, 2003). Banana workers in Costa Rica have been struggling since the end of the 1980s in a case against Dow for impotence and other health effects caused by exposure to the product DBCP. Although the courts have found in favour of the workers, the company has appealed and been responsible for the protracted action (Dinham and Malik, 2003).

TRAINING IN THE 'SAFE USE' OF PESTICIDES

One practical outcome of industry product stewardship has been the initiative entitled 'Safe Use' of pesticides. Three pilot projects in Kenya, Thailand and Guatemala were launched in 1991 through the industry association CropLife International (originally GIFAP, then Global Crop Protection Federation). The projects were a response to criticisms and concerns raised by governments, public interest groups, trade unions and farming communities regarding the effects of pesticide use on human health and the environment, and aimed to address the problem that farmers had insufficient information and training in pesticide use. Typically, such a project aims to train the distribution network, inform and educate end-users in the correct use of pesticides, improve poison control centres, educate school children and promote a positive perception of the industry. Having started with the three pilot projects, the regional CropLife centres in Asia-Pacific, Africa-Middle East, and Latin America are encouraging a scaling up of activities, and about 70 projects have been initiated. As indicated earlier, companies should now avoid the term 'safe use', as it does not adequately convey the importance of reducing hazards and risks.

In Guatemala, where large numbers of people had been poisoned by pesticides, companies introduced training through the local pesticide association, AGREQUIMA, which now collects on behalf of the government a 0.05 per cent levy on imported pesticides and administers these funds for training (Hurst, 1999). Industry literature indicates that the first phase trained 800 government extension agents, who then went on to train a further 226,000 farmers, 2800 schoolteachers and 67,000 schoolchildren, 700 pesticide distributor employees, 330 technical and sales people and 2000 physicians and health personnel (GIFAP, undated).

But studies of safe use programmes suggest that behavioural changes were often temporary, as the central concept of safe use is rooted in an assumption that pesticide problems are caused by irrational behaviour, while in reality the conditions facing many small farmers and agricultural workers mean their behaviour is perfectly rational (Hurst, 1999; Atkin and Leisinger, 2000; Murray and Taylor, 2001). For example, in Honduras a group of 15 young melon workers were poisoned after applying carbofuran with their bare hands before they ate lunch without washing. They were not provided with safety equipment or water for washing. Washing would have required leaving the field, losing their brief rest period or jeopardizing their employment (Murray and Taylor, 2001). This study recommended a wider set of actions to eliminate the most toxic pesticides, introduce safer products and alternative technologies; implement stricter administrative controls, and ensure all pesticide users have personal protective equipment. A seven-year in-depth study by the Syngenta Foundation concluded that:

> All available experience indicates that there are limits to the extent to which changes will be adopted within a generation. Even the best and most sustained efforts run into the paradoxical situation that not everyone who can adopt relatively simple modifications in behaviour will actually do so, even when it is shown that the changes are in the person's long-term best interest. Given that, any pesticide manufacturer that cannot guarantee the safe handling and use of its toxicity class Ia and Ib products should withdraw those products from the market (Atkin and Leisinger, 2000).

Nonetheless, company training projects are an important advance, although they should be obligatory, and costs need to be reflected in the product price rather than claimed from government levies or development aid funding. The industry association obtained funding for its training activities from a number of development agencies, including USAID, the European Union and the World Bank. Promoting 'safe use' through agricultural extension and local dealers is a form of advertising, and associates pesticides with safety, whereas public funds could be better utilized promoting IPM training, particularly through farmer field schools, where pesticides may be only one of many pest management methods used by farmers.

Some Corporate Strategies in Developing Countries

In addition to sales and marketing strategies, agrochemical companies are able to use their strong position to expand the use of their products in developing countries, even though other agricultural strategies may provide a more appropriate pest management solution for poor farmers. This section looks at the influences on the allocation of development funds, extending markets for hazardous products, and influences on policies to promote inappropriate products.

Agrochemical companies often advocate private–public partnerships, and seek access to development aid funding to promote their products. Increasingly, this is done in the context of IPM or 'safe use' training programmes, but the outcome can result in disadvantage to farmers. Documentation of one project in Senegal demonstrates how, with significant World Bank involvement, the French company Rhône Poulenc (now Bayer) increased pesticides use and associated risks (FAO, 2001a).

Pesticide use is high in the Senegal River Valley area, which is managed by the parastatal irrigation authority, SAED. A study by CERES/Locustox, a national centre for ecotoxicology co-sponsored by FAO, found that pesticide use increased by 4–5 times over the 15 years after 1984. Farmers typically used 13–15 toxic products, often mixed in cocktails, and these agrochemicals are perceived as beneficial 'medicines'. SAED advisers encourage calendar spraying rather than on an analysis of need. In tomatoes, 15 preventative applications involving five different products are recommended. Protective clothing is not available, and poverty and illiteracy is widespread.

In 1997, Rhône Poulenc was invited to demonstrate rational and safe use of pesticides to small-scale tomato farmers. SAED assumed that commercial competition would improve supply efficiency, the price to farmers of pesticides would fall, yields would increase, and training practices would be adopted. The World Bank was involved as a participant at the meeting making the strategy recommendations. It provided technical assistance to guide the choice of agrochemicals, and funded a delegation from the farmers' associations to visit Rhône Poulenc facilities.

The training lowered the number of different products used on tomatoes from 13–15 to seven, but increased the number of applications from 12 to 20. All applications recommended Rhône Poulenc products. Yields did increase, but did not necessarily improve farmer incomes because of the increased expenditure on pesticides. The Rhône Poulenc demonstration pack included pesticides not licensed for sale in Senegal and not permitted under a World Bank safeguard policy on pest management. The human toxicity index of the demonstration pack was double the index of the previously used pesticides. By contrast, IPM training in farmer field schools generally maintains or increases yields, while reducing input costs, with an overall positive impact on income. For example, the preliminary results of two years of agroecosystem training in vegetables in Senegal carried out by FAO in collaboration with CERES-Locustox have trained 800 farmers in farmer field schools. In all cases, farmers have reduced their use of pesticides and improved their income, and in one area farmers have fully replaced agrochemical pesticides with *Bt* and neem extract (Diallo et al, 2003).

Paraquat is a widely used herbicide first marketed in the 1960s by ICI (later Zeneca, now part of Syngenta). The product is out of patent, but Syngenta retains at least 50 per cent of sales (Dinham, 2003a). Although paraquat is a WHO Class II chemical (moderately hazardous), one teaspoon is fatal, and there is no antidote. Nonetheless, as a cheap herbicide, it is widely used in developing countries by small-scale farmers. Early formulations resembled coca cola, coffee or tea and home storage led to many cases of accidental consumption, although the company later introduced a dye, emetic and strong smell into formulations.

Syngenta indicates that paraquat is registered in over 120 countries, which shows the extent of usage in developing countries, and Asia and Latin America account for 45 per cent and 29 per cent of sales respectively. For workers on plantations and cropping systems where paraquat is used on a regular basis to kill foliage at the base of trees, paraquat can cause health effects ranging from unpleasant to serious: headaches, nausea, loss of finger- and toenails, skin rashes and ulceration, nosebleeds, eye irritation, excessive sweating (Wesseling et al, 1997b, 2001). The Malaysian government banned paraquat in 2002 because of its adverse effects on workers, particularly on oil palm plantations. Syngenta acknowledges problems with skin irritation and nail damage '*during occupational exposure, mainly in hand-held applications, as a result of unwashed spillages of commercial product, or from prolonged dermal contact with spray solution*', adding that the damage is '*indicative of inadequate standards of personal hygiene*' (Syngenta, 2002b). Some industrialized countries have banned paraquat (Austria, Denmark, Finland, Sweden, Hungary, Slovenia, Switzerland).

In 2003, the European Union approved continued paraquat use in Member States. Following this European decision, the Malaysian oil palm industry association took out two full page 'advertorials' (their description) in the national press protesting the ban and, drawing on information from Syngenta, challenged the Malaysian government to withdraw the ban on the grounds that the European Union had approved paraquat (*New Sunday Times*, Malaysia, 2003). This denies the important principle that governments must make pesticide registration decisions based on the assessment of risks under national conditions.

Endosulfan was developed by Hoechst, which later merged with the French company Rhône Poulenc to form Aventis (Bayer). The compound has been associated with both health and environmental problems in many countries. In Kerala, India, local people have been subject to 25 years of aerial spraying of endosulfan and other pesticides. Nearby villagers suffer both acute and chronic illnesses, with a high rate of birth abnormalities. Medical experts are persuaded of a clear link with endosulfan (Quijano, 2002). In West Africa, endosulfan was introduced in the 1999–2000 season after the cotton pest American bollworm, *Heliothis/Helicoverpa armigera*, became resistant to pyrethroid insecticides. The introduction was based on the recommendation of Projet Régional de Prévention et de gestion des Résistances de Helicoverpa armigera aux pyréthrinoïdes en

Afrique de l'Ouest (PR-PRAO), which covers several countries. The project was started by the national cotton research institutes, the agrochemical companies' Insecticide Resistance Action Committee, a French cotton company, and French research institutions (Ton et al, 2000).

In August and September 1999, reports began to surface of an alarming number of poisonings and fatalities. A government mission visited the major cotton growing areas, Borgou and Atacora, to evaluate the extent of poisoning. As no results were published, a local NGO established a team and visited Borgou province in the 1999/2000 season. Its investigation documented 147 cases of poisoning including ten deaths, with endosulfan largely responsible (Ton et al, 2000). The NGO mounted an investigation in the following season, and documented 241 poisonings and 24 fatalities, again predominantly from endosulfan (Tovignan et al, 2001).

An Aventis investigation first identified suicides as the cause of most fatalities, contradicting the NGO that found food and water contamination to be the major source. But the company later acknowledged that contaminated food accounted for the largest number of poisonings, although its recommendation was to introduce training in the 'safe use' of pesticides rather than cease use of endosulfan (Aventis, undated). An organic cotton project in Benin is achieving good results on yields, with improved farmer health and higher incomes as a result of less spending on pesticides (OBEPAB, 2002). Initial feedback from a cotton farmer field school IPM project in Mali found that pests could be controlled with the botanical pesticide neem rather than with four to six sprays of synthetic pesticides. Although yields were slightly lower, farmers' net revenue was 33 per cent higher because of the cost savings (Coulibaly and Nacro, 2003).

Despite overwhelming evidence of widespread farmer and farmworker poisoning, the pesticide industry continues to downplay the seriousness of such poisoning by suggesting a high degree of suicides and asserting that the vast majority of poisoning cases are not severe, and should not be a public health priority (CropLife, 2003). While agreeing to implement the revised FAO Code of Conduct, industry appears to consider acceptable the many health problems suffered by those using pesticides on a regular basis, which can include effects on the nervous system, impaired breathing, severe skin irritation, loss of finger- or toenails, temporary illness, vomiting, drowsiness, severe headache, nausea and blurred vision, among other symptoms. These illnesses often result in loss of labour time, and affect income.

CONCLUDING COMMENTS

Agrochemical corporations have played a major role in shaping modern agricultural production in both industrialized and developing countries. The products of their research and development dominate the agricultural input market. The industry is now highly concentrated into six research-based companies, with a large number of generic companies seeking to gain a greater foothold on sales. The health and environmental side-effects of many of these products have been acknowledged, and some have been removed from the market as a result. Nevertheless, many hazardous pesticides, and others with chronic health concerns, are freely available in developing countries. Workers and farmers who are not able to protect themselves are using these products under inappropriate conditions.

The major companies have signed up to the FAO Code of Conduct, and its implementation by governments, agrochemical and food industries is crucial to reduce the adverse effects of pesticides. Company projects are one essential element in implementation of the Code, but care must be taken to ensure that these projects do not encourage unnecessary pesticide use, that they are not associating pesticides with 'safety', and that the cost of the projects is reflected in the product price rather than drawing on public funds, and that least hazardous pest management strategies are prioritized. More assertive action may be needed in developing countries to find safer and more sustainable pest management solutions for poor farmers. Far more emphasis must be placed on a needs-based approach to pesticide application, and on safer and sustainable alternatives. The most important step companies could make would be to remove the most toxic pesticides from markets, in particular in countries where conditions are unsuitable for their use, and introduce safer products and technologies.

The widely held view of policy-makers that pesticides are essential to increase production is challenged by IPM training, and particularly through the farmer field school approach, which has shown that it is possible to maintain yields whilst reducing expenditure on pesticides. These strategies particularly benefit small-scale farmers in developing countries, and could make a significant difference to rural livelihoods. Governments, public research institutes and development agencies need to play a robust role in promoting the pest management strategies that address the needs of poor farmers and influence strategies in farming systems where agricultural workers use hazardous products without protection.

Chapter 5

Overview of Agrobiologicals and Alternatives to Synthetic Pesticides

David Dent

INTRODUCTION

The commercial and widespread availability of synthetic pesticides has transformed approaches to pest management, emphasizing the use of off-farm inputs and control as opposed to the management of pest problems. Some attempts have been made to substitute pesticides with 'agrobiologicals', the biological equivalents of synthetic pesticides. These include biopesticides based on bacteria, fungi, viruses and entomopathogenic nematodes and a range of other off-farm inputs that include pheromones and macrobiological agents such as predators and parasitoids. Other alternatives to chemicals include the use of pest resistant crop cultivars including transgenic crops and on-farm techniques such as crop rotations, intercrops, tillage systems, modification of planting dates and sowing densities, and overall improved habitat management. These options can be used individually or as part of integrated systems working at the pest, crop, farm or agroecosystem level. This chapter places pesticides use (both chemical and biological) in the wider context of pest management (including IPM – integrated pest management), and provides examples of where there has been widespread change and adoption of alternatives to chemicals.

FROM THE FIRST IPM TO TODAY

How long has integrated pest management been practised? Farmers and growers have integrated different methods for controlling pests for centuries as part of crop production processes – pests being just one of a number of constraints that farmers try to overcome in order to achieve desired yields. In the sense of the mix of options available, things have not changed dramatically for farmers over time. The options have always included genetic manipulation of crops and animals, environmental/habitat management and a variety of on-farm and off-farm inputs. However, it is the degree to which off-farm inputs have increased that has marked a major change from the first IPM to that practised today. Seed, fertilizer, pesticides and other pest control inputs such as pest monitoring devices, semiochemicals, even wild flower seeds for headlands are all purchased off-farm. These developments have been part of the industrialization of agriculture that occurred during the 20th century as governments sought to provide cheap food for growing urban populations. This was achieved by targeting investment in research and development, farm and input subsidies, and encouraging farmers to intensify production and industry to invest in the development of agricultural inputs.

Crop improvement through plant breeding became one of the major influencing factors affecting intensification. In terms of pest management, cultivars resistant to pathogen and insect pests were the first successes. In 1912 Biffen discovered that resistance of wheat to yellow rust was controlled by a major gene (Johnson and Gilmore, 1980) and Harlan, in 1916, demonstrated that resistance to the leaf blister mite, *Eriophyes gossypii,* was a heritable trait in cotton (Smith, 1989). These developments in plant breeding led the way to the discovery of other useful resistance genes and their incorporation into agronomically acceptable cultivars. The approach proved particularly successful and would have a major impact during the 1960s and 1970s, driving the highly successful Green Revolution though the breeding of high yielding varieties.

Subsidies were made available, making the production and use of pesticides more attractive to business and the farmers. It is not difficult to understand why – pesticides were and are effective (visibly so in many cases), are relatively low-cost, easy to use and versatile. They can be mass-produced, stored, distributed worldwide and sold through agricultural retailers for a range of pest species and crops. Chemical pesticides have been popular because they have suited the needs of farmers, industry and policy-makers as an efficient means of pest control helping to maintain productivity of high-input intensive cropping systems. Their use became institutionalized and farmers themselves became increasingly dependent on this single strategy (Zalom, 1993). Large farmers and industry wanted the benefits of the system despite increasing public concerns over safety and the environmental impact of pesticides (Morse and Buhler, 1997). However, the drawbacks of the solely chemical approach to pest management have become increasingly evident over time.

The resurgence of target pests, upsurges of secondary pests, human toxicity and environmental pollution caused by pest control programmes relying on the sole use of chemicals have caused problems in many cropping systems (Metcalf, 1986). The ecological and economic impact of chemical pest control came to be known as the pesticide treadmill, because once farmers set foot on the treadmill it became increasingly hard to try alternatives and remain in viable business (Clunies-Ross and Hildyard, 1992). In recent years, however, the availability of new active leads and the cost of meeting the increasingly stringent regulatory requirements has led to a decline of new product lines and the merger of a larger number of the major pesticide producers (see Chapter 4).

Pesticides influenced the development of IPM as a philosophy and in practice. It was the experience in the alfalfa fields of California's San Joaquin Valley of combining natural enemies, host plant resistance and use of chemicals pesticides in the 1950s against the spotted alfalfa aphid, *Therioaphis maculata,* that led to the integrated control philosophy of Stern et al (1959). At that time, as pesticide resistance and secondary pest outbreaks became an issue in the 1960s and 1970s, IPM was based largely on restricting pesticide use through the use of economic thresholds and the utilization of alternative control options such as biological products, biopesticides, host plant resistance and cultural methods (Thomas and Waage, 1996). During this period there was a proliferation in the development of new techniques and products to meet the growing need for alternative inputs to chemical pesticides. These included monitoring devices (insect traps), biopesticides such as *Bt*, semiochemicals (pheromones for mating disruption), insect predators and parasitoids, and the sterile insect technique.

In 1962, the gypsy moth pheromone was isolated, identified and synthesized. This inspired the search for semiochemical-based pest monitoring and control methods for the control of stored product pests, field crop pests, tree beetle pests and locusts. The first major success with sterile insect technique occurred in 1967 when the screwworm fly, *Cochlomyia hominivorax*, was officially declared eradicated from the US (Drummond et al, 1988). The first commercial release of *Bacillus thuringiensis*, a bacterium producing a product toxic to lepidopteran pests, occurred in 1972 (Burgess, 1981). This success encouraged research on biopesticides based on fungi, viruses and entomopathogenic nematodes.

The whole IPM concept was given greater prominence in the US by the large and influential Huffaker Project (1972–1979). This continued in the 1980s as the Atkinson project, focusing primarily on insect pest management in six crops: cotton, soybean, alfalfa, citrus fruits, pome fruits (apples, pears) and some stone fruits (peaches, plums) (Morse and Buhler, 1997). In this context, IPM was being developed and designed for intensive agriculture, undertaken by technologically sophisticated farmers.

A different model, however, came from south-east Asia with an emphasis on IPM training of farmers creating sufficient understanding of the interaction between natural enemies and hosts. IPM adoption across south-east Asia became political with various decrees and national policies – India and Malaysia declared IPM their official policy in 1985, and were followed by Germany and the Philippines in 1986, and Denmark and Sweden in 1987. At the 1992 United Nations Conference on Environment and Development, IPM was then endorsed as an effective and sustainable approach to pest management.

Sustainability has now become a major political driver for IPM. This is interesting because there has been a tendency within IPM to develop and employ control measures that provide only short-term solutions to problems. Such measures can work effectively only over a limited time-scale because, as soon as their use becomes widespread, pests will adapt and render them useless. Examples of this include prolonged use of a single insecticide, major gene plant resistance and *Bt* based insecticides. Such short-term measures may provide work for researchers and short-term economic gain for industry, but do not ultimately solve pest problems. However, it should be recognized that commercial companies are not necessarily in the business of alleviating the world's pest problems in a sustainable way, but rather providing solutions that will generate a viable income and maintain the longer-term prospects of the business. It is perhaps in this light that the latest technology available for use in IPM – transgenic crops – should be viewed.

The use of transgenic crops entered the agricultural production systems of a number of countries from the mid 1990s. The two most common traits in commercially cultivated crops are herbicide tolerance (to permit the use of broad-spectrum herbicides) and expression of *Bt* (to control lepidopteran pests). These have been incorporated mainly into maize, soybean, oil seed rape, beet and cotton. A number of small advances have been made with viral resistant crops, such as papaya in Hawaii and cassava in South Africa (Pretty, 2001; Nuffield Council on Bioethics, 2004). Most transgenic plants are grown in just four countries – the US, Canada, Argentina and China. One of the great challenges for the coming decades will be to explore how novel transgenic plants can be incorporated as safe and effective components of sustainable IPM systems (Hilbeck, 2002).

It is clear that the political landscape is changing in favour of more environmentally friendly and sustainable means of pest management. Consumers are now becoming a major driver in determining pest management practices, with retailers increasingly imposing good horticultural or agricultural practice standards on farmers (Parker, 2002). Opinions on pesticides have become polarized, with measures such as organic agricultural production gaining popularity. The agrochemical industry itself is not a unified body, with conflicts between producers of low-cost, 'generic' pesticides and newer, research-driven, proprietary products. Many of these newer agents, and most pesticides based on microbial organisms (biological pesticides or biopesticides), have orders-of-magnitude lower toxicity and greater specificity than those that first gave pesticides a bad public image. However, without appropriate application techniques, these more benign agents may be more costly or difficult to use than the older, broad-spectrum chemicals. Safe and rational use of chemical pesticides remains a necessity given the extent to which these products still dominate the pest control market.

PROBLEMS WITH PRODUCT REGISTRATION

Pesticide registration procedures and regulations provide a crucial means for both enabling and constraining the access and use of pest control products. At its extreme, this is about banning unsafe products such as those included in the WHO Class Ia products (e.g. aldicarb, lindane, chlordane and heptachlor), and the withdrawal of 320 pesticides across Europe in 2003. However, for the majority of products, it is about restricting their use to circumstances where any adverse effects can be minimized. However, such procedures can also be used as a constraint to the development and introduction of safer, more environmentally friendly alternatives.

EU legislation and prohibitive registration costs through the UK's Pesticide Safety Directorate (PSD) are now discouraging the development and commercialization of promising new biopesticide products. One reason is that the same regulations and evaluation criteria are applied to biopesticides as to new chemicals. This means that such biologicals are disadvantaged even though many are safer and provide longer-term, more sustainable pest control. The cost of registration is also an issue, with the cost of dossier evaluation by the PSD at £44,700 for biopesticides and £94,700 for chemical pesticides. These costs of registering biopesticides are disproportionate if the likely market size of these specialized niche products is taken into account. Because of their specificity (and hence environmental safety) the market size of a typical biopesticide is less than £6.4 million (excluding *Bt*), whereas imidacloprid (a relatively new insecticide product) has an estimated market size of £360 million (Bateman, pers. comm.).

In the US, Germany, Holland, Switzerland, Spain and France biopesticides enter a fast track and lower cost registration process that has aided the development of the biocontrol industry in these countries. The situation in the UK is preventing development of the industry and use of biopesticides. For instance, an application to the UK government for registration of a biopesticide based on the fungus, *Beauveria bassiana*, has been waiting for a licence for six years, even though this product is already licensed in Spain, Italy, Greece, Mexico, Argentina and the US, and is used as a matter of course on organic produce imported into the UK from these countries.

SUBSTITUTE SAFER PRODUCTS: BIOCONTROL AGENTS AND BIOPESTICIDES

Biological control agents include pathogens (bacteria, fungi, viruses) and entomopathogenic nematodes usually formulated as biopesticides, and insect predators and parasitoids. Biocontrol can be applied through introductions, augmentative releases, innundatively as biopesticides or through conserving existing field populations.

Conventional herbicides, fungicides and insecticides generally store well, have a relatively wide spectrum of activity and fast speed of kill, have relatively short persistence so need frequent applications, but have the potential for adverse environmental effects. By contrast, biological control agents tend to store poorly, have high target specificity and slow speed of kill, potentially long persistence through secondary cycling and hence need lower frequencies of application, but are environmentally friendly and present a low hazard to humans and livestock. However, there has been a tendency to develop and use some biocontrol agents and particularly biopesticides just as if they have the same properties as chemical pesticides.

The effectiveness of a biocontrol agent depends on two factors: its capacity to kill and to reproduce on pests (compounding its killing action) – in ecological terms, its functional and numerical responses. Currently, biopesticides based on viruses and fungi that have the potential for persistence and the compounding benefits of numerical responses, have been developed using the traditional chemical pesticide model involving quick kill, low persistence and frequent application. In this way, it is all the shortcomings of these biopesticides relative to chemicals that emerge, and few of the benefits. There may, therefore, be many opportunities to exploit the ecological benefits of biopesticides that are as yet little explored.

Biologicals cannot be successfully used against all pests. For instance, insects that feed directly on a harvestable product are considered low threshold pests because they cause damage at very low pest densities, such as *Cydia pomonella* on top fruit. The use of biocontrol agents for the control of these pests is generally not feasible because they are too slow acting to prevent damage – although infection by microbial agents will sometimes reduce feeding activity by an insect. However with high threshold pests that feed on foliage or stems and cause little yield loss at low pest densities are highly amenable to biological control, such as the red spider mite, *Panonychus ulmi*. Biocontrol agents are effective against a range of high threshold pests including aphids, whiteflies, stemborers, leaf miners, locusts and grasshoppers.

Hence, there are specific opportunities for the use of biological control agents, but as a group they do not provide a panacea as alternatives to chemical pesticides. Biological products currently represent just 1 per cent of the world market, and 80 per cent of that is taken just by *Bt*. Nonetheless, some commentators have estimated that biological control products could replace at least 20 per cent of chemicals, a market valued at US$7 billion. For this to be possible, however, the financial, regulatory and technical support to develop and expand the industry would need to be substantial.

RATIONAL PESTICIDE USE AND SAFE USE TRAINING

Rational pesticide use is the application of pesticides through optimization of their physiological and ecological selectivity. The physiological selectivity is characterized by differential toxicity between taxa for a particular pesticide and ecological selectivity refers to the operational procedures employed to reduce environmental contamination and unnecessary destruction of non-target organisms. Rational pesticide use (RPU) has been developed as a sub-set of IPM that combines four elements:

- selection of pesticides with low mammalian toxicity (preferably belonging to WHO toxicity Class III or lower, if used by non-specialist operators) and hence low impact on non-target organisms;
- accurate diagnosis of pest problems and forecasting of outbreaks;
- optimized timing of interventions to minimize pesticide use and maximize long-term efficacy;
- improved application of agents to maximize dose transfer to the biological target, reducing pesticide costs and minimizing residues and contamination to operators and the environment.

Practical measures to improve pesticide interventions include better diagnosis of pest problems and understanding of pest ecology. This enables growers to

make better informed pest management decisions (through participatory training and research). Pesticides also need to be developed with a greater specificity,
lower toxicity and ecotoxicity and have improved formulations, and safer and
more efficient pesticide application techniques need to be promoted through
applied research, database development and training.

A safe use project established by Novartis in 1992 in India, Mexico and
Zimbabwe involved communication campaigns addressing personal protection
and pesticide application, spray optimization, storage and disposal, pest and
beneficials identification, selection of pesticide, dosage and timing and improvement of farmer economics (Atkin and Leisinger, 2000). Overall it was believed
that the interventions did have a positive impact and a number of important
lessons were identified. Of greatest importance was that messages need to focus
on practical, basic, ready to use but effective options for farmers. A highly
technical approach is not needed to improve safety, and a small number of simple
changes made a big difference. In addition, the mix of communications media
used in each country was regularly refined during the project.

Effective training educates and informs, and raises the level of awareness of
the trainees and, under such conditions, it is possible to change behaviour.
However, research suggests that such changes in behaviour may be temporary
(Perrow, 1986; Spencer and Dent, 1991; Atkin and Leisinger, 2000). There is also
a concern that numbers trained may not necessarily equate to either changes in
behaviour or changes in the adoption of safer pesticide practices. Thus, the
possession of good knowledge about pesticides and how to use them safely may
not lead to actual safe practice. This has followed a vigorous debate about the
claimed and actual value of safe use programmes (Murray and Taylor, 2001).

Even though experience indicates that there is a limit to the extent to which
changes will be adopted within a generation (Atkin and Leisinger, 2000),
thousands of stakeholders (government extension agents, school teachers,
farmers, pesticide technical, sales and distribution staff, physicians and medics)
now have access to information that they did not have before. This represents an
improvement. However, one of the sobering messages to come out of such
studies is that social marketing campaigns have to be carried out on a sustained
basis – change cannot be maintained on time-restricted interventions only. There
is a pronounced need for continuing intervention to ensure persistent change and
no one has yet indicated either an ability or interest in supporting such long-term
costs.

HABITAT MANAGEMENT, ROTATIONS AND BIODIVERSITY

Conventional agriculture has had a major impact on soil structure, fertility, and
microbial and faunal diversity, resulting in an increase in root diseases unless
genetic resistance, soil fumigation and/or seed treatments are used (van Bruggen
and Termorshuizen, 2003). Although developed for a number of agronomic
reasons, rotational systems can be important in disease and insect pest suppression (Wolfe, 2002). A crop rotation may involve the use of agronomic techniques

such as catch cropping, cover cropping and green manures. The use of such techniques can lead to a higher diversity of the soil microbial flora and fauna and this in turn can have a beneficial impact on both soil health and plant health.

The length of rotation can be a key factor in determining whether or not a pest or pathogen can survive to reinfect a host crop species (van Bruggen et al, 1998). The length of the rotation has to extend beyond the time pathogens can survive without a host. A break of 2–4 years is usually sufficient to reduce the inoculum to a level that will allow the production of a healthy crop (Wolfe, 2002). However, one of the difficulties of evaluating the impact of rotations, cover crops and soil cultivation is that they tend to be highly location and cropping system specific. Nonetheless, a knowledge of soil food webs and their interactions with plants rises, and it is becoming increasingly possible to specify rotations, varieties and treatments more precisely that will minimize diseases, pests and weeds (Wolfe, 2002).

Organic farms tend to operate diverse crop rotations and combine these with incorporating livestock, use of both autumn- and spring-sown crops and incomplete weed control. However, crop protection in organic farming is generally not directed at controlling particular pathogens or pests but at the management of the whole environment so that plants are able to withstand potential attacks. The main practices that contribute to disease control are long, balanced rotations, organic amendments and reduced tillage, all geared towards maintenance of the soil organic content and fertility (van Bruggen and Termorshuizen, 2003). Integrated farm management takes a similar holistic approach to crop management within the overall context of the farming system. Farms that practise non-inversion tillage are likely to benefit not only soil fauna but also birds (Leake, 2002), while the introduction of management features such as field margins, hedgerows and set-aside can mitigate against single species dominance of a crop area.

Plant Breeding and Transgenic Crops

In the 1960s, the development of high yielding cultivars of wheat and rice heralded what has become known as the Green Revolution. At the beginning of the 21st century, plant breeding is undergoing another revolution with the introduction of techniques of genetic modification that permit the development of transgenic crops. This development represents one end of a continuum in the control of crop genetics that starts with farmer selection and the reuse of healthy seed taken from the most vigorous crop plants. In between is a broad range of plant breeding techniques that involve various degrees of control of genes associated with desirable characteristics. The traditional plant breeding methods utilized to develop resistance to pest insects, pathogens and plant nematodes involve a lengthy process by which appropriate characteristics are selected over many generations. This process has been of limited success with insect pests but more successful for pathogens.

The primary goal of conventional breeding is to produce higher yielding, better quality crops and, once that has been achieved, resistance characteristics can be incorporated, provided the method for introducing the resistance can itself be readily integrated into the breeding programme. Resistance, once it has been identified, is incorporated into plant material already having high yield potential and other favourable agronomic characteristics. For this to be possible, a good source of resistance is required and it must be controlled by simple inheritance, that is controlled by major genes, so that it can be easily incorporated using backcrossing breeding methods. A good source of useful characteristics in conventional breeding is the germplasm collections and pre-adapted plants of wild relatives of major crop species. By contrast, transgenic techniques make it possible to introduce novel non-plant genes or unrelated plant genes into crop cultivars.

The applications of biotechnology in agriculture are still in their infancy (Pretty, 2001; Persley and MacIntyre, 2002). Most current genetically modified plant varieties are modified only for a single trait, such as herbicide tolerance or insect resistance. The rapid progress being made in genomics may enhance plant breeding as gene functions and how they control particular traits are better identified. This offers the prospect for the successful introduction and breeding of more complex traits such as drought and salt tolerance.

The bacterium *Bt* produces more than 50 insect toxins that are protein crystals controlled by a number of *Cry* genes. The first isolation and cloning of a *Bt* gene was achieved in 1981, and the transformation of the endotoxin into tobacco first occurred in 1987, with the first field trials beginning in 1993 in the US and Chile (Vaeck et al, 1987). Transgenic *Bt* cotton was field tested in replicated trials in 1987 and in 1995 bulking of seed for commercial sale was undertaken (Harris, 1997). The first commercial cultivations were in 1996 in the US, since when the area of all transgenic crops worldwide has grown to 60 million hectares. Other crop and input trait combinations currently being field tested include viral resistant melon, papaya, potato, squash, tomato and sweet pepper, insect resistant rice, soybean and tomato, and disease resistant potato (Persley and MacIntyre, 2002).

Genetically modified herbicide tolerant crops (GMHT) can be grown under a regime of broad-spectrum herbicides applied during the growing season, providing good weed control (Persley and MacIntyre, 2002; Champion et al, 2003; Firbank et al, 2003). GMHT crops include cotton, soybean, maize, sugar beet and oil seed rape and GMHT vegetables will soon be available. The concern with such GMHT crops is that the broad-spectrum herbicides used on a commercial scale on some crops may be more damaging to the agricultural landscape and farmland biodiversity than some of the more selective herbicides they replace (Firbank et al, 2003).

Of course, there remain many concerns about the environmental effects of some transgenic crops, such as gene transfer to non-GM crops. However, the larger issue is whether or not there exists alternative crop development approaches that will ensure equally high yields, resistances to specific pests, tolerance to local environmental conditions (e.g. salinity) and, in addition, contribute to greater sustainability. Over the past century plant breeding and other management aspects of crops have been gradually taken away from farmers and researched, developed and sold as off-farm inputs. However, addressing problems in agriculture requires a move from the pathosystem to ecosystem.

The implication is that the level of detail and understanding that is required in order to make pragmatic decisions (by the farmer) is not that supposed by researchers (corporate or otherwise) attempting to manipulate a single pest in a single crop (Dent, 2000). Integrated farm management is moving towards lower inputs and a return to the use of cultural control methods. On this basis alone, there will be a future need for seed that is adapted to local needs and local pest problems. Participatory plant breeding/varietal selection and crop improvement represents the start of this trend (Altin et al, 2002).

THE CUBAN BIOCONTROL INDUSTRY – A CASE OF CHANGE

With the breakdown in trade relations with the former USSR in 1990 and the continued economic and political blockade of Cuba by the US, pesticide imports to the island declined by more than 60 per cent (Altieri and Nicholls, 1997; Funes et al, 2002). In 1990 the Cuban President declared the start of the 'Special Period in Peacetime' that introduced policy reforms to enable the island's agricultural and economic productivity to be rebuilt. As a result, Cuba moved from high-external input to low-input and organic agriculture, including the implementation of biological, control-based, integrated pest management approaches throughout the country. The process included the break-up of state farms into smaller units under more direct management by producers; the creation of a national network of small laboratories producing a variety of biocontrol agents, botanical pesticides and bio-fertilizers; legalisation and promotion of private sector farmers' markets; widespread development of urban agriculture; and a new emphasis on farmer-to-farmer and farmer-to-extensionist exchanges, on-farm research and agroecological training for producers and scientists alike.

The Cuban government's IPM policy focused on biological control in its search for techniques that would enable biologically sophisticated management of agroecosystems (Rosset and Benjamen, 1994). Earlier, Cuba had experienced several decades of biological control. This was mainly with mass-reared parasitoids, particularly *Lixophaga diatraeae*, used since 1968 for the control of the sugar cane borer in almost all sugar cane in the country, and Trichogramma in the 1980s against lepidopteran pests in pasture, tobacco, tomato and cassava, and predatory ants for control of the sweet potato weevil, *Cylas formicarius*. Based on previous experience of success with biocontrol, the Ministry of Agriculture significantly accelerated and expanded the production of natural enemies to replace the lost pesticide imports. Key components of the strategy were the Centres for the Production of Entomophages and Entomopathogens (CREEs), where the artisanal production of biocontrol agents takes place. By 1994, 222 CREEs had been built throughout Cuba and were providing services to cooperatives and individual farmers.

The main products produced by the CREEs include *Bt* for lepidopteran and mosquito control, *B. bassiana* for coleopteran pests and particularly weevils attacking sweet potato and plantain, *V. lecanii* for whitefly control (*Bemisia tabaci*), *M. anisopliae* for a range of insect pests and Trichoderma spp. for soil borne pathogens particularly in tobacco. Production of *Nomuraea rileyi* and *Hirsutella thomsonii* are also undergoing production scale-up. However, there are a number of obstacles to the uptake and effective use of these products. Quality standards are often poor and production output is limited by resources such as a regular power supply, and producers and extensionists still tend to be unfamiliar with biocontrol agents, which limits their effectiveness.

Other changes have been introduced as part of an integrated approach. Whereas monoculture was predominant in Cuba, since the start of the Special Period there has been an increase in intercropping. The most widespread example is the maize/sweet potato intercrop for reducing sweet potato weevil and armyworm infestations, enabling high productivity without pesticides. The Cuban experience is perhaps unique in that the severity of the reduction in availability of chemical pesticides has driven the process of change.

DOMINICAN REPUBLIC ORGANIC PRODUCTION – A CASE OF CHANGE

The continuing and growing demand for organic produce is increasing the opportunities for non-chemical pest control inputs. The highest values of total organic food sales are in the US (2000 figures) (US$8 billion), followed by Germany (US$2.1 billion), the UK (US$1 billion) and Italy (US$1 billion). Overall the US and European markets are much the same size. Although these markets are only a relatively small proportion of the total food markets in these countries, demand is continuing to grow and exceeds domestic supply. Organic production and export sales of these crops from the Caribbean to the US and Europe is increasing and, hence, the market for non-chemical control products is growing in order to meet the pest control needs of these organic producers.

The main market for biopesticides in the Caribbean is currently being driven by the organic production systems for commodities in the Dominican Republic. Organic production figures available from the Dominican Republic indicate that the principal organic crops grown are banana, cocoa, coffee, orange, mango, lemon, coconut sugar and pineapple with 80 per cent of produce exported to Europe. The total export of produce was in the order of 53,000 tonnes, worth US$20 million. Organic food production is now spreading more widely across the Caribbean, providing a new niche market for biopesticides.

The use of biological control began in the early 1970s when parasitoids were released into cabbage fields to control the diamondback moth (*Plutella xylostella*). However, the supplies were too infrequent and in numbers that were too small to have an impact and the effort was abandoned. Nonetheless, the country now has a thriving organic industry and there is a commitment to research on the development of biopesticides. Currently, retailers are selling the imported biopesticides Turilav based on *B. thuringiensis*, Beauveril and Brocaril based on *B. bassiana*, Biostat based on *Pycealomyces lilacinus*, Destructin based on *M. anisopliae*, and Vertisol based on *V. lecanii*. In Jamaica and Trinidad, biopesticides make up 5.6 per cent and 8.3 per cent respectively of the total pesticides on sale in the retailers surveyed compared to 28 per cent in the Dominican Republic. The situation with regard to the percentage of insecticide based biopesticides is even more marked. Jamaica and Trinidad have only 3–4 products each (all *Bt*) compared with the Dominican Republic where retailers have seven different *Bt* product formulations and five fungal based products for insect control (37.5 per cent of the total insecticide products available from the retailers). In addition, Dominican retailers had biopesticide products for sale for use against nematode and pathogen pests. The market for organic produce is driving the demand for biocontrol based solutions for pest and disease management.

This experience of organic agriculture in the Dominican Republic holds valuable lessons for the Caribbean region. Agriculture, particularly bananas (it is the largest exporter of organic bananas to Europe), is of great importance. However, given the increasing pressures and demands of an open market, the agriculture sector is now facing new challenges. These relate to the need to adopt sustainable production methods and high-quality products, as well as an urgent need to improve competitiveness and extend participation in global markets. The increasing interest in tropical fruit and the growing market for organic products in the US and Europe has resulted in an increasingly important role for exported organic products. Organic banana production first began in 1989; it now involves some 2500 smallholder farmers and is being seen to have a major impact on poverty alleviation in rural areas. The growth of organic production has been sustained by developing high levels of knowledge of organic production measures and awareness of market requirements among producers. The future market potential is considered promising, especially for newly developing European markets. Direct trading links established for organic products between producers and commercial organizations have been seen to bring increased benefits to the producers.

Organic products exported from the Dominican Republic now amount to around 20 per cent of all fruit and vegetable exports and a wide range of crops are produced organically. Increasing revenues for small- and medium-scale farmers, who generally tend to be located in areas of rural poverty, is of crucial importance and can be achieved through an increase in high value export opportunities. The socio-economic benefits of organic production are seen in the high level of involvement of small- and medium-scale farmers and the agrarian reform associated with this, such as the formation of cooperatives, which provide important opportunities for rural development and mutual advancement.

CONCLUDING COMMENTS

Many agrobiologicals represent safe and effective alternatives to chemical pesticides, but systems of registration and regulation tend not to favour them.

IPM requires the availability of a range of options to farmers in order to ensure long-term control of pests, diseases and weeds. Pest management can be made safer by: eliminating the most hazardous products (e.g. banning WHO Class I products and introducing tougher regulatory constraints through registration procedures); substituting with safer biocontrol agents and biopesticide products; implementing administrative controls that emphasize training and education in the safe use of existing products and improving agroecological knowledge; making available personal protective equipment only as a measure of last resort – research has shown that it often gives a false sense of safety when, in fact, residues are still reaching the user (Fenske, 1993). This last step should only be taken after the most hazardous pesticides have been eliminated, IPM and less hazardous categories of chemicals and substitutes are available and educational activities instigated.

Chapter 6

Farmer Decision-making for Ecological Pest Management

Catrin Meir and Stephanie Williamson

INTRODUCTION

As global agribusiness and the world's food systems undergo restructuring and concentration, so crop production systems are having to change with respect to the provision of farm inputs, the relationships between trading enterprises and individual farmers, and the type of produce demanded by consumers (Reardon and Barrett, 2000). As a result, farmers in both developing and industrialized countries are increasingly faced with rapid and profound changes in production technologies, processing and purchasing systems and market requirements (Williamson, 2002). These changes mean that to remain competitive in global markets, farmers require new management skills and knowledge (Blowfield et al, 1999; Hellin and Higman, 2001). Amongst these, sound decision-making about pest management strategies and pesticide use is particularly critical, even for farmers growing mainly subsistence crops for local consumption, in view of rising production costs, increased competition and consumer concerns about food quality and safety.

This chapter reviews what is known about farmer decision-making for pest management and why it is important to understand decision-making processes if farmers are to be successfully motivated to reassess their approaches to pest management, as well as become more aware of alternatives to pesticides. We describe farmer perceptions, the external influences on farmer decision-making, some of the training and agricultural extension methods that aim to influence farmers' pest management knowledge and practices, and review several different cases from developing countries.

UNDERSTANDING FARMERS' DECISION-MAKING

A good understanding of farmer decision-making is particularly important in pest management in view of the increasing need for farmers to adopt IPM (integrated pest management), rather than rely only on pesticides. However, IPM does not just require a farmer to be aware of and decide to use a new pest management technique. It requires more input to the decision-making process by farmers than does the application of pesticides (Matteson et al, 1994). Farmers need more knowledge of their agroecosystems, and the skills to select and manage the most appropriate pest management methods. Pest management problems are often complex, requiring information about many factors. This complexity is compounded by the fact that farmers usually have incomplete information about both the problem and potential techniques to manage it (Norton, 1982). Much pest management research has therefore aimed to develop decision tools and expert systems that can be used to make sound decisions on behalf of farmers (Norton and Mumford, 1993).

In general, evaluations of farmer decision-making in pest management have concentrated on what happened (or not) as a result of an intervention, not on why (or why not). Over 20 years ago, Tait (1983) observed that research tended to concentrate on the results of farmers' decision-making, not the process. This is still true today: useful information on farmers' thinking and decision-making processes remains scarce. Yet such information is important: pest management decisions are subjective and depend very much on the individual farmer's knowledge, goals, resources, risk averseness and values.

Norton and Mumford (1993) proposed a basic decision model based on four key influences: the pest problem (including level of attack and the damage caused), control options available, farmers' perceptions of the problem, and of the availability and effectiveness of control options; and farmers' objectives including monetary goals and attitude to financial risks, health hazards and community values. These factors influence farmers' assessment and evaluation of the problem, choice of control method and the outcome of that method, which in turn influences future pest levels. The importance of farmers' perceptions was further emphasized by Heong and Escalada (1999), who proposed a pest belief model to explain the use of pesticides. Their model includes perceived benefits of a certain action in reducing the perceived susceptibility or severity of the pest attack, along with any perceived negative aspects of a particular action. To these perceptions, they added the influence of an individual's belief in the effectiveness of spraying pesticides and their belief in the threat of crop loss. The resulting combination of these perceptions and beliefs determine farmers' spray behaviour.

Others have also highlighted the importance of farmers' perceptions in pest management, but rather than proposing models, have emphasized the importance of dedicating more research effort to understanding farmers' particular points of view (Bottrell, 1984; Bentley, 1992a, 1992b; Paredes, 1995; Gómez et al, 1999, Meir, 2004).

WHAT FARMERS KNOW AND THINK

Understanding farmers' perceptions is clearly essential if we are to understand farmers' pest management decision-making. A wide range of factors can influence farmer perception of pest management and we have identified eight key issues (Box 6.1). In the past, these issues have often been largely ignored: most researchers and extensionists can recount anecdotes about specific IPM technologies that remained unadopted because important farmer perceptions, practices or resource constraints were not considered. More recently, however, there is evidence of a welcome shift towards studying farmer perceptions as an integral part of pest management research and implementation (Motte et al, 1996; Adipala et al, 2000; Ebenebe et al, 2001).

Understanding what farmers know and think requires careful analysis. Distinctions have often been made between farmer 'knowledge' and 'beliefs', with outsiders referring to ideas which coincide with scientifically accepted facts or theories as 'farmers' knowledge', and anything else as 'farmers' beliefs'. However, farmer knowledge dismissed by outsiders as being mere beliefs has not infrequently turned out either to be empirically valid or to be based on scientific facts which were not immediately obvious to the outsiders in question (Bentley et al, 1994). Moreover, farmers' apparent lack of knowledge of phenomena such as plant diseases (especially as expressed in scientific terms) does not necessarily mean that they lack management strategies for these problems (Fairhead, 1991). Farmers' knowledge may also be expressed in very different terms to those that outsiders would use, since knowledge tends to be constructed according to what is important to local people and the culture in which they live. To understand and appreciate farmers' knowledge, outsiders need not just to learn what farmers know, but how to look at the world through farmers' eyes.

Box 6.1 *Eight issues affecting farmer perceptions and their pest management decisions*

1   Expectations, needs and desires (e.g. ability and need to minimize risk; desire to keep fields 'clean'; need to conform to specifications laid down by external agents)
2   Experience and perception of biophysical conditions (e.g. past experience and current expectations of climatic conditions; past experience of losses associated with pest problems)
3   Knowledge, belief and experience of pest management strategies (e.g. knowledge of pest causing the perceived problem; experience of efficacy and cost of pest control methods)
4   Perception of pesticides (e.g. perception of the efficacy and necessity for pesticides; personal exposure to the consequences of pesticide misuse)
5   Availability of resources (e.g. timely availability of and access to pest control inputs; comparative and opportunity cost of labour; availability of and perceived trustworthiness of advice)
6   Market related factors (e.g. potential market prices; market demands in terms of quality or specifications)
7   Opinions and information from others (e.g. of family, friends and neighbours; influence of pesticide advertising)
8   Others' actions (e.g. neighbours' pest management strategies and their perceived success; strategies of farmer leaders)

Nonetheless, farmers do not know everything. They rarely know the causal agent of pest problems, especially for plant diseases (Sherwood, 1995; Sherwood and Bentley, 1995; Paredes, 1995), and may well confuse insect- with disease-caused damage (Kenmore et al, 1987; Meir, 1990). Farmers often assume that any insect in their crop causes damage, not being aware of the existence of natural enemies (Bentley, 1992a), and feel that insect damage will automatically result in crop loss, not realizing that, in some cases, plant compensation may mean yield loss that is, in fact, minimal (Heong and Escalada, 1999).

Bentley (1992b) has proposed that farmers know more about things that are easily observed and culturally important, and less about things which are difficult to observe or not perceived as important. This analysis provides a useful guide as to what may influence farmers' decision-making. It is also important to bear in mind that cultural and/or social factors may mask who actually makes the on-farm decisions, and therefore whose knowledge is important. For example, women are often assumed to have no input into pest management decisions, especially if they have no presence in the field, whereas in reality they may be important decision-makers in terms of allocating cash or labour inputs (Rola et al, 1997).

Farmer perception of risk has a particularly strong influence on pest management decisions. Heavy losses, particularly those that occurred in the recent past, can influence farmers to spray more pesticides in subsequent seasons, rather than basing their decisions on actual pest populations or on the returns from spraying (Mumford, 1981; Waibel, 1987; Heong and Escalada, 1999). When farmers' decisions to apply pesticides are taken on the basis of worst losses experienced or unrealistic expectations of pesticides' abilities to control pests, they will often be uneconomic (Mumford, 1981; Heong and Escalada, 1999).

Perceived crop losses and the effectiveness of pesticides are not the only influence on farmers' decisions to use pesticides. Farmers generally like pesticides because they are easy to apply and usually less labour intensive than alternatives (Rueda, 1995). They also like it when pests are killed quickly. The decision to apply pesticides, and the choice of which to apply, is often reinforced by what farmers' neighbours do. In the case of herbicides, many farmers feel that the appearance of clean, weed-free ground denotes a good farmer (Meir, 1990; Castro Ramirez et al, 1999). Pesticide producers and retailers tend to reinforce farmers' tendencies to see pesticides as the sole answer to pest problems, which has all the more effect due to the absence of alternative information (Heong and Escalada, 1997, 1999). However, Gomez et al (1999) have observed that farmers' fear of loss due to pests tends to decrease as they become aware of a range of management techniques, rather than feeling they have to rely only on applying pesticides. Personal experience of health problems related to pesticide use can also lead to the reduction or elimination of pesticides (Meir, 1990).

Farmers' pesticide decisions may, in addition, be affected by outside influences, either from the private or governmental sector. Cotton growers in South India, for example, are almost entirely reliant on local moneylenders for inputs of credit. Agrochemicals are closely tied in to this credit system, leaving farmers with little choice in their pest management techniques, even if they want to use

IPM approaches (Verma, 1998). During the Green Revolution era, national government research and extension systems removed farmers' decision-making power through direct state intervention in pest management via calendar spraying regimes and enforced control methods. The advent of more participatory extension methods has led some governments to recognize that farmers' knowledge and involvement is more important than the transfer of pre-packed technologies, and in some cases to start devolving decision-making power back to farmers (Fleischer et al, 1999). Local government influence can be important too: the mayor of one district in the Philippines banned all local advertising of synthetic pesticides in order to support an IPM training programme (Cimatu, 1997).

Consumer demand and market requirements also exert an increasingly weighty influence on pest management decisions, even on farmers in developing countries. The competitive and cosmetic demands of many African vegetable markets can lead farmers to apply pesticides several times a week, often in ignorance of (or ignoring) stipulated safety periods before harvest (Williamson, 2003). However, emerging price premiums available for pesticide-free or organic produce can be a powerful motivating factor for farmers, leading them to change their pest management decisions and even to generate demand for specific IPM methods (Williamson and Ali, 2000).

In practice, however, there are always difficulties in adopting new systems. In South Africa, citrus growers are starting to adopt IPM, mainly to stay in business as synthetic pesticide costs increase. However, this change is also in response to the deregulated export market, which enables national exporters to differentiate between IPM and conventional produce. Fewer export markets now accept non-IPM fruit and there is a current competitive advantage with higher prices for IPM produce, which provides incentives for farmers to change their practices (Urquhart, 1999). However, IPM production in citrus requires more intensive management and more time in administration, and some farmers are unwilling to take on these extra burdens.

IMPACT OF OUTSIDE INTERVENTIONS ON FARMERS' DECISION-MAKING

All IPM programmes deliberately seek to influence farmers' pest management practices and some specifically aim to improve their decision-making. This section looks at some examples of interventions that have resulted in farmers altering their management practices, and examines the relative merits and successes of these approaches.

The dominant paradigm for connecting the results of agricultural research with pest management in the field was for many years the transfer of technology model. This concentrated on transferring results developed mainly on research stations to farmers in the field with the help of extension staff. It is perceived by many as having considerable shortcomings (van Huis and Meerman, 1997;

Röling and de Jong, 1998). It has failed to provide pest management techniques appropriate for the highly varied biophysical and socio-economic environments in which many farmers grow their crops (Rowley, 1992). It has also failed to acknowledge and benefit from farmers' experience and abilities (Hagmann et al, 1998), concentrating instead on providing farmers with simplified and generalizable recommendations. As a result, technology transfer approaches have had limited impact on improving farmers' decision-making, and hence on farmers' ability to cope with variations in environmental and economic factors (Visser et al, 1998).

At the same time as technology transfer models came under criticism, farmer participatory approaches to training and research started to be used by a range of agricultural development institutions. The aim was to empower farmers as active subjects in the process of making crop production more sustainable, replacing their conventional role as passive recipients of information and inputs. Learning, as a process of knowledge construction and interpretation by individuals interacting with their ecological and cultural environment, is replacing the earlier extension theory of pouring relevant facts and messages into the empty vessels of backward farmers (Pretty and Chambers, 1994). Ter Weel and van der Wulp (1999) provide an excellent review of these research and training issues in relation to IPM.

The following sections outline five examples of interventions in farmer decision-making that have aimed to change farmers' behaviour, fill gaps in farmers' knowledge, and improve farmers' decision-making capacity.

DEMONSTRATION PLOTS IN CONJUNCTION WITH HOLISTIC MODELLING: COCOA, INDONESIA

This programme, run by the Indonesian estate crops government agency in collaboration with Imperial College, London, provides smallholder Indonesian cocoa growers with crop protection and production messages for the effective management of the cocoa pod borer moth (CPB) (Mumford and Leach, 1999). CPB control problems had added considerably to the cocoa production crisis, with low yields exacerbating poor management. A critical issue was that the interactions of factors in CPB management are too complex for farmers to understand or assess themselves within a short time span (for instance, measuring weekly or average percentage of pods infested by CPB gives unreliable or misleading information). Furthermore, the pest problem was a new one in a relatively new crop, so farmers lacked the experience to develop their own solutions to the problem.

The project combined research and demonstration plots over several seasons, and used a simulation model of cocoa production, pest dynamics and pest management costs and effectiveness. This allowed the likely impact of different harvesting regimes and management practices to be tested, and sound production advice to be drawn up. Demonstration plots and analysis showed that complete, frequent and regular harvesting every seven days was the best strategy in terms of highest gross returns, compared with farmers' practice of monthly harvesting. Farmers' initial reaction to the main recommendation was that this would involve more work. But the improved yield and income from the recommendation plots was impressive, and the returns far exceeded the extra labour required. Farmers were able to appreciate this through cost–benefit analyses of the demonstration plots in which they participated. As a result of their trust in these results, they more readily accepted further advice on strict plantation floor hygiene.

PERSUADING FARMERS TO TRY A SIMPLE HEURISTIC FOR RICE IN THE PHILIPPINES AND VIETNAM

Asian rice farmers often apply insecticides at the wrong time or for the wrong targets, particularly foliage feeders such as leaf-folder caterpillars (Heong et al, 1995). The approach developed by the International Rice Research Institute (IRRI) and communications researchers at the University of Los Baños to tackle farmers' perceptions involved persuading farmers to try out a 'rule-of-thumb', also known as a heuristic (Heong and Escalada, 1997). The heuristic, which had first been proven on the research station, was this: *'Leaf-folder control is not necessary in the first 30 days after transplanting.'* The goal was to see if using the advice, which was in conflict with their current practice, influenced a change in farmers' pest management perceptions and practice. Farmers who participated in half-day training sessions reported higher yields, a reduction in pesticide applications, and a shift of the first pesticide application towards later in the season. Conducting a simple experiment to test a heuristic generated from research was concluded to be an effective way to improve farmers' pest management decision-making. Farmers were felt to be more likely to participate in an experiment if they perceived the problem to be of economic significance and the source of conflict information to be highly credible.

In Vietnam, the heuristic was modified to the *'...first 40 days after sowing'* (Heong et al, 1998). A mass media campaign with this message included distribution of over 20,000 leaflets, drama slots on radio, and billboard poster publicity. More than two years later, the number of insecticide sprays had dropped from an average of 3.35 to 1.56 per season and those who did not use insecticides at all increased from 1 per cent to 32 per cent of farmers. The proportion of farmers who believed that leaf folders could cause yield losses fell from 70 per cent to 25 per cent. Cost savings on insecticides and labour was the most important incentive for farmers to stop early season spraying, as cited by 89 per cent of farmers surveyed. The mass media approach was readily adopted by extension services in 15 Vietnamese provinces, covering 2 million farming households in the Mekong delta.

COMPLEMENTING FARMERS' KNOWLEDGE: MAIZE, BEANS AND VEGETABLES IN HONDURAS

The aim of a novel Natural Pest Control Course in central Honduras was to complement farmers' existing knowledge by enabling them to learn about insect reproduction and about natural enemies, and to stimulate their inventiveness with these new ideas (Bentley, 2000; Meir, 2000). The course was based on the premise of listening to and valuing farmers' knowledge and experience, and building new ideas on this base. The trainers made extensive use of practical sessions and inductive questioning to bring out farmer knowledge and make a bridge between existing knowledge and new ideas developed from research. This enabled the course concepts to be built on farmers' existing knowledge, which was then combined with new observations prompted by the training sessions, and small pieces of information added by the trainers.

A total of 512 natural control techniques were adopted, adapted, invented or reinforced by 100 farmers (interviewed up to four years after training) as a result of the course, with techniques involving natural enemies being the most popular. Farmers said that the new techniques had increased their productivity and improved the farm environment. Some 40 per cent of the techniques directly replaced pesticide use (mostly in WHO Classes Ia and Ib). Many others were used by farmers who could not afford pesticides. The training helped many farmers become established users of natural control, with sufficient confidence to continue experimenting with these approaches for themselves as well as to demand further information. They applied the concepts they had learned to other pest problems and developed new solutions, continuing the process of learning through observation and experiment. Moving to large-scale use of natural control was not a one-off event for farmers, however, but a process that took time and required continued support and encouragement throughout for its success to be optimized.

PARTICIPATORY RESEARCH AND EXTENSION: COFFEE, VEGETABLES, BANANAS AND PLANTAIN IN NICARAGUA

Since 1995, the CATIE-IPM Program has educated extension staff to work with farmers to strengthen their decision-making in pest management, with an emphasis on the active participation of women and children (CATIE-INTA, 1999; Gómez et al, 1999; Padilla et al, 1999). A critical part of the IPM training is to help farm families understand the considerable ecological and biophysical variability in their cropping systems and to improve their evaluation and management skills.

Farmer training is based around discussion of farmers' knowledge and current practices, together with group experimentation with cultural and biological control options (research- or farmer-derived), and joint evaluation of results including cost–benefit assessment. Sessions are linked to critical periods in crop growth stages and focus on key pests and diseases at each stage. The programme works mainly with smallholder farmers but also includes similar training for farm managers and workers on large private and cooperative estates. The groups establish pest and disease monitoring via regular field scouting methods, and select options from a range of different practices to solve their local pest and soil problems. Impact evaluation confirmed that the process had strengthened farmers' knowledge and analytical capacity while developing their experimentation skills and increasing their dissemination of IPM ideas to other farmers. Economic data from seven coffee estates, for example, indicated that yields improved by 28 per cent, coffee berry borer management costs were reduced by 17 per cent, and borer damage to export beans dropped from 7 per cent to 0.8 per cent.

DECISION-MAKING AND EXPERIMENTATION SKILLS FOR INTEGRATED VEGETABLE MANAGEMENT IN SOUTH-EAST ASIA

The farmer field school (FFS) has now become the best-known approach to IPM training. The FAO Community IPM training programme in rice pioneered the development of farmer-centred discovery-learning and participatory methodologies in the early 1990s to enable farmers to understand the agroecology of pest, natural enemy and crop interactions and the negative impacts of pesticides (Vos, 1998, 2000, 2001; van den Berg, 2001; van den Berg and Lestari, 2001). As a result, rice FFS farmers drastically reduced their reliance on insecticides and the programme has since expanded to cover a wide range of cropping systems in three continents (FAO, 2001d). FFSs continue to develop: some FFS programmes have evolved towards field experimentation by farmers and farmer field school rice curricula have been adapted to the needs of vegetable farmers to help them learn about integrated pest, crop, nutrient and water management.

In FFS, the field is the primary classroom for training farmers and extension staff, building on four key tenets: (i) grow a healthy crop; (ii) observe fields weekly; (iii) conserve natural enemies; (iv) help farmers understand ecology and become experts in their own fields. Farmers collect data in the field, analyse it in groups along with trained facilitators, and then decide on any action to be taken. Through weekly agroecosystem analysis of a sample of plants, they compare their current practice with plots where integrated crop management methods are used. Direct observation is complemented by exercises to demonstrate or visualize important processes, such as plant compensation for mechanical damage, insect predation and parasitism, causal agents of plant diseases and their spread, and the action of pesticides on pests and beneficial organisms.

Vegetable FFS programmes have incorporated a strong element of participatory action research with FFS-trained farmers to develop and adapt location-specific strategies. This is particularly important where off-the-shelf systems did not exist, for example, in soybean and sweet potato. Since 1999, FFS Participatory

Action Research (PAR) groups in Vietnam have focused on soil-related diseases causing major yield loss in tomato, beans, cabbage, cucumber and onion. PAR use the disease triangle framework to assess host–pathogen–environment interactions, identifying which management options are applicable, which are already used by farmers and which are incorrectly applied, for example, poorly composted material in which disease spores have not been killed. In cucumber, PAR research groups increased yields by 80 per cent and reduced the incidence of blight, so enabling farmers to reduce pesticide use from nine to five sprays per season.

THE RELATIVE MERITS OF MESSAGE-BASED AND LEARNING-CENTRED APPROACHES

Although these examples of interventions had different objectives and were designed to fit a variety of different circumstances, they can be broadly grouped into two categories: message-based and learning-centred. Both categories differ from a conventional technology transfer approach in that they concentrate on enabling farmers to experience, or at least see, the effects of new pest management methods or strategies for themselves, and how and why these work. Both approaches have also been effective at changing farmers' pest management practices but there are important differences between them. Message-based interventions are an evolution of the transfer of technology model, concentrating on the pest management message, or the end result of farmers' decision-making: changes in farmer behaviour. Learning-centred interventions, on the other hand, represent a radical change in the approach to extension, since they focus on affecting farmers' decision-making processes, rather than any single end result. Unlike message-based interventions, the success of learning-centred approaches is not judged on the adoption of technologies, but on whether farmers are subsequently able to make more informed decisions, and to benefit from them (Visser et al, 1998).

Learning-based approaches exhibit some or all of the following principles:

- Farmers learn by doing: learners find out or discover the new ideas for themselves in a hands-on process; these learning processes are much more effective than conventional teaching methods, especially for adults (Rogers, 1989).
- Farmers are in charge of the learning process and are able to adapt it to their own conditions, incorporating their own criteria and maximizing the relevance to them of what they learn (Visser et al, 1998).
- Farmers learn about the ecological principles underlying pest management for themselves, through guided experimentation and discussion (van de Fliert, 1993; Vos, 1998); old information is reconstructed through the learning activities, and integrated with new information and ideas (Meir, 2004).
- Farmers gain skills and experience in applying their new knowledge to different pest problem situations (Hamilton, 1995); this is a crucial part of the

learning process, since integrating and applying principles to real life situations is a skill in itself (Röling and de Jong, 1998).

- Farmers are motivated and empowered by the process of the training and by the nature of what they learn, with the exchange of ideas and experience by farmers and the opportunity to learn formal science and carry out research giving them prestige in their own and others' eyes (Kimani et al, 2000).

Learning-based approaches are clearly more of a process, compared with message-based approaches, which can be viewed as more of an event. Each approach has advantages and disadvantages (Table 6.1), and different pest management problems may require one or the other approach, or a combination of the two. Message-based approaches are more appropriate where there is a simple message to communicate, where the message produces reliable, readily observable results relatively quickly, and has a clear economic benefit in farmers' eyes, and where the techniques do not require a deep understanding of the crop ecosystem, or high levels of skill to apply effectively. They can also be effective where there is a pest management crisis requiring urgent action, and pest–crop–market interactions are too complex for farmers to assess effectively. By contrast, learning-centred approaches may be more appropriate where there is no simple message to communicate, when the pest problem is complicated (requiring manipulation of a number of techniques and hence a deeper understanding of the crop ecosystem), and when farming conditions are heterogeneous, so that farmers need to be able to adapt techniques to meet their needs.

Clearly, interventions may combine messages with conceptual learning and improved knowledge of pests and beneficials. One example is IRRI's expansion of the heuristics experimentation approach to enable rice farmers to assess the importance of crop compensation for 'whitehead' damage by stem borers (Escalada and Heong, 2004). Another is the cotton Insecticide Resistance Management (IRM) training programme in India, based on farmers scouting for pests, simple threshold levels and recommended insecticide rotation. This was demonstrated in village participatory trials in 24 villages across four states during 1998–1999 (Russell et al, 2000). In all the areas, the quantity of insecticide active ingredient was reduced by at least 29 per cent and net profit rose by US$40 to US$226 per ha on these farms, compared with farmers not involved in the scheme.

However, it seems clear that learning-centred approaches can have a much broader effect. A comparative assessment of the IRM described above with an FFS programme based in India, (Williamson et al, 2003) showed that both approaches expanded decision-making from a single point (i.e. spray pesticide) to a process (assess conditions, choose option, take action, review progress). However, the FFS training provided farmers with new decision tools and a wider range of control options than the IRM training. FFS farmers reported how they decided on timing and choice of control techniques based on the relative abundance of pests and natural enemies observed and on the level of pest and disease incidence.

The focus on enabling farmers to learn for themselves which characterizes learning-centred interventions may help farmers not just to adapt existing

Table 6.1 *Advantages and disadvantages of message-based and learning-centred interventions*

Message-based Interventions	
Advantages	**Disadvantages**
• may be quicker to have an effect on farmers' pest management decision-making • may require less investment than learning centred approaches • may provide short-term solutions to pest management crises requiring immediate action, or to certain pest problem situations • can reach larger numbers of farmers	• contain messages that may not be effective over time • generally require the development of a message for each different pest problem situation • do not necessarily improve farmers' pest management in other crop/pest combinations • do not usually contribute to long-term improvements in farmers' abilities to make better decisions

Learning-centred Interventions	
Advantages	**Disadvantages**
• enable understanding of principles that allow farmers to adapt their pest management techniques to change • provide farmers with the tools to make better informed decisions and to continue learning • may motivate and empower farmers to do more/request information from outsiders • can make significant contributions to farmers' abilities to make better decisions, to farmer empowerment and to increased farmer self-reliance • may contribute to resolving problems of lack of communication between agricultural professionals and farmers • provide increased opportunity for farmer/outsider working partnerships	• can take longer than message-based approaches • can be more expensive in terms of initial investment than message-based approaches • requires more initial investment than message-based approaches • requires the assimilation of new skills and difficult role changes for formally qualified professionals

technologies but to contribute to the development of new ideas. Ooi (1998) describes one Indonesian FFS rice farmer's experimentation with dragonfly perches for brown planthopper control, which has now been taken up by neighbours on 40 ha of paddy. However, in certain specific situations, such as for that of the cocoa pod borer example earlier, or the difficult problem of black Sigatoka disease in bananas, the potential of participatory approaches to develop adequate control strategies has been questioned (Jeger, 2000).

Learning-centred approaches can also contribute to the process of overcoming economic and political constraints to farmer adoption of new pest management technologies – constraints that can be crucial, but are often neglected (Norton et al, 1999). The farmer empowerment and group support gained during the learning process can give farmers the confidence to demand change, and hence contribute to overcoming these constraints. The CATIE-IPM programme in Nicaragua, for example, generated significant demand from smallholder farmers for biopesticides (Williamson and Ali, 2000), while FFS programmes in certain countries have fostered new marketing opportunities for organic and 'ecological' produce, community initiatives in group savings, resource management and even influenced political arenas at local and national levels (FAO, 2001d).

The comparative cost effectiveness of the two different approaches is the subject of debate. Experiential learning-based programmes can appear quite expensive in terms of gross training costs compared with conventional extension activities, but their proponents argue that the breadth of benefits, and their long-term nature mean that the investment costs are more than repaid (Mangan and Mangan, 1998; Neuchâtel Group, 1999). An independent socio-economic evaluation of a Kenyan pilot FFS by ISNAR (Loevinsohn et al, 1998), for example, used a conservative financial analysis to calculate that the benefits accruing to FFS farmers, plus the farmers they interact with, would repay the initial project investment in 1–2 years. However, the impact and value for money of FFS programmes has been questioned by the World Bank (Feder et al, 2004), while other practitioners favour the use of simpler, more message-based or information delivery interventions (Clark, 2000) which tend to require fewer human or financial resources. Direct comparisons are clearly difficult, especially since few extension approaches in pest management, whether message-based or learning-centred, have been adequately assessed in terms of real costs and benefits.

The success of large-scale expansion of learning-centred interventions will depend very much on issues of fiscal sustainability, as well as quality control (Quizon et al, 2000; Praneetvatakul and Waibel, 2002). It is therefore of particular note that FFS in East Africa are now moving towards self-financing mechanisms (Okoth et al, 2003), whilst the FIELD Alliance in Indonesia, Thailand and Cambodia is an independent evolution from earlier FAO programmes, made up of NGOs, Ministry of Agriculture staff, farmer trainers and farmers' associations, aiming to strengthen farmer and rural community movements via action research and building networks (FIELD, 2003).

Diffusion and dissemination are also contested issues between the two approaches. The results of some message-based interventions may appear to diffuse without the need for much outside intervention (Heong and Escalada, 1997). However, this may be due more to the nature of the message required for such interventions (e.g. easily observed results, simple to implement, no deep understanding of the underlying reasons required) than to the nature of the intervention. The extent to which diffusion of the results of learning-centred interventions occurs is as yet unclear. Some studies have found that the ideas do diffuse to untrained farmers (Loevinsohn et al, 1998), whilst other studies have found diffusion rates have been low (van de Fliert, 1993; CIP-UPWARD, 2003).

WHY SHOULD WE INVEST IN NEW IPM APPROACHES

A variety of different programmes have independently arrived at a learning-centred approach as an effective way of enabling farmers to perceive the need for changing their pest management practices. It seems clear that technology transfer thinking cannot deal effectively with the heterogeneous and dynamic field conditions of most farming systems, whether smallholder soybean cultivation in Indonesia (van den Berg and Lestari, 2001) or Californian nut orchards (Schafer, 1998). A new approach is needed, with IPM being conceived more as a methodology and less as a technology (van Huis and Meerman, 1997). As well as better equipping farmers to deal with change, learning-centred interventions are part of a new trend to devolve more power to farmers. This can be particularly important for women, who traditionally may have little decision-making power even in tasks for which they are responsible, but who can make better decisions once they have gained the tools and information to do so (van de Fliert and Proost, 1999; Loevinsohn et al, 1998; Rola et al, 1997).

The success of learning-centred interventions is going to depend significantly on investment in enabling crop protection professionals to appreciate and understand farmer decision-making, and on follow-up support to ensure the learning process continues beyond the initial intervention (Meir, 2000). Some management research projects are beginning to assess farmers' perceptions and practices (Paredes, 1995; Meir, 2004), and in some cases this is becoming routine in research and development planning, conducted by cross-disciplinary teams (Boa et al, 2000; Chikoye et al, 1999). Evaluation of IPM training programmes needs to include more rigorous analysis of decision-making and information sharing systems before and after intervention as standard practice, to provide an important indicator of the sustainability and growth of the processes initiated, in addition to the equity aspects of who participates and who benefits (Nathaniels et al, 2003a, 2003b; van den Berg et al, 2003).

IPM practitioners and crop protection professionals should also learn from experiences in farmer decision-making in agricultural and natural resource management groups (Pretty and Ward, 2001), such as the Local Agricultural Research Committees in the tropics (Braun et al, 2000; van de Fliert et al, 2002), the Landcare movement in Australia (Campbell, 1994), and livestock farmer discussion groups in New Zealand (Riddell, 2001). Implementation of learning-centred approaches for pest and crop management training and farmer-centred research requires different institutional support and policy environments from those required by technology transfer, and this should be viewed as a long-term investment in human and social capital.

Chapter 7

The Human and Social Dimensions of Pest Management for Agricultural Sustainability

Niels Röling

INTRODUCTION

As a social scientist who has spent his working life in an agricultural university, I am thoroughly familiar with, and have deep respect for, the instrumental, 'hard science' perspective on agricultural sustainability. Such a perspective deals with causes and effects. It is the basis for a diagnosis of environmental problems, and also for monitoring and evaluation, whether or not we make any headway in dealing with them. In studies of concerted efforts to deal with environmental problems, research by natural scientists has often played a key role in creating understanding of the problems that people experience.

A typical example of this type of instrumental thinking is the OECD Pressure State Response (PSR) model. Pressure can justifiably be seen as the *causes* for environmental conditions in agriculture to change. Pressure is the *driving force* for that change. The effect of this pressure on the environmental conditions in agriculture is the *state*, and the action that is taken to do something about the change in the state is the *response* (Parris, 2002).

This chapter takes a different, hopefully equally justifiable, stance. It looks at *pressure* as the creative moment in knowledge-based action when the realization dawns that we got it wrong. In other words, I take a constructivist or cognitivist stance, and assume that people and communities are doomed to live by their wits. We have become the most successful species because we are not specialized. Having eaten the fruit of knowledge, we were chased from paradise. Like all other species, we must survive by creating and capturing opportunities in the ecosystem on which we depend. But the way we do this, and the way we deal with the imperatives of the ecosystem, are not given. We must painstakingly construct them through knowledge-based action.

The fact that we are doomed to bring forth realities that allow us to take effective action in our domain of existence (Maturana and Varela, 1992) implies the possibility that we get it wrong. A constructivist perspective is not necessarily unrealistic or relativistic. Thomas Kuhn (1970) made 'getting it wrong' an essential element in the process of scientific development. Scientists build a model or theory of the world, based on a certain paradigm, that is on assumptions about the nature of reality (ontology), and about the nature of knowledge (epistemology), and on a methodology (Guba and Lincoln, 1994). As science evolves, the 'body of knowledge' it develops becomes more coherent and consistent within the paradigm. Evidence that does not 'fit' tends to be ignored. However, eventually, it is no longer possible to ignore the lack of correspondence with the changing context. The elaborate structure collapses and this creates room for a shift in paradigm and for new, more useful, ways to construct reality. A new lifecycle, post-normal science, can begin (Funtowicz and Ravetz, 1993; Holling, 1995; Hurst, 1995).

This chapter is based on the assumption that we live not in an epoch of change, but in a change of epoch (Da Souza Silva et al, 2000). We have successfully built a technology and an economy that allowed a sizeable proportion of humanity to escape much of the misery that comes with the proverbial 'vale of tears' to which we had been banished. The rest of humanity is bent on making the same 'great escape'. However, in the process of co-evolving our aspirations and the technologies to satisfy them, we have transformed the surface of the earth. We have become a major force of nature (Lubchenco, 1998). We have taken on the management of the earth and not made a good job of it. We are beginning to realize that we might be getting it wrong, that we might be jeopardizing the very capacity of the earth to be coaxed into generating human opportunity. We are becoming worried by the extent to which our use renders the land dry, barren, degraded and lifeless. Those who have not yet made the great escape are beginning to realize that there might be nothing left for them. We are facing a second fall from paradise, but this time it is not the archangel, but we ourselves who banished us.

I am not saying this is true, but that we are increasingly beginning to believe that it is true. And with that comes the realization that we must turn around our huge global human apparatus for satisfying our ever-expanding aspirations. We can no longer afford the single-minded pursuit of control and growth for the sake of satisfying the demands of an ever-larger aggregation of individual preferences. Progress might crash. That is the *pressure*.

This chapter will work with this definition of pressure. It is structured as follows. First, it will reiterate the indispensability of a constructivist perspective for mobilizing the reflexivity and resilience required during a change of epoch. Second, it will provide a theoretical underpinning for the human predicament of having to juggle coherence and correspondence. Third, it will further analyse the nature of pressure using that theory. Fourth, it will try to analyse pressure in terms of the nature of human knowledge and its inadequacy, given the task we are beginning to perceive. Finally, I will present a brief case that hopefully illustrates what I am talking about. The challenge is not in dealing with land but in how people use land. *Pressure* is the realization that we got it wrong. The *state* refers to the coherent human apparatus that gets it wrong. And the *response* must deal with ourselves, with restoring correspondence between that human apparatus and our domain of existence, the land. The focus is, therefore, on the interface between humans and the land and pays only fleeting attention to issues such as power, struggle, negotiation, competition and similar subjects.

THE CONSTRUCTIVIST PERSPECTIVE

Although this section might be superfluous for some readers, I believe it is useful to spell out my understanding of the constructivist assumptions about the nature of knowledge (epistemology) because it is so central to my understanding of pressure.

Most of us have been trained to think in the following terms. The world around us exists, whether we are there to see it or not. It is. We must get to know that world so that we can control it for our purposes. Luckily, we can build objective knowledge by using scientific methods that eliminate bias. Thus we can build a body of true, or, since Popper (1972) potentially falsifiable, knowledge. Every dissertation or other bit of sound research is expected to add to the body of human knowledge. We talk of proof, of validation, of evidence, of cause and effect, and even of the 'end of science', i.e., the moment when we know everything there is to know (Horgan, 1996).

It took me years to accept wholeheartedly a radically different perspective and even now I sometimes 'fall back'. The new perspective holds that there is no way, in the sense that it can be explained even by positivist, reductionist science, by which the external environment can be projected on the brain for us objectively to know. Even a frog does not bring forth *the* fly, but at best *a* fly, and even then not *any* fly, but a fly that can be caught. That need to eat flies presumably is the reason that the frog brings forth flies to begin with (Maturana and Varela, 1992; Damasio, 2003b). Given the way brains work, organisms are doomed to either perish or bring forth realities that allow them to take effective action in their domain of existence. Evolution and learning are two important mechanisms by which organisms learn to construct, create and grasp livelihood opportunities. In the case of humans, constructing reality, or learning, is a social affair, involving language, accumulation and transfer of knowledge and culture, conflict, investment, and building institutions and organizations (Berger and Luckman, 1967).

Different people build different realities, partly depending on the objectives that define what is taken to be 'effective action'. In the end, every individual has a different perspective. There is not one truth, but multiple truths. Even the US Supreme Court had to admit that there is no way by which one can determine which expert witness is right. An expert witness is someone who believes he/she is right. The extent to which he/she is right is determined by the number and type of people he/she is able to convince. Or as Mahayana Buddhism puts it: 'An objective world is a manifestation of the mind itself'. At a societal level, what is taken to be true is the outcome of social interaction, including negotiation, attraction, persuasion, signification and so on.

Doomed to construct we might be, but that does not mean every construction is equally effective. Constructivism or cognitivism does not have to be relativism. In my view, we might get it entirely wrong. We can build a cosy coherent reality world, in which our values, theories, perceptions and actions are mutually consistent. But this reality world can become divorced from its domain of existence; for example, it can fail to correspond to ecological imperatives. Then we are in for surprises, for totally unexpected feedback. The effect ('state') is that our cosy reality world has become obsolete. We might not be aware of it, we might ignore it, and elites might have the power to maintain a community's lifestyle long after it has become unsustainable (Pain, 1993). But in the end, coherence will have to give way to the need to rebuild correspondence. That is the *pressure*.

The key to survival is resilience, the ability to note discrepancy, to adapt the reality world to feedback, and to relinquish the institutions, organizations, power positions and interests that we had built around obsolete reality worlds. The key asset we have in survival is rapid, deliberate learning. We used to call this 'science'. But when it insists that it builds truth, science forfeits the claim to be a survival mechanism. It becomes a liability. For example, I am convinced that neo-classical economics, with its arrogant reification of the market, is a serious threat to human survival, a pressure of the first order. It is a blinding insight that reduces resilience now that we are beginning to realize we might have got it wrong. A body of 'true knowledge' is a stumbling block in building the know-ledge required for a change of epoch.

Constructivism is pretty radical. Yet acceptance of the constructed nature of truth and reality is a necessary condition for resilience. It is indispensable in a change of epoch. Survival implies that we deliberately deal with the construction of human knowledge and institutions in the process of building effective and adaptive action.

COGNITION: DEALING WITH THE DILEMMA BETWEEN COHERENCE AND CORRESPONDENCE

Based on a great many authors, it is possible to build a credible argument that cognition is the key manifestation of life (Röling, 2002a). There is not an organism that is not capable of perceiving its surroundings, assessing this perception against some criterion, and taking action on the basis of that assessment and on some understanding of how the context is structured (Maturana and Varela, 1992; Capra, 1996, 2002). I have settled for the illustration shown in Figure 7.1 as the best representation of the key elements of cognition, or perhaps it is better to speak of knowledge-based action.

Figure 7.1 shows a cognitive agent in its context or domain of existence. The cognitive agent is intentional, expressed in values, objectives, criteria or emo-tions. It might not always be led by goals, but its emotions or criteria allow it to judge its outcomes. 'Emotion is in the loop of reason all the time' (Damasio, 2003a). 'Outcomes' result from the ability of the cognitive agent to perceive its context. Depending on the sophistication of its design, a cognitive agent might have anything between a hard-wired response or a flexible 'theory', that is beliefs

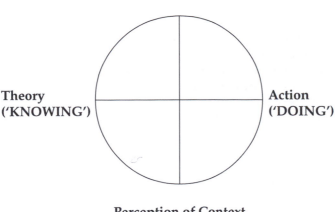

Values, Emotions, Goals ('WANTS')

Theory ('KNOWING')

Action ('DOING')

Perception of Context ('GETS')

Figure 7.1 *The elements of knowledge-based action*

Source: Adapted from Kolb, 1984; Maturana and Varela, 1992; Bawden, 2000

about the world and how it functions. Finally, a cognitive agent is able to take more or less effective action in its context with a view to affect its outcomes. The context, or domain of existence, also contains other agents, and action therefore includes interaction. This chapter focuses on context in terms of ecological imperatives and on how people deal with the surprises emanating from them.

Virtually all organisms can learn or adapt the coherent set of elements of cognition to changing circumstance. Such adaptation might take the form of genetic adaptation or evolution. A typical example is a genus of soil bacteria in Southern France where Bordeaux Mixture has been sprayed as a fungicide in vineyards for 150 years. It is now impossible to use copper water pipes in these areas because the bacteria have adapted to be able to 'eat' copper. Proper learning, however, is the adaptation of the set of cognitive elements by the living organism itself. Mal-adaptation does have to lead to death or extinction. Learning is a useful attribute in rapidly changing circumstances. Deliberate learning is indispensable when the organism itself has become the main cause of change, if not threat to its own survival.

The elements of cognition tend to *coherence*. Action tends to be based on theory about the context, and on perception of the context as assessed against intentionality. In that sense, action is the emergent property of the (collective) mind. But the cognitive agent also seeks outcomes that, in the end, ensure continued structural coupling with its domain of existence. That is, it seeks *correspondence* with its context (Gigerenzer and Todd, 1999). We could say that effective action is the emergent property of a *cognitive system* that not only comprises the elements of the cognitive agent but also its domain of existence or context (Figure 7.1).

Coherence and correspondence do not necessarily add up. If fact, increased coherence might lead to reduced correspondence, as we have seen when we discussed Kuhn's (1970) scientific revolutions. Yet the cognitive agent is doomed to seek coherence and correspondence at the same time. The interesting bit is how this dilemma is played out.

The OECD Pressure, State, Response Model can be seen in this light. *Pressure* (cause or driving force) arises from an assessment of the perceived context against some criteria, values or objectives that leads to an undesirable outcome. The effect or *state* is the ensuing disarray of the cosy coherent reality world. Learning can involve:

- re-building theory to allow identification of the causes of the undesirable outcome as the basis for a response or action to change it to a more desirable one;
- adapting intentionality to render palatable those outcomes that are perceived as unchangeable;
- adapting perception to be better able to assess the context (new indicators and standards, monitoring procedures, agreed information systems);
- developing new ways of acting and technologies to deal with the causes of the undesirable outcomes;
- mutually re-aligning the changed elements to build coherence.

In many ways, human communities and societies can be seen as *collective* cognitive agents. Through their common concerns, division of labour, their interdependence, reciprocal mechanisms and organic solidarity, people form collectives with shared interests, values, objectives and criteria. They build knowledge through slow painstaking cross-generational learning of what works and what does not. They develop indicators and standards, as well as shared surveillance mechanisms to perceive the environment and assess it against criteria. They develop an ability to take collective decisions and concerted, distributed action. What collectives learn becomes solidified in institutions, rules and regulations that constrain behaviour (North, 1990) or at least provide preferred behaviour (Pretty, 2003).

Studying the dilemma between coherence and correspondence becomes really interesting where such collective cognitive agents are concerned. In the first place, the individuals making up the collective are not always willing and/ or able to submit to collective discipline, even if it is in their own long-term interest. The social dilemma literature has made us all too aware of the mechanisms involved (Ostrom, 1998).

But second, it is of great interest to study the mechanisms by which collective cognitive agents build shared intentionality, theory, perception and action, and establish coherence among these elements. Such a study becomes even more interesting, bordering on acquiring survival value, when we are dealing with a change of epoch, that is when we realize that collectively we might have got it entirely wrong, that we are in danger of losing correspondence and therefore must redesign the entire collective cognitive agent. That implies designing new

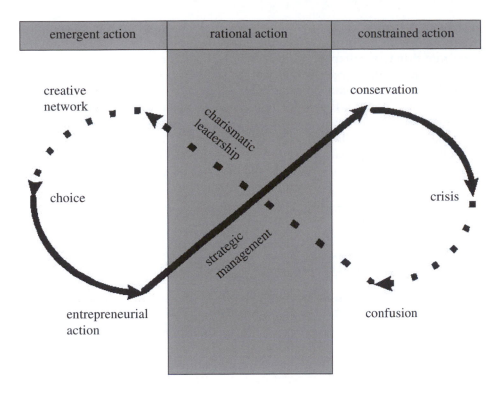

Figure 7.2 *The cycle of succession, crisis and renewal*

Source: Based on Hurst, 1995, in turn based on Holling, 1995

mechanisms and standards for perceiving the environment, new theories to understand what is happening, new criteria that deal with adaptation to environmental imperatives, new ways of acting to 'improve' our outcomes and new institutions. Much of what we invested in before becomes obsolete. Thus we are talking about an unprecedented social upheaval.

In order to use a theory of collective cognitive agents in crisis, we need some further understanding of crisis. Holling (1995) and Hurst (1995) have provided the model of the 'lazy eight' to deal with dynamic situations. The model was originally developed by Holling to explain the ecological cycle and later adapted by Hurst to explain the dynamics of organizations. My use of the model in this chapter draws on Van Slobbe's (2002) analysis of the crisis in Dutch water management, and on work for the EC-financed SLIM Project (Jiggins et al, 2002) (Figure 7.2).

The 'lazy eight' provides a perspective on ecological or organizational dynamics by suggesting phases that have specific attributes. The lifecycle phase under strategic management, called succession in ecology, deals with growth, consolidation, increasing inter-connectedness and investment. It is the phase of increasing coherence among the elements of the collective cognitive system.

Increasing coherence and conservation eventually mean reduced resilience and ability to respond to signals from the changing environment. In the end, the elaborate structure collapses. That is the crisis. This crisis leads to search for a renewal, a phase of re-organization, learning and experimentation under charismatic leadership that can eventually lead to a new lifecycle. The interesting aspect of this dynamic is that different social mechanisms, such as different types of management, learning styles, ways of arranging human affairs can be hypothesized to mark the different stages (Jiggins et al, 2002).

While single loop learning (more of the same but doing it better and on a larger scale) seems characteristic for the lifecycle phase, double loop learning (doing things differently as a result of adapting the cognitive elements) seems more likely to take place during the renewal phase (Maarleveld, 2003). We are now ready to look at *pressure* again.

IS THE LIMIT DESCENDING FROM THE SKY?

The enormity of the *pressure* we could be facing is only slowly dawning on us. We have created a society that is totally geared to generating wealth and satisfaction. Industrial nations have virtually abolished religion because it is not really needed any longer to pacify suffering and to cope with uncertainty. We have learned to control many of the factors that cause suffering. We feel we know enough not to have to worry about uncertainty. Instead, we devote our collective energies to the never-ending pursuit of satisfying our ever-expanding 'preferences', as the economists like to call them. Some trend-setters create a new, more attractive, lifestyle and, before you know it, it has become a 'living standard', an indispensable element in what is defined as a reasonable income. People who never considered themselves poor before and who proudly nourished the traditions of their ancestors become poor peasants overnight as a result of exposure to the living standards of the rich.

Most of our institutions and forms of organization are now devoted to the pursuit of wealth. We have banks, corporations, insurance firms and stock exchanges. Our education systems prepare people to function in that pursuit. Our democratically elected politicians pay most attention to the indicators of what we collectively find most important: economic growth, employment and incomes. If growth stagnates to 1 per cent or less a year, we are in crisis. Governments begin to fall and misery strikes our households. We have established a system of global communication that ensures a continuous flow of advertisements and lifestyle demonstrations designed to expand our insatiable preferences. Our main science, the body of knowledge that helps us theorize about society, is economics. It postulates ever-increasing preferences and individuals in selfish pursuit of them. This science has become powerful to the extent that it has become 'common'. Most people take it for gospel truth. Most media pay a great deal of attention to the indicators and standards by which we assess the economy. The market has become our most important institution and most of us believe that one can best leave 'free market forces' to design future society.

Yet it is very obvious that a society with ever more people, each with ever-expanding preferences, must, in the end, hit a brick wall. It is a temporary product of historical circumstance that can only last for some time. I think few people will disagree with me on that score, it is just unimaginable to think about the implications. It does not fit with anything we know or are capable of. So we ignore the brick wall while the going is good.

Working as a rural sociologist in Nigeria in the late 1960s, I became friends with Klaas K., the expert who headed FAO's Fertiliser Programme in the country at the time. Its slogan was 'Fertiliser is the spear point of development'. Klaas encountered strange local behaviours however. Instead of embracing fertilizers as the best thing ever, local village chiefs came to him and asked him to remove the fertilizers he had brought and stop the programme in their villages. Fertilizers had caused strife and fighting. The reason was that local villagers thought that fertilizer was a 'medicine', a magic concoction that allowed people who used it to pull soil fertility from surrounding fields to their field. Klaas had encountered what anthropologists call 'the image of the limited good' (Foster, 1965). The 'Good' is inherently limited. If the one gets more, the other gets less.

Of course, we thought this a ridiculous idea, although quite understandable for poor people in small communities who did indeed face circumstances in which it is useful to consider the good as limited. Now they had become part of modern society. Everybody could have a better life. The cake could grow in size. The rich might get richer, but also the poor would prosper in the process. The sky is the limit, and all that.

Looking back on the episode now, I think that it is likely that the Nigerian farmers in question no longer hold the image of the limited good. But it is equally likely that they still are not using fertilizers. They probably cannot afford them. With their production techniques, they cannot compete against cheap imported foodstuffs and must continue to rely on their now degraded self-provisioning system. But they have adjusted their intentionality to allow them to cope with frustration and poverty. They might, for example, have become members of 'The Holy Ghost and the Latter Day Saints' so as to seek escape from this world. But, if I know my Nigerian farmers, they probably also have re-invented their theories and ways of perceiving the situation so as to be able to grasp opportunity whenever it arises.

Global society is revisiting the image of the limited good. The famous saying of George Bush Sr., 'The US lifestyle is not negotiable', is an admission that there is not enough for everybody. The country is taking active steps to safeguard its lifestyle by developing its capacity to protect its sources of cheap energy.

And fertilizers? No one would dare to call them a spear point of development any longer. The worry is that crops are showing decreasing yield responses to the same fertilizer dosage. The externalized costs of using fertilizers and pesticides in terms of disrupting aquatic systems and polluting drinking water are becoming increasingly obvious. With the Green Revolution we have done the easy thing (Castillo, 1998). We have greatly increased production on the land on which conditions could easily be controlled (e.g., through irrigation). The next 'Great Leap Forward' is to come from increasing production in less favourable, more diverse, more drought-prone areas, especially in Africa. But so far, no second Green Revolution has been forthcoming. Farmers in those areas cannot afford agrochemical inputs. And, also, highly productive areas in the US and Europe are in the process of becoming under-utilized in terms of agricultural production because of market forces (Blank, 1998; Röling, 2002b). Yet we do not know how we are going to produce the food for a world population that will go on increasing for another 40–50 years, even if the rate of its growth is now decreasing, especially where many do not have the purchasing power to operate in the global market. And so the CGIAR calls for more investment in agricultural research ('cutting edge science') to produce the food we shall require. Let us look at this crazy, confused world a little more systematically.

The question we must ask is, are we really facing an anthropogenic ecocrisis? One of the problems with the lazy eight (Figure 7.2) is that it is very difficult to know when or whether we have hit the end of the lifecycle phase. As a result, very different assessments are being made of the conditions we are in. Plausible arguments mounted by credible scientists lead to very different conclusions. Perhaps that is the best indicator of crisis at the moment. Below I give a number of these assessments to illustrate the great diversity.

Lomborg (2000) claims that we should continue to focus on growth, and that there is nothing wrong with our environment. In fact it is getting better all the time. The areas under forest are increasing, pollution is diminishing, and so on. There is no crisis, no *pressure* at all. Lomborg became an adviser to the Danish Government. Recently, convincing evidence of his selective use of information has reduced his initial influence and the endorsement of his work.

Others claim that it is very hard to prove objectively that we are facing a crisis. Climate change might, after all, be a purely 'natural' phenomenon. Until we know for sure, there is no reason to worry. This approach has been replaced by the 'precautionary principle' by people who fear that it might be too late if we wait until we know for sure.

How can so many be so short-sighted, so totally blinded by their short-term interests? We are living in 'risk society' (Beck, 1992). We have entered an age of 'post-normal science' (in the sense of Kuhn) in which great uncertainty exists with respect to issues with very high stakes, that is with respect to issues that threaten the future of organized human society (Funtowicz and Ravetz, 1993). The ecologist Holling (1995) comes to the conclusion that human society with its focus on linear growth is inconsistent with the cyclical nature of ecosystems (Figure 7.2). Instead of seeking to enhance target variables and suppress every-thing else, we should engage in 'adaptive management' to tease out human opportunity through probing, learning, experimentation, and so on. In fact, it is Holling and his students who, as ecologists, insisted that the only way out for humans is social learning of adaptive management (Gunderson et al, 1995). As scientists they choose to operationalize this social learning by creating simulation models of ecosystems that ecologists can use to inform politicians of what is happening in complex ecosystems so that politicians can take adaptive decisions.

Lubchenco (1998) used her inaugural address as President of the American Association for the Advancement of Science to emphasize the fact that we are entering 'the age of the environment'. This gives science a new social contract, now that the cold war is over. Humans have become a major force of nature; they are transforming the surface of the earth, using a large proportion of the available fresh water resources, and so on. Science will have to tell people 'how it is', what is happening to the globe. In other words, Lubchenco, like other scientists, is inclined to give science an important role in two elements of the cognitive agent: perception of the context and theory building (Figure 7.1). The World Resources Institute (2000) together with a host of international organizations has tried to make a careful assessment of the state of the earth's ecosystems. It comes to the conclusion that the web of life is 'frayed' but that all is not lost. But the trend is not in the right direction on most indicators. The report led to a call for more research, that is more and better perception of the context.

Capra (2002) has developed *'a conceptual framework that integrates the biological, cognitive and social dimensions of life that enables us to develop a conceptual framework to some of the critical issues of our time'*. Capra sees two global networks emerging, one the rise of global capitalism, and the other the creation of stable communities based on ecological literacy and ecodesign. The former is concerned with electronic networks of financial and information flows, the latter with ecological networks of energy and material flows. The goal of the former is to maximize the wealth and power of elites; the goal of ecodesign is to maximize the sustainability of the web of life:

> *These two scenarios ... are currently on a collision course. We have seen that the current form of global capitalism is ecologically and socially unsustainable. The so-called 'global market' is really a network of machines programmed according to the fundamental principle that money-making should take precedence over human rights, democracy, environmental protection and any other issue. However, human values can change* (Capra, 2002).

And so we come to a new and more precise definition of pressure. The pressure is not to change ecosystems, to develop new more sustainable technologies, or to invest more in science so that we better understand environmental impacts. The real pressure is to change the intentionality of our nested collective cognitive agents (Figure 7.1). Capra (2002) feels that there are many signs that such a shift in values has already started (e.g. the ecology movement and the rise of feminist awareness). This conclusion, although optimistic to a degree, also emphasizes the enormity of the pressure we are dealing with.

Focus on Agriculture: The Pressure To Get Off The Treadmill

Agriculture provides a dramatic example of the kind of pressure we are facing. It is called the agricultural treadmill. It provides a 'blinding insight' for which we have no alternative for the time being. Typically, it is based on the assumptions of economics with respect to human rationality. The treadmill works as follows (Cochrane, 1958):

- many small farms all produce the same product;
- because not one of them can affect the price, all will produce as much as possible against the going price;
- a new technology enables innovators to capture a windfall profit;
- after some time, others follow ('diffusion of innovations'; Rogers, 1995);
- increasing production and/or efficiency drives down prices;
- those who have not yet adopted the new technology must now do so lest they lose income (price squeeze);
- those who are too old, sick, poor or indebted to innovate eventually have to leave the scene; their resources are absorbed by those who make the windfall profits ('scale enlargement').

This is a very coherent and well known story indeed. And policy based on the treadmill has very positive outcomes. For one, the advantages of technological innovation in agriculture are passed on to the customer in the form of apparently cheap food. For example, an egg in Europe still has the same nominal value as in the 1960s. We now spend only about 10 per cent of household income on food and beverages, and of this only a fraction goes to the primary producer (Pretty, 2002). The very structure of agriculture makes it impossible for farmers to hold on to rewards for greater efficiency (Hubert et al, 2000). Meanwhile, labour is released for work elsewhere. One farmer can now easily feed 100 people. When the treadmill runs well at the national level in comparison with neighbouring countries, the national agricultural sector improves its competitive position. Furthermore, an important advantage is that speech making farmers do not protest against the treadmill. They only profit from it. A farmer on the treadmill can only make a good living if he is ahead of the pack. Unlike industrial workers, farmers collectively, usually, do not claim rewards for greater labour productivity. A final advantage is that the treadmill will continue to work on the basis of relatively small investments in research and extension. These have a high rate of return (Evenson et al, 1979).

All in all, it is very understandable that policy-makers have grasped the treadmill as the fundament for agricultural policy. It represents market forces in optimal form. According to the World Trade Organization (WTO), we must work towards a global treadmill. For example, the 4 million small farmers in Poland must leave the scene quickly so that Polish agriculture can become 'competitive'. A competitive agriculture is the key slogan, and also applies for global agriculture. The treadmill is a very coherent and alive model and a good example of what Capra (2002) calls global capitalism.

However, the treadmill has a number of negative aspects that are increasingly less acceptable to society at large:

- Not consumers but input suppliers, food industries and supermarkets capture the added value from greater efficiency. Large corporations are well on their way to obliterating competition in agriculture. Only farmers are squeezed.

- The advantages of the treadmill diminish rapidly as the number of farmers decreases and the homogeneity of the survivors increases. The treadmill has a limited lifecycle as a policy instrument.
- Eventually, the treadmill is unable to provide farmers with a parity income. That becomes clear from the subsidies we must give our farmers. We want to reorient that flow of subsidies, but do not as yet have a good alternative. At the time of writing, the European Commissioner for Agriculture was working on policy reform. In the meantime, recent research shows that 40 per cent of farmers in the Netherlands are already engaged in a deepening and broadening of farming and off-farm work (Oostindie et al, 2002).
- The competition among farmers promotes non-sustainable forms of agriculture (use of pesticides and hormones, loss of biodiversity, unsafe foods). The treadmill is contradictory to nature conservation, drinking water provision, landscape conservation and other ecological services.
- The treadmill leads to loss of local knowledge and cultural diversity.
- A global treadmill unfairly confronts farmers with each other who are in very different stages of technological development, and have very different access to resources. Although the costs of labour in industrialized countries are many times those in the developing countries, labour productivity in agriculture in the North is still so much greater that small farmers in developing countries do not stand a chance (Bairoch, 1997). The global treadmill prevents them from developing their agriculture and denies them purchasing power at the same time. This effect is further acerbated by subsidies paid to farmers in industrialized countries.

The treadmill leads to short-term adaptations that can be dangerous for long-term global food security. There is, for example, the possibility, however much disputed, of the disappearance of arable farming from the Netherlands. In the US, some now speak of the 'Blank Hypothesis', which suggests that domestic agriculture will disappear by 2030 because food can be produced more cheaply elsewhere (Blank, 1998). The new American subsidies might prevent this for a while. But it does become evident that the treadmill does not support the contribution to global food security of the most productive agricultural areas in the world. There are those who say that organic agriculture cannot feed the world. I think it is more appropriate to say that one cannot feed the world as long as the treadmill is in operation.

I conclude that within the self-imposed boundaries of treadmill thinking there is no way to solve some of the more important challenges that now confront us. To further reduce the fraction of our incomes that goes to primary production, we make ever-greater externalized costs. The market simply fails where sustainability and world food security are concerned. Nevertheless, we blindly hold on to an idea that worked well in a specific historical phase now that we have entered a totally different phase. The treadmill does not fit our post-normal age. We have to re-invent agricultural economics. In agriculture, the major pressure is for us to get off the treadmill and to imbed land use in other social and economic mechanisms.

IMPLICATIONS FOR CROSSDISCIPLINARITY

It is obvious from the preceding sections that inappropriate values and intentionality are key *pressures* to worry about. However, humans are also creatures that have to live by their wits. They need good reasons; they need narratives that make sense as a minimum condition for change. Another condition is a sense of reciprocity in making sacrifices (Ridley, 1995). Reaching that condition perhaps is the greatest liability in building a sustainable society. Now that it is becoming obvious that there is not enough for everybody to have an American lifestyle, industrial countries are digging in to protect their lifestyles, thus creating a permanent cause of war and sustainability. This chapter does not address that issue. It does try to examine the kind of narrative we might need to build a sustainable society. Natural science and economics obviously are part of the problem in that they have shown us how to get to a situation in which we have to worry about the consequences of our success as a species. This asks for a critical analysis of our domains of knowledge. I distinguish three ideal types.

1 *Using Instruments.* This ideal type deals with the manipulation of cause/effect relationships through the use of instruments. It focuses on technology, and its pathological expression is the assumption that there is a technological fix for every human problem. The future can be planned and experts have answers. The use of instruments can also be applied to humans in what is called social engineering. Policy instruments, especially regulation backed up by force, whether legitimated or not by parliamentary procedure, is expected to yield expected results.

2 *Exchange of Values.* The basic tenet behind this ideal-type is that the individual pursuit of utility more or less automatically leads to the achievement of the common good through the invisible hand of the market. Society is an aggregation of individual preferences. Based on methodological individualism, this ideal type ascribes people with certain motivations (rational choice theory) and deducts normative models for action from this assumption. Hence we are dealing with an axiomatic domain of knowledge. When the market fails, this ideal type recognizes two options: regulation and fiscal policy, that is the manipulation of incentive structures through compensation, subsidies and fines.

3 *Learning.* Where the tenets of the two previous ideal types are widely shared, this third ideal type does not feature much in the public discourse. It deals with the reasons why people act and shape their networks the way they do. This ideal type does not ascribe motives to people but focuses on human sense making, cognitive conversion and transformation, concerted action based on trust, negotiation and reciprocity. Given that humans have become a major force of nature, and that human behaviour is becoming a major cause

of the human predicament, this third ideal type perhaps deserves more attention, as it is likely to provide the levers for learning our way to a sustainable future.

These three domains are compared using the four essential characteristics of knowledge-based action presented in Figure 7.1 (see Table 7.1). Each of the three knowledge domains is coherent. Each 'owns' a part of the domain of existence with which it tries to correspond. What does not fit is externalized to the other domains. Each domain demands a considerable amount of formation before a secondary school graduate becomes a professional within the domain. A well known example is the observation that economics students behave more selfishly in tests than others (Ridley, 1995; Gandal and Roccas, 2002). It appears to be a result of professionalization. The example also shows how closely values are mixed up with science.

Table 7.1 *Three domains of knowledge*

	Using instruments	Using incentives	Learning
Predicament, problem perceived, success perceived ('GETS')	Lack of control over causal factors	Competition, scarcity, poverty	Lack of control over ourselves, disagreement, lack of trust, conflicting interests
Dynamics ('KNOWING')	Causation, use of 'instruments', power, hierarchy	Rational choice in satisfying preferences, struggle for survival, market forces, exchange of values.	Interdependence, learning, reciprocity, tendencies toward coherence and correspondence
Values, emotions, goals, purposes, wants ('WANTS')	Control over biophysical and social resources and processes	Win, gain advantage, satisfaction,	Convergence to negotiated agreement, concerted action, synergy
Basis for effectiveness ('DOING 1')	Technology, power differential, use of instruments	Strategy, anticipation, exchange	(Facilitation of) awareness of issues, conflict resolution, agreement, shared learning
Policy focus ('DOING 2')	Hard systems design, regulation, (social) engineering	Fiscal policy, market stimulation, compensation	Interactive policy making, social process design, facilitation, stimulation

Although each knowledge domain is formed according to its own self-imposed 'discipline', the domains are highly dependent on each other. It is very clear that there is no sustainable society that is not based on a high degree of sophisticated instrumental knowledge. There is also no society that does not have an economy. Finally, there is no sustainable society unless it can reflect on itself and the way it relates to its ecological base.

Two issues arise. In the first place, the third domain has been largely under-developed, as I have already observed. It is an increasingly important 'third way of getting things done'. It is high time for society to transform its current self-image that is largely based on rational choice theory and to become imbued with an understanding of constructivism and knowledge-based action. This is a *'pressure'* of the first order.

In the second place, it is not at all clear how the three domains, each with its own coherent 'reality world', optimization perspectives and logic, can be made to be complementary and synergistic, as the easy talk about 'the triple bottom line', or 'the three Ps' (people, planet and profit) would suggest. To be sure, the three domains have found ways to deal with each other. For example, natural scientists tend to work within a framework of market forces. The technology they develop must be economically feasible. Technological change occurs in a context dominated by the treadmill. Economists assume that natural scientists can produce a new technology when a problem, such as the depletion of fossil energy, arises. After all, technological innovation is an essential cornerstone of their growth-based paradigm. The protagonists of the third domain, finally, have delineated their domain by stating that they occupy themselves with problems for which the market fails and technological fixes are not available. In all, we all get along splendidly, but a coherent narrative that can sustain a sustainable society is still far away. That is another *pressure* of prime importance.

APPLICATIONS FOR PEST MANAGEMENT

During his involvement in the battle against the outbreak of Spruce Budworm in the forests of New Brunswick in Canada, Miller (1983, 1985) observed his colleagues and developed the scheme illustrated in Figure 7.3 to classify them. I found it useful to discuss different paradigms with my agricultural colleagues.

The four quadrants represent four paradigms, based on different ontologies (holism versus reductionism) and epistemologies (positivism versus constructivism). Miller's colleagues in Quadrant I had an instrumental and hard science perspective on problems that they isolated in order to better deal with them. Their reaction to the Budworm was spraying with pesticides. Scientists in Quadrant II had an equally hard and instrumental perspective, but they tended to contextualize problems in a more holistic manner. Hence they analysed the Budworm outbreak from an ecosystem perspective and advocated IPM through the use of natural enemies, and so on. Not many of Miller's colleagues fitted Quadrant III. The ones who did, considered the Budworm problem as embedded in human reasons and defined the 'pressure' very much as I have done in this

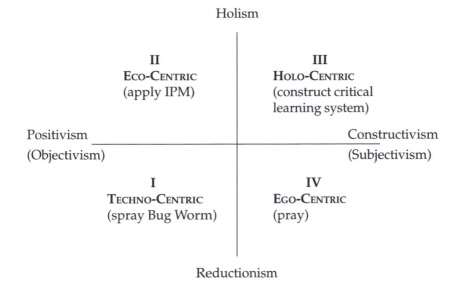

Figure 7.3 *Four expert reactions to the spruce budworm based of different paradigms*

Source: Based on Miller 1983, 1985; Bawden, 2000

chapter. For them, the Budworm problem could not be explained and dealt with unless one took into account human intentionality, and unless one considered human learning as a key ingredient in the solution. There was a good reason for taking this stance. The main cause of the outbreak was that paper mills had planted vast tracts of land with the same species of tree and created optimal conditions for pest and disease outbreaks. Miller did not have any colleagues in Quadrant IV. He reserved it for a spiritual, if fatalistic, response, hence 'pray'.

Where Quadrant I implies a focus on component issues and problems to deal with sustainability, and Quadrant II a focus on sustainability as a property of agroecosystems, Quadrant III implies that sustainability is the emergent property of human interaction, that is an emergent property of a soft system (Bawden and Packam, 1993). In similar vein, Quadrant IV could imply that sustainability is grounded in a spirituality that is based on the intuitive grasp of a reality beyond discursive cognition.

Miller's quadrants should not suggest a rank order from negative to positive. In fact, a strong claim can be made that dealing with sustainability requires attention to all four quadrants, or at least to the first three. Good, sound, reductionist research is required to identify species, pheromones, etc. Similarly, it is very useful to take an agroecosystem perspective and to analyse the inter-action among organisms in their ecosystem context as it is understood by biology, hydrology, etc. Finally, it is useful to consider hard systems as sub-systems of soft systems and to construe the human reasons and interactions that lie at the basis of a resource dilemma. Research of agricultural research (Van Schoubroeck, 1999;

Tekelenburg, 2002) has shown that successful work is based on approaches that combine all three. The operationalization of Quadrant III requires not only participatory technology development, but also participatory exploration of the context. The research must be contextualized in terms of resource dilemmas and multiple stakeholder processes so as to ground the research in the needs of the intended beneficiaries and ensure that the results can be used given the prevailing institutional and policy conditions. Giampietro (2003) calls this paying attention to the 'pre-analytic choices'.

Even if the sequence of Quadrants from I through IV does not reflect a rank order of desirability, one can say that scientists who are 'stuck' in Quadrant I are totally blind to the thinking required in II and III and will reject it as unscientific. Likewise, type II scientists are impervious to type III thinking. Therefore, it *is* desirable for scientists, even if their work is limited to Quadrant I problems, to expand their thinking to include at least also Quadrant III. What I observe in my university is that most scientists not only have embraced ecosystems thinking (i.e. Quadrant II) with a vengeance, but that many have realized that Quadrant III thinking is a necessary condition for the professionalism underpinned by their discipline. Thus a hydrologist will realize that the irrigation engineers she trains will not be able to do a professional job unless they know about irrigators' associations, and power struggles between top- and tailenders. Likewise, an entomologist training crop protection professionals nowadays will provide them with a thorough training in community IPM and farmer field school curriculum development and management. In Wageningen, we call this beta/gamma knowledge, that is the mix of science and social science required to make things 'work'. It should be remarked here that, in Wageningen, irrigation, crop protection, forestry, soil science and so on, largely on their own, invented the social science dimensions of their disciplines, while the social scientists themselves were still totally absorbed in their own disciplines.

Entomologists and social scientists engaged in research of farmer field schools and community IPM only shifted as a result of the tremendously creative work by FAO's 'IPM in Rice Programme in Asia', and especially Indonesia (see Chapter 8; Pontius et al, 2002; CIP-UPWARD, 2003; Jiggins et al, 2004). Especially for those, like myself, who had been struggling to find alternatives to the perennial practice, grounded in the agricultural treadmill, of 'transferring' knowledge based on Quadrant I and II thinking to farmers, the development of the farmer field school and especially community IPM was a life-saver. Of course, the old guard does not easily give up. Thus World Bank economists (Quizon et al, 2000) considers farmer field schools a *'fiscally unsustainable form of extension'* and fail totally to grasp the significance of moving technology development and farmer learning into Quadrant III.

Summing up, I conclude that *pressure* can no longer usefully be seen only in terms of the intractability of some natural resource. A natural resource becomes problematic as a resource dilemma, that is because multiple stakeholders make competing claims on it, or because the same stakeholder has multiple conflicting uses for it. To put it another way, in an era when Homo rapiens (Gray, 2002) has become the predominant creature on this planet, the future cannot be designed only on the basis of technical or economic knowledge. It must be based on a thorough understanding of ourselves and especially on designing institutions that reciprocally limit our potentially ever-expanding aspirations. In the end, Homo rapiens does not require more technical fixes to make pesticides more efficient (i.e. a Quadrant I solution) and by a few years extends the lifecycle of the agricultural treadmill. We only deserve the title 'sapiens' if we manage to think in the terms of Quadrants III and IV.

Chapter 8

Ecological Basis for Low-toxicity Integrated Pest Management (IPM) in Rice and Vegetables

Kevin Gallagher, Peter Ooi, Tom Mew, Emer Borromeo, Peter Kenmore and Jan-Willem Ketelaar

INTRODUCTION

This chapter focuses on two case studies primarily arising from Asian based integrated pest management (IPM) programmes. One case study provides an in-depth analysis of a well researched and widespread rice-based IPM, while the second study focuses on an emergent vegetable IPM.

The powerful forces that drive these two systems could not be more different. Rice production is a highly political national security interest that has often justified heavy handed methods in many countries to link high yielding varieties, fertilizers and pesticides to credit or mandatory production packages and led to high direct or indirect subsidies for these inputs. Research, including support for national and international rice research institutes, was well-funded to produce new varieties and basic agronomic and biological data. Vegetable production, on the other hand, has been led primarily by private sector interests and local markets. Little support for credit, training or research has been provided. High usage of pesticides on vegetables has been the norm due to lack of good knowledge about the crop, poorly adapted varieties and a private sector push for inputs at the local kiosks to tackle exotic pests on exotic varieties in the absence of well-developed management systems.

However, other pressures are now driving change to lower pesticide inputs on both crops. Farmers are more aware of the dangers of some pesticides to their own health and their production environment. The rise of Asian incomes has led to a rise in vegetable consumption that has made consumers more aware of food safety. The cost of inputs is another factor as rice prices fall and input prices climb. More farmers are producing vegetables for urban markets, so driving competition to lower input costs as well. Highly variable farm gate prices for vegetables make farmers' economic decisions to invest in pesticide applications a highly risky business. Research on vegetables is beginning to catch up with rice, allowing for better management of pests through prevention and biological controls. IPM programmes in both crops aim to reduce the use of toxic pesticide inputs and the average toxicity of pest management products that are still needed whilst improving the profitability of production.

INTEGRATED PEST MANAGEMENT IN RICE

This chapter has been prepared to provide a conceptual guide to the recent developments in rice IPM within an ecological framework. It is not a 'how to' guide but rather a 'why to' guide for IPM programmes that are based on ecological processes and work towards environmentally friendly and profitable production. We provide a broad overview of IPM practices in rice cultivation including its ecological basis, decision-making methods, means of dissemination to farmers and future needs to improve these practices. The breadth of pest problems, including interaction with soil fertility and varietal management are discussed in depth. Although the main focus is on Asian rice cultivation, we also provide examples of rice IPM being applied in other regions.

IPM in rice has been developing in many countries since the early 1960s. However, much of the development was based on older concepts of IPM including intensive scouting and economic thresholds that are not applicable under all conditions (Morse and Buhler, 1997) or for all pests (e.g. diseases, weeds), especially on smallholder farms where the bulk of the world's rice is grown and which are often under a weak or non-existing market economy. During the 1980s and 1990s, important ecological information became available on insect populations that allowed the development of a more comprehensive ecological approach to pest management, as well as greater integration of management practices that went beyond simple scouting and economic threshold levels (Kenmore et al, 1984; Gallagher, 1988; Ooi, 1988; Graf et al, 1992; Barrion and Litsinger, 1994; Rubia et al, 1996; Settle et al, 1996).

Since then, an ecological and economic analytical approach has been taken, for management that considers crop development, weather, various pests and their natural enemies. These principles were first articulated in the Indonesian National IPM Programme, but have expanded as IPM programmes have evolved and improved. Currently programmes in Africa and Latin America now use the term integrated production and pest management (IPPM), and follow these principles: grow a healthy soil and crop; conserve natural enemies; observe fields regularly (soil, water, plant, pests, natural enemies); and farmers should strive to become experts. Within these principles, economic decision-making is still the core of rice IPM but incorporates good farming practices as well as active pest problem-solving within a production context.

IPM in rice seeks to optimize production and to maximize profits through its various practices. To accomplish this, however, decision-making must always consider both the costs of inputs and the ecological ramifications of these inputs. A particular characteristic of Asian rice ecosystems is the presence of a potentially damaging secondary pest, the rice brown planthopper (BPH), *Nilaparvata lugens* (see Box 8.1). This small but mighty insect has in the past occurred in large-scale outbreaks and caused disastrous losses (IRRI, 1979). These outbreaks were pesticide-induced and triggered by pesticide subsidies and policy mismanagement (Kenmore, 1996). BPH is still a localized problem, especially where pesticide overuse and abuse is common, and therefore can be considered as an ecological focal point around which both ecological understanding and management are required for profitable and stable rice cultivation. BPH also becomes the major entry point for all IPM educational programmes since it is always necessary to prevent its outbreak during crop management. Other pests that interact strongly with management of inputs are rice stemborers and the various diseases discussed below.

A major issue when considering IPM decision-making is one of paths to rice production intensification. In most cases, intensification means the use of improved high yielding varieties, irrigation, fertilizers and pesticides – as was common in the Green Revolution. However, two approaches to intensification should be considered. The first is input intensification in which it is important to balance optimal production level against maximizing profits and for which higher inputs can destabilize the production ecosystem. The second route to intensification is one of optimizing all outputs from the rice ecosystem to maximize profits. In many lowland flooded conditions, this may mean systems such as rice-fish or rice-duck that may be more profitable and less risky, yet require lower inputs (and often resulting in lower rice yields). In areas where inputs are expensive, where the ecosystem is too unstable (because of drought, flood) to ensure recovery of input investments, or where rice is not marketed, then such a path to intensification may be more beneficial over time. However, such a system has a different ecology due to the presence of fish or duck, and therefore will involve a different type of IPM decision-making.

ECOLOGICAL BASIS OF RICE IPM

IPM in much of Asian rice is now firmly based on an ecological understanding of the crop and its interaction with soil nutrients and crop varieties. We present below an ecological overview of our current understanding of how the rice ecosystem operates during the development of the crop.

The rice ecosystem in Asia is indigenous to the region and its origins of domestication date back 8000 years to the Yangtze Valley in southern China (Smith, 1995), and more widely some 6000 years ago (Ponting, 1991). Cultivation practices similar to those of today were reached by the 16th century (Hill, 1977). This period of time means that rice plants, pests and natural enemies existed and coevolved together for thousands of generations. Rice ecosystems typically include both a terrestrial and an aquatic environment during the season with regular flooding from irrigation or rainfall. These two dimensions of the rice crop may account for the extremely high biodiversity found in the rice ecosystem and its stability even under intensive continuous cropping – and contrasts with the relative instability of rice production under dryland conditions (Cohen et al, 1994).

Box 8.1 *The brown planthopper (BPH)*

The brown planthopper (BPH), *Nilaparvata lugens* Stål (Delphacidae, Homoptera), is an insect that has been associated with rice since the crop was grown for food in Asia. This insect is known to survive well only on rice and in evolutionary terms has co-evolved with the rice plant.

Rice fields are invaded by macropterous adults. Upon finding a suitable host, female BPH will lay eggs into the stem and leaf stalks. Egg stage lasts from 6 to 8 days. Nymphs resemble adults except for size and lack of wings. There are five nymphal stages. The complete lifecycle lasts 23 to 25 days. When food is suitable, the next generation of adults are often brachypterous or short winged. Both nymphs and adults prefer to be at the base of rice plants. BPH feeds by removing sap from rice plants, preferably from the phloem.

Usually, populations of BPH are kept low by the action of a wide range of natural enemies indigenous to tropical rice ecosystems in Asia. Outbreaks reported in the tropics during the 1970s were associated with regular use of insecticides. The more effective the insecticide, the faster the resurgence of BPH populations which led to a large-scale dehydration of rice plants, a symptom known as 'hopperburn'. Insecticides removed both BPH as well as their predators and parasitoids. However, eggs laid inside the stem are relatively unharmed by spraying and, when these hatch, BPH nymphs develop in an environment free of predators. In unsprayed fields, the population of BPH did not increase to any significant level, suggesting the importance of biological control. Today, farmers learn about predators by carrying out experiments and when they discover the role of these natural enemies, they are less likely to use insecticides. In Indonesia, Presidential Decree 3/86 provided the framework and support for farmers to understand and conserve natural enemies and this has in turn helped rice fields in Indonesia to be relatively free of BPH in the last ten years. This has coincided with an extensive programme to educate farmers based on the farmer field school model.

The irrigated rice systems in Africa, Americas and Europe also include this aquatic and terrestrial element within which high levels of biodiversity are also found.

Insects

Studies by Settle and farmer research groups in Indonesia (Settle et al, 1996) show that flooding of fields triggers a process of decomposition and development of an aquatic foodweb, which results in large populations of detritus-feeding insects (especially Chironomid and ephydrid flies). These insects emerge onto the water surface and into the rice canopy in large numbers, very early in the growing season, providing critical resources to generalist predator populations long before 'pest' populations have developed (Figure 8.1).

This is quite different from the usual predator–prey models taught in most basic IPM courses and provides a mechanism to suggest that natural levels of pest control in tropical irrigated rice ecosystems are far more stable and robust than purely terrestrial agroecosystems. This stability, however, was found to be lower in rice landscapes that are subject to long (more than three month) dry seasons and where rice is planted in large-scale synchronous monocultures, as well as in areas where farmers use pesticides intensively. Increased amounts of organic matter in the soil of irrigated rice fields, by itself a highly valuable practice for sustainable nutrient management, has the additional advantage of boosting both populations of detritus-feeding insects and insect predators, and thereby improving natural levels of pest control (Settle et al, 1996).

A second consideration for rice IPM is the ability of most rice varieties to compensate for damage. The rice plant rapidly develops new leaves and tillers early in the season replacing damaged leaves quickly. The number of tillers produced is always greater than the number of reproductive tillers allowing for some damage of vegetative tillers without effecting reproductive tiller number. The flag leaf contributes to grain filling but the second leaf provides photosyn-thates as well, while lower leaves are actually a sink that compete with the panicle. Finally, photosynthates appear to move from damaged reproductive tillers to neighbouring tillers so that total hill yield is not as severely impacted as expected when a panicle is damaged by stemborers.

Thus, early season defoliators (such as whorl maggot, case worms and armyworms) cause no yield loss up to approximately 50 per cent defoliation during the first weeks after transplanting (Shepard et al, 1990; Way and Heong, 1994) although higher damage occurs when water control is difficult. As early tillering is also higher than what the plant can ultimately support reproductively, up to 25 per cent vegetative tiller damage by stemborers ('deadhearts') (caused by *Scirpophaga* spp., *Chilo* spp. and *Sesamia* spp.) can be tolerated without signifi-cant yield loss (Rubia et al, 1996). Significant damage (above 50 per cent) to the flag leaf by leaffolders (*Cnaphalocrocis mdeinalis* and *Marasmia* spp.) during pan-icle development and grain filling can cause significant yield loss, although this level of damage is uncommon where natural enemies have been conserved (Graf et al, 1992). Late season stemborer damage (whiteheads) also causes less damage than previously expected such that up to 5 per cent whiteheads in most varieties does not cause significant yield loss (Rubia et al, 1996; Way and Heong, 1994).

The conspicuous rice bug (*Leptocorisa oratorius*) is another major target for insecticide applications. However, in a recent study involving farmers and field trainers at 167 locations, van den Berg and Soehardi (2000) have demonstrated that the actual yield loss in the field is much lower than previously assumed. The rice panicle normally leaves part of its grain unfilled as if to anticipate some level of loss (Morrill, 1997). Numerous parasitoids, predators and pathogens present in most rice ecosystems tend to keep these potential pests at low densities (Barrion and Litsinger, 1994; Loevinsohn, 1994; Shepard and Ooi, 1991; Ooi and Shepard, 1994; Matteson, 2000).

Thus, under most situations where natural enemies are conserved, little yield loss is expected from typical levels of insect pests. Up until recently, insecticide applications for early defoliators, deadhearts and whiteheads often led to lower natural enemy populations allowing the secondary pest, rice brown planthopper (*Nilaparvata lugens*), to flare up in massive outbreaks (Rombach and Gallagher,

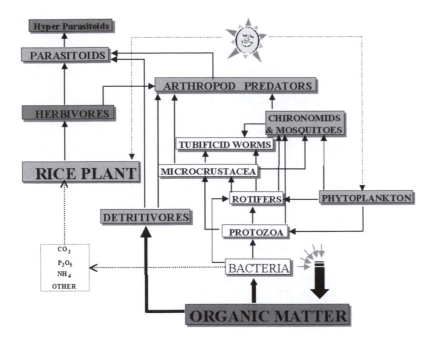

Figure 8.1 *Hypothesized flow of energy in tropical rice ecosystems. Organic matter drives the development of high early-season populations of predators through parallel pathways: (1) micro-organisms (zooplankton and phytoplankton) are fed on by filter-feeders (mosquitoes and midges), and (2) organic matter directly feeds detritus-feeding insects (Diptera larvae, Collembola, and some Coleoptera larvae). Each of three pathways dominates at different times of the season: micro-organism/filter-feeders early-season; plant/herbivore mid-season, and detritivores post-harvest. The pattern of interaction leads to consistently high populations of generalist predators early in the growing season, and low and stable populations of herbivores later in the season*

Source: after Settle and Whitten, 2000

1994). Work by Kenmore et al (1984) and Ooi (1988) clearly showed the secondary pest status of brown planthoppers. Although resistant varieties continue to be released for brown planthopper, the highly migratory sexual populations were found to have high levels of phenotypic variation and highly adaptable to new varieties. Although wrongly proposed to be 'biotypes', it was found that any population held significant numbers of individuals able to develop on any gene for resistance (Claridge et al, 1982; Sogawa et al, 1984; Gallagher et al, 1994). Huge outbreaks have not reoccurred in areas where pesticide use has dropped, due either to changes in policy regulating pesticides in rice or due to educational activities (Box 8.2).

A few minor pests are predictable problems and therefore should be considered for preventive action with natural enemies, resistant varieties, or specific sampling and control. These include black bug (*Scotinophara* spp.), gall midge (*Orseolia oryzae*), and rice hispa (*Dicladispa* spp.) which are consistently found in

---

Box 8.2 *Predators of BPH: hunting spiders*

Predators are the most important natural enemies of BPH. Together with parasitoids and insect pathogens they keep populations of BPH down. An important group of predators commonly found in rice fields is the spiders. Of particular importance are hunting spiders, especially *Lycosa pseudoannulata*. This is often found near the water level, the same area where BPH feed. A lycosid is known to feed on as many as 20 BPH per day. Its voracious appetite makes it a very important natural enemy of BPH. However, there are often questions asked about this predator.

A common one is: What will the spider feed on in the absence of BPH? Like other spiders, *Lycosa* and *Oxyopes* do not depend entirely on BPH for its food. There are many flies in the field that provide the bulk of the food for spiders. Studies in Indonesia have shown the importance of 'neutrals' in supporting a large population of predators in the rice fields. Spiders are found in rice fields before planting and they survive on these 'neutrals'. During the dry season, rice field spiders are known to hide in crevices or in grasses around the field. Like all predators, spiders are very susceptible to insecticides and so sprays or granular applications into the water will destroy these beneficial arthropods, thus allowing BPH to multiply to large numbers.

---

certain regions; thrips (*Stenchaetothrips biformis*), whereas drought causes leaf-curling that provides them a habitat; armyworms (*Mythimna* spp. and *Spodotera* spp.) in post-drought areas that are attracted by high levels of mobilized nitrogen in the rice plant, and panicle cutting armyworms cause extreme damage.

Green leafhoppers (*Nephotettix* spp.) are important vectors of tungro (see below) but by themselves rarely cause yield loss. White-backed planthoppers (*Sogatella* spp.) are closely related to brown planthoppers in terms of population dynamics and are not usually a major yield reducing pest. Rice water weevil (*Lissorhoptrus oryzophilus*) introduced from the Caribbean area in North America and north-east Asia is a problem pest requiring intensive sampling (Way et al, 1991) that deserves greater research on its natural enemies. In upland ecosystems, white grub species and population dynamics are not well studied and are difficult to manage. Way et al (1991) provide an overview of insect pest damage dynamics, while Dale (1994) gives an overview of rice insect pest biology.

Diseases

The need to grow more rice under increasingly intensive situations leads to conditions that favour diseases. High planting density, heavy inputs of nitrogen, and soil fertility imbalance result in luxuriant crop growth conducive to pathogen invasion and reproduction. This is made worse by genetic uniformity of crop stand that allows unrestricted spread of the disease from one plant to another, together with continuous year-round cropping that allows carry over of the pathogen to succeeding seasons. Reverting to the less intense, low yield agriculture of the past may be out of the question, but a thorough understanding of the ecological conditions associated with the outbreak of specific diseases may lead to sustainable forms of intensification. We briefly describe the specifics for three major diseases of rice, namely, rice blast, sheath blight and rice tungro disease.

Blast (*Pyricularia grisea, Magnaporthe grisea*) occurs throughout the rice world but is usually a problem in areas with a cool, wet climate. It is a recognized problem in upland ecosystems with low input use and low yield potential, as well as in irrigated ecosystems with high input use and high yield potential (Teng, 1994). Fertilizers and high planting density are known to exacerbate the severity of infection. Plant resistance is widely used to control the disease, but varieties often need to be replaced after a few seasons because pathogens quickly adapt and overcome the varietal resistance. Recent work by IRRI and the Yunnan Agricultural University demonstrated that the disease can be managed effectively through varietal mixtures (Zhu et al, 2000; see Box 8.3).

Box 8.3 *Diversity defeats disease*

Glutinous rice is highly valued in Yunnan, China, but like many varieties that have been 'defeated' by rice blast, it cannot be grown profitably without multiple foliar applications of fungicide. Rice farmers, guided by a team of experts from IRRI and Yunnan Agricultural University, have successfully controlled rice blast simply by interplanting one row of a susceptible glutinous variety every four or six rows of the more resistant commercial variety. This simple increase in diversity led to a drastic reduction of rice blast (94 per cent) and increase in yield (89 per cent) of the susceptible variety. The mixed population also produced 0.5–0.9 tonnes more rice per ha than their corresponding monocultures, indicating high ecological efficiency. By the year 2001, this practice had spread in over 100,000 ha of rice in Yunnan, and is being tried by other provinces.

Varietal diversity creates an entirely different condition that affects host pathogen interaction. To begin with, a more disease-resistant crop, interplanted with a susceptible crop, can act as a physical barrier to the spread of disease spores. Second, with more than one crop variety, there also would be a more diverse array of pathogen population, possibly resulting to induced resistance and a complex interaction that prevents the dominance by a single virulent strain of the pathogen. Finally, interplanting changes the microclimate, which may be less favourable to the pathogen.

Sheath blight (*Rhizoctonia solani*) is a problem during warm and humid periods and is also aggravated by dense planting and nitrogen inputs above 100 kg ha^{-1}. No crop plant resistance is known for sheath blight. A number of bacteria (*Pseudomonas* and *Bacillus*) isolated from the rice ecosystem are known to be antagonistic to the pathogen. Foliar application of antagonistic bacteria at maximum tillering stage appeared to effect a progressive reduction of disease in the field over several seasons (Du et al, 2001). Incorporation of straw and other organic matter, with its effect on soil fertility, pH and possibly on beneficial micro-organisms may reduce sheath blight incidence in the long term.

Rice tungro disease caused by a complex of two viruses transmitted by the green leafhopper (*Nepthettix virescen*) is a destructive disease in some intensively cultivated areas in Asia where planting dates are asynchronous (Chancellor et al, 1999). Overlapping crop seasons provide a continuous availability of host that enables year round survival of the virus and the vector. Controlling the vector population with insecticide does not always result in tungro control. Synchronous planting effectively reduces the disease to manageable levels. When and where planting synchrony is not possible, resistant varieties are recommended. In addition to varieties with a certain degree of resistance to the vector, varieties highly resistant to the virus itself became available recently. Farmers should also employ crop or varietal rotation, and rogue intensively.

Fungicidal control of blast and sheath blight is increasing in many intensified rice areas. It is extremely important that these fungicides be carefully screened not only for efficacy as fungicides but also for impact on natural enemies in the rice ecosystem. One example is the release of iprobenfos as a fungicide for blast control. Iprobenfos is an organophosphate that was originally developed for brown planthopper control and is highly toxic to natural enemies. Its use in the rice ecosystem is likely to cause ecological destabilization and consequent outbreaks of brown planthopper. Fungicides should also be carefully screened for impact on fish, both to avoid environmental damage in aquatic systems and to avoid damage to rice–fish production.

In general, clean and high quality seed with resistance to locally known diseases is the first step in rice IPM for diseases. An appropriate diversification strategy (varietal mixture, varietal rotation, varietal deployment, crop rotation) should counter the capacity of pathogens to adapt quickly to the resistance of the host. Management of organic matter has to be geared not only towards achieving balanced fertility but also to enhancing the population of beneficial micro-organisms.

Farmers in Korea who face heavy disease pressure can learn to predict potential outbreaks using educational activities that combine various weather and agronomic input parameters with disease outcomes. Computer based models are also being commercially sold to predict disease potential based on meteorological monitoring. With increasing nitrogen applications, however, greater disease incidence can also be expected.

Weeds

The origin of puddling for lowland rice cultivation is thought to have been invented to create an anaerobic environment that effectively kills several weeds including weedy and red rice. In most IPM programmes for lowland rice, weed management has therefore been a closely considered part of agronomic practices during puddling and later during aeration of the soil with cultivators. At least two hand weedings are necessary in most crops, and considered in many countries economically viable due to low labour cost or community obligations to the land-less, who are then allowed to participate in the harvest. With raising labour costs, decreasing labour availability and more effective herbicides, this situation is rapidly changing to one of using one or two applications of pre- or post-emergence herbicides. As in the case of fungicides, it is critical that these herbicides do not upset natural enemies, fish or other beneficial/non-target organisms in the aquatic ecosystem including micro-organisms (see Figure 8.1). In the case of upland rice, similar changes are rapidly occurring, although better dry land cultivators are already developed for inter-row cultivation as an alternative to herbicides.

Non-herbicide, but low labour weed management methods are also emerging from the organic agriculture sector. The International Association of Rice Duck Farming in Asia supports research and exchanges among mostly organic farmers. In rice–duck farming, a special breed of duck is allowed to walk through the field looking for food that is either broadcast or naturally occurring, and the action of walking up and down the rows is sufficient to control most weeds. In Thailand, mungbean and rice are broadcast together with some straw covering in rainfed rice fields. When the rains come, both crops germinate. If there is abundant rain, the mungbean will eventually die and become part of the mulch, but if the rain is insufficient for the rice then the mungbean will be harvested.

No-till, no-herbicide combined with ground cover from winter barley straw or Chinese milky-vetch is being used in South Korea in both conventional and organic systems. Organic farmers in California use a water management system in which there is a period of deep (30 cm) flooding followed by complete drying – the rice can cope with the changes, but young weeds cannot. A widely adopted method in Central Thailand involves growing rice from ratoons. After harvest, the stubble is covered with straw and then irrigated, which allows the rice plant to emerge. This method not only controls weeds effectively but also increases organic matter and requires no tilling.

However, for the majority of rice cultivation, labour saving often means moving towards direct seeded rice and thus more weed problems. Red rice (weedy off-type of rice) is already the key pest in most of the Latin American direct seeded rice production areas. It seems clear that more direct seeding will lead to more herbicide use in rice production. Yet herbicide resistance is also sure to emerge eventually and there are obvious health and environmental costs associated with some herbicides. Thus it is important that IPM for rice weeds be improved and considered in the broadest terms (e.g. promoting modern rice varieties that are red in colour among consumers may be part of the solution to red rice problems). Crop rotations are only feasible in some areas, while simple line sowers or tractor sowing in rows combined with manual or tractor cultivation may provide some solutions for lowland and upland rice.

Genetically modified herbicide-resistant rice will eventually be on the market, but Asian consumer preferences may not favour these varieties. However, the resulting increase in herbicide use could have obvious adverse effects on the aquatic systems that are associated with most rice production. In addition, a major problem of herbicide resistant rice is the possibility of the transfer of gene resistance to weedy rice, although such transfers would not occur to wild grass species. Use of herbicide resistant rice in monocropping could also create long-term, serious problems of glyphosate resistance in weed species previously susceptible to the herbicide. The ecosystem level interactions of herbicide resistant rice will need careful assessment prior to their use.

Community Pests

Insects, diseases (with the exception of tungro virus) and weeds in rice eco-systems are generally managed with decisions on individual farms or plots. However, some pests, particularly rats, snails and birds, require community-level planning and action. Management of these pests requires facilitation of community organizations not generally supported by extension services with the possible exception of some multi-purpose cooperatives and water-user associations.

Numerous species of rats occur in rice fields and can cause considerable damage. Rats migrate from permanent habitats to rice fields as food supply changes throughout a yearly cycle, with rice plants the most preferred after the panicles have emerged. Some natural enemies of rats, particularly snakes, are harmed by pesticides and are often killed by farmers, thus resulting in more rats. The most effective management strategies are to ensure baits are appropriate to the species present, and then carry out continuous trapping along feeding routes, fumigation or digging of rat holes, and establishing early season bait stations using second generation anticoagulant baits (although more toxic zinc phosphide and repackaged and unlabelled aldicarb is still commonly seen, but strongly discouraged in most countries due to deaths of children and small livestock). Community programmes can include educational activities on rat biology and behaviour (Buckle, 1988), and an emphasis on action during the early season vegetative stage is considered the key to rat management (Buckle and Smith, 1994; Leung et al, 1999). An innovative owl habitat programme in Malaysia has been successful in increasing owl populations to control rats in rice and planta-tion crops.

The Golden apple snail, *Pomacea canaliculata*, was originally introduced to rice growing areas as an income generating activity for a caviar look-alike, given its brightly pink coloured egg clusters. It has since become widespread from Japan to Indonesia and is now one of the most damaging pests of rice. It was introduced without appropriate tests in any country, even though it is on the quarantine lists of several countries. The snail feeds on vegetation in aquatic environments, including newly transplanted rice seedlings up to about 25 days when the stems become too hard. Without natural enemies and with highly mobile early stages that flow with irrigation water, the golden snail spreads rapidly. Pesticides are often used before transplanting or direct seeding, mainly highly toxic products such as endosulfan, organo-tin products and metaldehyde. These products have serious health implications and also cause the death of potential fish predators and natural enemies early in the season (Halwart, 1994). The use of bamboo screens as inlets to fields to inhibit snail movement is reported as the first line of snail defence. Draining fields that have several shallow ditches where the snails will congregate allows for faster collection or ease of herding ducks into the fields to eat the snails. In Vietnam, snails are reported to be collected, chopped, cooked and used as fish food to such an extent that they are now a declining problem.

Birds can be very damaging, especially when occurring in large flocks. The Red-billed Quelea, *Quelea quelea*, in sub-Saharan Africa and various species in Asia are known as consistent problems in rice ecosystems. In most Asian countries and in Chad, netting is used to trap large numbers of birds for sale as food. Mass nest destruction is also possible for some species. In Asia, these methods have effectively reduced pest bird populations to very low numbers. In Africa, the capture method may bring benefits to local people in terms of income or a good protein addition to the diet, but the impact on pest bird populations has been small. During the ripening period in north-east Asia, some fields are protected by being covered with bird nets. Reflective ribbons or used video or cassette tape are widely used to scare birds in Asia. Sound cannons and owl or hawk look-alikes are also used in many countries, although some birds become quickly habituated to mechanical devices. Use of poisoned baits, and the destruction of bird nesting habitat, are discouraged both because they are seldom effective and also because of the potential negative effect on non-target species in adjacent aquatic environments.

DOES IPM WORK FOR RICE FARMERS?

Although there is a large amount of grey literature (see www.community ipm.org) related to rice IPM impact among farmers, there is little peer-reviewed published data. This is in part a reflection of the financial and technical difficulty of conducting these studies. Longitudinal studies in agriculture are notoriously difficult due to seasonal changes. Latitudinal studies (comparisons across sites) are also difficult due to the fact that finding an identical IPM and non-IPM control is rarely possible given the diversity of ecological and social conditions. Nonetheless, such evidence as does exist indicates considerable benefits for rice IPM farmers.

The first, and perhaps strongest indicator, is the greatly reduced incidence of brown planthopper. Wide area outbreaks accompanied with massive losses have no longer been experienced during the past 15 years since IPM programmes have become widely implemented in both policy and field training. In most cases, changes in policy involved removal of pesticide subsidies, restrictions on outbreak-causing pesticides, and investment in biological research and educational programmes for decision-makers, extension workers and farmers. These policy changes most often came about as a result of successful small-scale field trials. The FAO Inter-Country Programme for Rice IPM in south and south-east Asia, headed by Peter Kenmore, brought policy-makers in contact with researchers and farmers who could explain from their own experience the ecological basis of farming with IPM methods. The banning of 57 pesticides and the removal of pesticide subsidies known to cause brown planthopper outbreaks in 1987 in Indonesia by the former President Suharto came about after cabinet officials were brought into a dialogue with both senior Indonesian and IRRI scientists and farmer groups who had shown the outbreak effects of the pesticides and their ability to produce high rice yields without these pesticides (Eveleens, 2004).

The second indication comes from case study literature (FAO, 1998). Table 8.1 gives a typical result found across hundreds of communities surveyed in rice IPM programmes. This shows the key changes in practices, especially the common outcome of investing less in pesticides and more in fertilizers (including P and K). Other large-scale studies provide similar data, although a recent study in Vietnam notes an increase in the use of fungicides (FAO, 2000). The authors have noted that with higher levels of fertilizers (as would be found in Vietnam) such increases in fungicide are predictable. This data also reveal the multidisciplinary aspect of rice IPM in that it encourages farmers to look beyond the pest complex into the multiple parameters for achieving a profitable high yielding crop.

Table 8.1 *Financial analysis of ten IPM field school alumni and ten non-alumni farms from impact assessment in Lalabata, Soppeng, Ujung Pandang, South Sulawesi, Indonesia (FAO, 1998)*

	IPM Alumni (Rp. 000 ha⁻¹)	Non-alumni (Rp. 000 ha⁻¹)
Ploughing	105	84
Planting	113	102
Weeding	49	47
Harvest	67	59
Seeds	18	21
Urea	80	96
SP36	30	12
KCl	25	12
ZA	41	0
Pesticides	7	28
Irrigation	25	25
Total costs	560	501
Yield (kg ha⁻¹)	6633	5915
Returns	2786	2485
Income	2226	1983
Difference	+243	

Note: farm gate rice price Rp. 420/kg
Source: FAO, 1998

GETTING IPM INTO THE HANDS OF FARMERS

'IPM is not for farmers but is by farmers' is often noted in IPM programmes. Getting IPM into the hands of farmers, however, is not always easy. Several methods have been developed with various levels of information and completeness. Most agricultural extension services now recognize the importance of natural enemies and are quick to point out the need to conserve them, even though their co-promotion of various insecticides, fungicides and herbicides is at odds with this apparent awareness of natural enemies. Work by Heong and others from the Rice IPM Network (Heong et al, 1998; Heong and Escalada, 1999;

Huan et al, 1999) has developed interesting radio messages to get the word out on a large scale, for example, that early spraying of insecticides during the first 40 days of the crop is not only unnecessary but increases the risk of higher pest populations later in the crop. The radio messages are accompanied by field-based plant compensation participatory research groups in many cases (Heong and Escalada, 1998). This programme has been effective in increasing awareness of the adverse effects of insecticides on natural enemies and the role of plant compensation in recovering without yield loss from early season pest damage and has resulted in reduced early insecticide sprays.

Study groups of various types are now common in many rice systems. There are reports from organic agriculture, rice–duck groups, Australian rice farmer association and many others. The FAO Community IPM Programme in Asia (Matteson et al, 1994) has promoted study groups now called 'farmer field schools' under which structured learning exercises in fields ('schools without walls') are used to study both ecosystem level dynamics transferable to other crops (predation, parasitism, plant compensation) as well as specific rice IPM methods. Already, more than 1.5 million farmers have graduated from one or more season long field schools in Asia over the past decade with good cost effectiveness as an extension methodology (Ooi et al, 2001).

Community-based study groups, study circles, field schools and other approaches are now being integrated with wider community-based organizations, such as IPM clubs, water-user groups, women's organizations and local farmer unions (Pretty and Ward, 2001). With the large-scale training and visit style extension programmes generally being phased out in most countries, it will be necessary for local communities to become organized in ways in which they can increasingly cover their own costs for experts. Primary school programmes on IPM are also emerging in Thailand, Cambodia, the Philippines and other countries as part of environmental education curriculum related to Asian rice-culture. Such programmes as farmer field schools in many countries or Landcare in Australia and the Philippines are providing innovative models in community-based study and action.

The future of IPM in rice in Asia, if not globally, should see the phasing out of all WHO Class Ia, Ib and II products, while phasing in production methods that allow for whole ecosystem approaches. Organic pest management (OPM) alongside the rapid expansion of certified organic rice production is certainly an area fertile for research and training in addition to modernized IPM approaches.

VEGETABLE PRODUCTION IN TROPICAL ASIA

Vegetables are an important part of the diet, adding valuable nutrients that would otherwise be insufficiently available in staples such as rice or maize. New production areas are continuously being opened up, sometimes at the expense of rice land, to meet the demands for vegetables, particularly crucifers, carrots, potatoes, tomatoes and beans. Many vegetable crops perform best under cool temperatures found in higher altitudes, but increasingly vegetable production is expanding into the lowlands with the release of new heat-tolerant varieties mainly bred in Asia. But the achievement of good yields, particularly in the warm humid lowlands, is often constrained by pests and diseases. Most vegetables are heavily sprayed and in many places poor horticultural practices exacerbate these crop production constraints. In general, a lack of skills among vegetable producers and limited or no access to sources of information on new and ecologically sound crop production practices provide a clear rationale for why much of the intensified vegetable production in Asia is currently facing serious problems.

PROBLEMS ASSOCIATED WITH THE INDISCRIMINATE USE OF SYNTHETIC PESTICIDES

The indiscriminate use of synthetic pesticides in intensified vegetable production in tropical Asia is a serious problem (Shepard et al, 2001). Pest problems in tropical vegetable production occur frequently and are often acute. Yields are highly variable while farm gate prices vary considerably on both a daily and seasonal basis. Compared with rice, the riskiness of vegetable production provides a stimulus for farmers to rely on preventive pesticide applications. For some vegetable crops the average frequency of application of chemical pesticides is 10–20 times per season (see Table 8.2), with up to 80 applications per season for *brinjal* (eggplant) production in parts of South Asia (e.g. in Bangladesh).

Table 8.2 *Average number of pesticide applications per season for selected vegetable crops in Cambodia*

	Cucumber	Yard Long	Mustard Bean	Cabbage	Radish
Number of pesticide applications per crop cycle	7.5	9.1	5.7	12.1	10.0

Source: adapted from van Duuren, 2003

The application of cocktails of pesticides by vegetable farmers is also a common phenomenon, particularly in Cambodia and Indonesia. Farmers mix insecticides with fungicides and herbicides in an effort to make them more effective. These cocktails commonly include banned or restricted, and often highly toxic, insecticides such as DDT, endosulfan, chlordane, sodium cyanide, methyl parathion, mevinphos, methamidophos or monocrotophos.

A most recent survey among 332 vegetable producers in Cambodia indicated that 55 per cent of farmers interviewed were using WHO Class Ia pesticides. This figure increases by another 18 per cent when farmers who are using Class Ib pesticides are included. Thus, an alarming 73 per cent of interviewed farmers were frequently handling highly and extremely toxic pesticides under conditions that are far from those that can possibly be considered safe (see Table 8.3).

Table 8.3 *Proportion of vegetable farmers (n = 360) using protective clothing during pesticide applications*

Clothing and protective gear	Proportion of farmers using each element during application (%)
Long sleeved shirts and long pants	82
Cotton mask to protect from inhalation	64
Traditional scarf wrapped around head	52
Boots	38
Gloves	8
Rain coat	3
Protective glasses	2
Ordinary clothing	8

Source: adapted from van Duuren, 2003

The health hazards to farmers and their families are serious (Murphy et al, 1999; Sodavy et al, 2000).

In Cambodia and Laos, labelling of pesticide products is often inappropriate as labels are usually in foreign languages (Thai, Vietnamese). This situation is aggravated by high illiteracy rates in the rural population. Original products are often re-packaged and contain no label at all (EJF, 2002). For example, a recent study conducted in Cambodia revealed that only 8 of 77 pesticide traders said they could read foreign labels on the pesticides they sold, whilst 97.5 per cent of the pesticides were labelled in a foreign language (CEDAC, 2000). The extensive use of synthetic pesticides results in a range of unsustainable production practices arising from undesirable externalities. The frequent applications of pesticides most often causes a resurgence of pest populations because of the destruction of natural enemies. Resistance of target pests against pesticides has become a serious problem for many important vegetable pest problems, such as diamondback moth (*Plutella xylostella*) on crucifers and Fruit and Shoot Borer (*Leucinodes orbonalis*) on eggplants.

However, greater awareness among consumers in urban communities of the dangers of pesticide residues on vegetables has created a growing demand for vegetables free of residues. Governments in many developed Asian countries have now established maximum residue levels (MRL) regulations for pesticides on imported vegetables. Clearly, the consequences of international trade restrictions related to residues of toxic pesticides on vegetable produce can no longer be ignored. Nonetheless, vegetable production in tropical and subtropical Asia remains in a 'crisis phase' (Lim and Di, 1989; Shepard and Shepard, 1997), requiring urgent attention to safeguard the production of healthy food and producers' livelihoods across Asia.

VEGETABLE IPM: ECOSYSTEM CONSIDERATIONS AND THE NEED FOR 'INFORMED INTERVENTION'

What is it that drives the frequent use of pesticides in intensified vegetable production in most of Asia? The main factors seem to be the high risk of crop losses, the acute occurrence of serious crop pests, and the heavy promotion of pesticide use by the private sector. In addition, the low level of ecological literacy and wider low-level education of farmers, particularly in a country such as war-ravaged Cambodia, further explains the rampant use of pesticides. However, it is important to understand that there are some major differences between the need for human intervention in ecosystem management when comparing vegetables to rice in tropical Asia.

While many locally consumed vegetables are native to tropical Asia, most vegetables produced in Asia for local consumption and for market supply are exotics. Many popular cash crops such as tomatoes, crucifers and potatoes were relatively recently introduced to Asia from temperate regions. Similarly, many important pest and disease problems are exotic, such as the diamondback moth, which was introduced from Europe into Asia without its naturally existing complex of natural enemies. If not swiftly and adequately managed, crop protection problems can lead to serious cosmetic damage and total crop failure. Vegetable ecosystems are much less stable compared with paddy rice ecosystems.

Any rice farmer field school (FFS) alumni farmer would be able to explain and demonstrate in their field that there are several highly effective predators and/or parasitoids indigenously available as integral components of a well-functioning rice ecosystem. As a result, pest populations are well regulated and rarely reach critically damaging levels. Management interventions are therefore rarely needed. In fact, human intervention with applications of pesticides causes more problems than it solves. The basic message that IPM farmer field schools spread focuses on ensuring that farmers do not intervene when pest problems occur, and rather let nature play out its well regulated population dynamic games. The IPM management strategy to be employed in Asian rice production can thus be characterized as 'informed non-intervention'.

But the situation is very different in vegetable ecosystems in tropical Asia, as these are much more designed to prevent and manage pest and disease problems from causing serious crop loss. The crop protection strategy therefore employed in intensified tropical vegetable production is based on informed (and pro-active and preventive) intervention (Whitten and Ketelaar, 2003).

MAKING VEGETABLE IPM WORK: THE NEED FOR FARMERS TO BECOME IPM EXPERTS

With the urgent need to address problems associated with the indiscriminate use of pesticides, the FAO Inter-Country Programme for Vegetable IPM in south and south-east Asia has carried out applied research, extension and farmer education activities to promote and support IPM in vegetables. Based on impact assessment of farmer training work conducted by this FAO Programme and its associated National IPM Programmes in a variety of crops in several Asian countries, IPM trained vegetable farmers can now avoid excessive and inappropriate use of pesticides (Larsen, 2001; Lim and Ooi, 2003).

Farmers who undergo season-long discovery based training in farmer field schools become ecologically literate, and so can understand much better how ecosystems function and what is the likely impact of their management decisions. By being better able to identify field problems and assess their potential impact on yields, farmers can considerably reduce the use of pesticides in vegetable production and limit any remaining applications to those situations where human intervention is necessary As a result of FFS training, farmers can also make better decisions on which pesticides to purchase, how and when to apply them, and how to avoid cocktail formulations.

These IPM trained farmers are then better placed to access new information and to adapt and adopt novel options that reduce further dependency on pesticides. The potential for better understanding and improved access of farmers to biocontrol interventions, such as the employment of viral and fungal pathogens and the introduction of parasitoids for pest management, is considerable. But it requires proactive action from a range of different stakeholders (research, private sector, extension workers and farmers) so as to ensure that functional biocontrol can reach its full potential. This would further assist the process of eliminating toxic products from vegetable production.

In Cambodia, the National IPM Programme, with UN Food and Agriculture Organization (FAO) support, is currently implementing an FFS-based farmer training programme in several major vegetable growing areas in the country. Farmers who have undergone training are now running training for other farmers and are actively experimenting with growing crops with lower inputs of pesticides. When interventions are indeed needed, farmers now prefer to resort to biocontrol, using the biopesticide, *Bacillus thuringiensis*, which has no adverse effects on farmers' and consumers' health.

Local NGOs, such as Srer Khmer, are actively supporting a self-sustained and multiple season IPM learning process and are facilitating the mobilization of IPM farmer groups and their associations. IPM farmer clubs are also increasingly becoming interested in embarking on the production of organic vegetables for the local niche markets. With active support from the government and NGOs, formal education efforts are underway to give school children access to ecosystem education, using the discovery-based learning methodologies employed in the FFS for adults. The hope is that this will lead to greater ecological literacy among Cambodia's youth and future farmers.

FUTURE NEEDS IN RICE AND VEGETABLE IPM

There is still much room for improvement for IPM. Indeed, the ecological view of rice and vegetables presented here must be given greater support by international and national scientists and policy-makers to widen economic and ecosystems benefits already being realized by some farmers. A new CD-ROM produced by the International Rice Research Institute (IRRI) is beginning to bring together basic rice information in an accessible format, while the World Vegetable Centre in Taiwan has developed a web-based study programme. Both programmes could be helpful in training extension staff but still remain distant from farmers. Other major challenges remain. Post-harvest pests are still a problem and deserve greater research on non-toxic management methods, and environmentally friendly methods of controls for all types of pests, especially weeds and fungal pathogens, are required to reduce the pressure on natural resources.

Some countries are calling for major changes. South Korea has banned pesticide use in Seoul's watersheds and is promoting organic agricultural investments to ensure both clean water and high levels of production. Other communities are moving away from grain maximization to diversification such as rice–fish–vegetable culture as a response. This is expected to increase as demand for more profitable non-grain products increases and nitrogen use is reduced to lower environmental impacts and the incidence of expensive-to-control fungal pathogens. However, IPM development is required in more countries. These programmes should ensure that educational systems (both formal and non-formal) are responding to the future needs of reducing the environmental impact of agriculture whilst improving yields. IPM is clearly a major aspect of this education.

There is a need to phase in new plant protection methods and products including subsidizing the commercialization of locally produced products such as pheromones, attractants, natural enemies, pest-exclusion netting (for insects and birds), high-quality seed, improved disease resistance and balanced soil fertility products. High foreign exchange costs for imported pesticides and increasing consumer awareness of the social costs arising from pesticides and inorganic fertilizers can be expected to drive rice IPM system development. The trend will be towards lower impact and local production of environmentally friendly pest management. A significant redefinition of IPM to exclude Class I and most Class II products could be a most important step to revitalize private sector, research and extension IPM activities.

Chapter 9

Towards Zero Pesticide Use in Tropical Agroecosystems

Hans R. Herren, Fritz Schulthess and Markus Knapp

INTRODUCTION

Agricultural production in tropical agroecosystems is greatly impacted by pests (arthropods, pathogens, nematodes and weeds) with the result that the use of synthetic pesticides has been on the rise. This is particularly true for cash and horticultural crops that have a significant economic return. Recently, however, the use of pesticides is being restricted on crops destined for export, following the introduction of new maximum residue levels in industrialized countries.

For the purposes of this chapter, we will define 'pesticides' as including all synthetic chemical insecticides, acaricides, fungicides, nematicides and herbicides. However, botanicals, semiochemicals (mostly used in trapping devices) and biopesticides come under a different category that we will not exclude from the farmers' pest management tactics toolbox. These pest management agents are usually specific (except in the case of pyrethrum), are applied on a 'need' basis and do not have adverse effects on people, animals and the environment. We also emphasize that there may be instances where there is a need to intervene to stop an outbreak that could not be prevented through agronomic or habitat management practices.

Plant health suffers further from several serious scientific constraints that must be addressed for the implementation of a 'zero pesticide' scenario. At the individual organism level, it is the behaviour of the pest and how to manipulate it; at the population level, it is the understanding of the pest population fluctuation and its causes; and at the system level, it is the lack of know-how and training of implementers for pest management knowledge and technology integration at the policy, strategy and tactical intervention levels.

In this chapter, we analyse two case studies on the practicalities of eliminating synthetic pesticides from the 'ecological' IPM toolbox without jeopardizing the quality and quantity of food production, while at the same time improving farmers' revenues and the sustainability of their production systems. We also include in our analysis the animal disease vector management component, as this cannot be taken in isolation of what is happening elsewhere in the agricultural production system. An ecosystem and geographic information system approach

(ES/GIS), linked to an adaptive management system and community–farmer participation (Meffe et al, 2002), will need to be introduced as the basis for decision-making on any intervention, whether preventative or curative. However, achievement of zero pesticide use in the tropics may be constrained as much by policy and market issues as it is by scientific or technical considerations.

KEY ISSUES IN ECOLOGICALLY SOUND PEST MANAGEMENT IN THE TROPICS

We identify six key issues for pest management decision-making that need the attention of farmers and policy-makers, as well as the scientists who are trying to assist the farmers with the implementation of a zero pesticide use strategy. These are: (i) education and information availability; (ii) economic environment and imperatives; (iii) agricultural production systems; (iv) availability and affordability of alternative pest management tools and implementation strategies; (v) market requirements, consumer education; and (vi) policy environment.

Education and Information Availability

Farmers have, over the years, developed a body of knowledge on different pest problems that affect their crops and have selected for plants that have a certain level of tolerance or resistance against them. They have also explored their environment to identify products that may be helpful in combating some key pests. Such methods do work fairly well in an environment where the acceptable damage level is greater than zero and on indigenous crops and indigenous pests. On introduced crops, there has been a slow adaptation of local pests, and usually farmers have had ample time in selecting cultivars that show appropriate levels of resistance. When it comes to introduced pests, however, the problem commonly overwhelms farmers, and so government-level interventions are needed. This is also true for migratory pests such as locusts, armyworms and birds.

In general, farmers in the tropics lack basic agricultural training. International development agencies and local governments have made some efforts, but as a whole, very few farmers have attended vocational school or courses. In addition, the level of input from local extension services is often inadequate to satisfy the knowledge needs for modern agricultural practices of farmers. With the exception of south-east Asia, farmer field schools (FFSs) are reaching only a fraction of the farmers who need help.

It is our view that lack of access to education and information is as much a constraint to increasing production and productivity as are the seeds, pest management tools and other key agricultural inputs. We suggest that a component to implement a zero pesticide strategy in tropical agroecosystems is dependent on access to specialized agroecosystem management training and provision of decision support tools in the framework of an adaptive management system, where the experience of previous seasons is integrated into the production cycles.

Economic Environment and Imperatives

Farmers in the tropics have two main difficulties that act against them: one is economic and the other environmental. On the economic side, tropical developing countries all have to manage several problems that greatly affect the way pest management is done. Because of a widespread lack of integrated pest management (IPM) policies, or lack of enforcement, pesticides that are normally banned often appear in local markets and are used without restrictions. These are mostly old and potent products that are cheap and often do not work. Only when new regulations on export crops are introduced does the utilization of these products become an issue, while in most domestic markets, maximum residue regulations, if any, are usually not enforced.

Poverty is another key constraint, and is responsible for zero pesticide use in some areas. But the lack of education and training under these conditions leads to low yields and high losses due to pests. It is clear that what is needed is a low-input, low-cost, sustainable strategy to control agricultural pests. The transition from present day pesticide-based agriculture to a zero pesticide approach comes at some cost, as the system has to recover its resilience, often compromised by over-use of broad-spectrum pesticides. Where only synthetic pesticides have been used, and often also synthetic fertilizers, there is a need to rebuild soil organic matter and fertility as the key elements of a sound pest management strategy (as a healthy plant can afford some pest damage).

For farmers in the average tropical zone, pests are a very difficult problem to tackle without know-how and inputs. Governments, therefore, need to take a much more proactive role in training, technology diffusion and provision of decision support tools. These will, however, only have the necessary impact if farmers do adopt them, the condition being that whatever investment in time or money is made will be rewarded in the end. This will require further government intervention at harvest time, with provision for storage, processing and some price guarantee, as is the case in most developing countries. Excess agricultural production for the market will only be produced if the investment can be recouped.

Tropical Agricultural Production Systems

In many parts of the tropics, the agricultural environment is difficult and unpredictable, with not only indigenous and site-specific pests (bio-types), but also many exotic introductions and migratory species. The tropical environment allows pest populations to breed without the annual breaks and attendant mortalities found in temperate zones. On the positive side, tropical systems allow for the year-round presence of beneficial organisms, thus avoiding the high population variation both of pests and their antagonists, although pest infestation and damage show distinct seasonal peaks. Despite this, pest management issues are usually more recalcitrant and difficult to handle in the tropics. Apart from climatic considerations, the small size of farms and high crop diversity often actually favour natural control mechanisms. However, the trend towards a simpler cropping system, usually maize- or cassava-based, is working against a pesticide-free strategy. New developments in the area of genetically modified pest-resistant and herbicide-tolerant crops are also leading towards mono-cropping systems with no ground cover (Herren, 2003). These developments will certainly lead to increased need in terms of active pest management activities of the curative type, rather than promoting the preventative, agroecological approach.

There is no doubt that more labour-efficient food, feed and fibre production systems are needed, with increased productivity and sustainability. These, however, will be of a type that is dictated by the need for diversity and resilience in terms of pest management strategies, as prevention is better than cure. More research on tropical agroecosystems as a whole, that will accommodate and integrate the large body of research results already available on the individual and population levels, is therefore a must. This approach will provide the needed information for a decision support system that will help tropical farmers in their pest management intervention choices.

The tropical environment is not only very taxing on the crop pests and diseases, it also seriously affects human and animal health, and thus the whole agricultural process. The impact of animal diseases transmitted by arthropod vectors such as ticks and tsetse flies is enormous, and one of the main reasons for the low productivity of many African agricultural systems. The loss is not only quantifiable in terms of milk and meat, but even more so in terms of lost draught power and organic fertilizer. The heavy morbidity and death toll inflicted by mosquito-borne malaria is a further factor in causing low agricultural productivity. In spite of their very many serious side-effects, pesticides have been and are being used to control the vectors of these major tropical diseases. Here too, there is a great need to develop new vector management strategies based on an adaptive systems approach.

AVAILABILITY AND AFFORDABILITY OF ALTERNATIVE PEST MANAGEMENT TOOLS AND IMPLEMENTATION STRATEGIES

New zero pesticide strategies have already been developed and implemented in several instances. Organic cotton production is well established in Africa and India, as is the maize-based system that we at the International Centre of Insect Physiology and Ecology (ICIPE) call the 'push–pull' approach for maize stem-borers and striga weed in Africa. In terms of affordability, there are two issues: one is at the production level under any pesticide-alternative strategy, the other the costs of the investment in terms of time and money. In the pesticide-use option, decisions are made on the bases of direct, short-term and on-farm cost–benefit analysis. Here, 'on-farm' means leaving out any indirect costs that may accrue due to externalities arising from the production, storage and destruction of unused products, and the deleterious effects on the environment. There is a need for research to develop new production methods that will lower the direct production costs and increase the availability of such biological products. When it comes to management practices such as push–pull, the costs are higher in terms of staff time to plant the intercrop and harvest in good time than in the purchase of products, which requires actual cash.

There are a group of chemicals, the semiochemicals, that have yet to be developed to their full potential. These are natural or synthetic products that are copies of naturally produced substances that affect insect behaviour. Such synthetic products have a key role to play in arthropod population monitoring and also in their control. For example, they offer viable and readily implementable alternatives in tsetse, fruit fly and locust control. In the past, they have been rather out of reach for the average farmer, but new research at ICIPE shows that semiochemicals can be produced very competitively and locally in the tropics.

Often, the issue in implementing sound pest management lies less with the individual tactics available than with the strategy. One important aspect is the integration of the knowledge from the individual, population and ecosystem levels into an implementation action plan that will include the policy, strategy and tactical levels. This can be defined as 'knowledge and technology integration and delivery'.

Market Requirements, Access and Consumer Education

Farmers will be willing to invest in more pest management practices than the quick fix approach when markets are ready to acknowledge the extra effort through some financial reward for quality. What is often seen as a lack of production and low quality is also a function of lack of demand and appropriate prices. In general, farmers in the tropics have a difficult time accessing urban markets, and usually only do so via several middlemen, which increases the product price at the consumer end (restricting the market size) but lowers the price at the producer end (reducing the incentive to produce more and at better quality levels).

A zero pesticide strategy will also require that there is a demand for products produced under such a label. In the tropics, local consumers are often not aware of the options in terms of product quality, such as the benefits of organically against conventionally produced products with the support of synthetic pesticides and fertilizers. A consumer education campaign thus has to go hand-in-hand with any introduction of a zero pesticide strategy, to ensure that farmers have a further incentive to stick to such a strategy, even though it may at first be more elaborate and demanding than the quick fix offered by the synthetic chemical approach.

Policy Environment

Science and technology, market forces and environmental and health considerations generally work in support of a zero pesticide strategy to manage pests in agricultural systems. But these considerations will not lead to full implementation if the policy environment is not conducive. Guidelines and regulations are needed not only to promote adoption but also to make it the solution of choice because of the potential private and public benefits. It is important that on a national and regional level, policies that promote zero pesticide strategies are formulated, implemented and enforced. A notable past success is the IPM policy introduced in Indonesia, which was followed by a substantial reduction of pesticide use in rice production. The UN Food and Agriculture Organization (FAO) has also developed IPM guidelines, but their adoption, implementation and enforcement by the member countries has been weak to date.

The major interests of the agrochemical industry are obviously being targeted by the IPM guidelines, and even more so by the introduction of organic production, where there is a zero chemical tolerance. The interests of the agrochemical industry, understandably, differ substantially, although, even here, there is room for the production of bioproducts on a large scale. At present, with only one exception, the case of *Bacillus thuringiensis* (*Bt*), bioproduct production has not been seriously considered for market share reasons.

Zero tolerance, however, does not mean zero intervention, and there are many possible avenues to allow active intervention on a curative basis should the preventative approach not yield the necessary results. At the policy level, there is a need to recognize that there are alternatives available, and that, with some assistance from governments and donors, biological methods can yield formidable results (Herren and Neuenschwander, 1991; Zeddies et al, 2001). What is now needed in terms of policy support is to include zero pesticide use as the ultimate goal in pest management approaches, with commensurate investment in supporting research, capacity building and product development and production areas.

Regulations must also be in place to establish standards for products produced under a zero pesticide strategy, with guidelines for certification together with a certifiers' accreditation process. In an environment where a zero pesticide strategy is being rewarded appropriately by the market system and supported by policies and their strict implementation, farmers are likely to adopt new technologies readily.

CASE STUDY 1: LEPIDOPTERAN CEREAL STEMBORERS MANAGEMENT

Background

In sub-Saharan Africa (SSA), average yields of maize are around 1.2 t ha^{-1}, which is far below the 6.1 ha^{-1} obtained in breeding trials (CIMMYT, 2001). Major constraints to maize production are lepidopteran stem- and cob-borers, which reduce yields by 10–70 per cent (Polaszek, 1998). In West Africa, the most important species are the pyralid *Eldana saccharina* Walker, the noctuids *Sesamia calamistis* Hampson and *S. botanephaga* Tams & Bowden and the pyralid cob-borer *M. nigrivenella* (Endrödy-Younga, 1968; Schulthess et al, 1997; Buadu et al, 2002). In Cameroon and Central Africa, the noctuid *Busseola fusca* Fuller is the predominant species across all altitudes (Cardwell et al, 1997). By contrast, in East and Southern Africa, *B. fusca* is a mid-altitude/highland pest, whereas in the lowlands the exotic crambid *Chilo partellus* (Swinhoe) predominates (Polaszek, 1998).

The search for control of stem- and cob-borers has been a prime concern of agricultural researchers in Africa since the 1950s (Polaszek, 1998; Kfir et al, 2001). Various control strategies have been tried, some with partial or local success, but all have limitations and none have provided a complete solution. Chemical control using systemic insecticides provides only protection against early attacks, but not against borers feeding in the cob (Sétamou et al, 1995; Ndemah and Schulthess, 2002). Development of maize varieties resistant to borers was the first approach used by international agricultural research centres (IARCs), however only moderate resistance has been achieved so far.

Thus in the 1990s, IARCs started to look for alternative solutions. First, country-wide surveys and farmer questionnaires were conducted to determine the extent of losses in maize production due to pests, and farmers' perceptions of these losses. Multivariate analyses of the survey data generated hypotheses on the interactions among physical components of the cropping system such as edaphic and crop management factors with biotic components of the system. The hypotheses were tested in selected benchmark sites, on-farm participatory trials, on-station, or in the lab or greenhouse, using controlled experiments. This approach led to several novel approaches to combat stemborers.

Habitat Management

With the exception of *C. partellus*, all stemborer species attacking maize are indigenous to Africa, and these pests evolved with wild grasses and sedges (Bowden, 1976; Schulthess et al, 1997). In addition, small-scale farmers tradition-ally intercrop maize with non-hosts or other cereals in order to obtain greater total land productivity; in many cases, pest densities are known to decrease in diversified systems (see Risch et al, 1983; van den Berg et al, 1998). Consequently, any attempt to control these pests must take into consideration the close link between the pests and natural habitats. The latter includes alternate hosts and associated crops, as well as the physical and chemical properties of soils, all of which affect the plant, its pests and thereby the natural enemies (Schulthess et al, 1997; Khan et al, 1997a, 1997b; Ndemah et al, 2002). Several IPM technologies based on manipulation of the wild and cultivated habitat of maize pests and their natural enemies have been developed by IARCs.

Trap plants: Surveys in roadside fields in several countries in western Africa showed that the higher the wild grass abundance around the field, the lower the pest incidence on maize (Schulthess et al, 1997). Oviposition preference and life table studies revealed that some wild grass species are highly attractive to ovipositing female moths, although mortality of immature stages is close to 100 per cent versus 70–80 per cent on maize (Shanower et al, 1993; Schulthess et al, 1997). In addition, wild habitats were shown to play an important role in maintaining stable parasitoid populations during the off-season, thereby lowering pest incidence in crop fields during the growing season (Schulthess et al, 2001; Ndemah et al, 2002, 2003). Planting border rows with grasses led to an increase in parasitism, a decrease in pest densities and increased yields (Khan et al, 1997a, 1997b; Ndemah et al, 2002).

Push–Pull: Using a similar approach, ICIPE in collaboration with Kenyan national programmes and Rothamsted-Research developed the 'push–pull' technologies, which are especially suited for mixed crop–livestock farming systems where stemborers and striga are major biotic constraints to maize production. These 'push–pull' strategies involve trapping stemborers on unsuitable trap plants (the pull) such as Napier grass (*Pennisetum purpureum*) and Sudan grass (*Sorghum vulgare sudanense*) and driving them away from the crop using repellent intercrops (the push) such as molasses grass, *Melinis minutiflora* Beauv., and the legume *Desmodium uncinatum* (Jacq) DC (Khan et al, 2002). Molasses grass also increases stemborer parasitism by the braconid larval parasitoid *Cotesia sesamiae* (Cameron) (Khan et al, 1997b), while *D. uncinatum* inhibits striga by causing suicidal germination (Khan et al, 2002). All of these plants are of economic importance to farmers in eastern Africa as livestock fodder, thus contributing to increased livestock production (milk and meat) by availing more fodder and crop residues. These plants have shown great potential for stemborer and striga management in farmer participatory on-farm trials with more than 1500 farmers in Kenya and Uganda (Khan et al, 2002). The 'push–pull' technique has been shown to increase maize yields by 18–35 per cent on average in ten districts in Kenya and three districts in Uganda.

Mixed cropping: In SSA, intercropping maize often led to a reduction of pest densities of 50–80 per cent, especially if the associated crop is a non-host. The mechanisms involved were reduced host finding by the ovipositing female moth; unsuitable hosts acting as trap plants; and increased parasitism or increased mortality due to starvation and/or predation of migrating larvae (van den Berg et al, 1998; Chabi-Olaye et al, 2002; Schulthess et al, 2004).

Management of soil nutrients: Survey work and lab and field trials conducted in Benin showed that increasing soil nitrogen favours both plant growth and survival and fecundity of stemborers (Sétamou et al, 1993; Sétamou and Schulthess, 1995). Thus, although pest densities were higher, yield losses decreased with increasing N dosage applied. This was corroborated by results from crop rotation trials in Cameroon, which included legumes (Chabi-Olaye, IITA-Cameroon, unpubl. data). Other nutrients such as silica and potassium have a negative effect on survival and fecundity of *S. calamistis* and *E. saccharina* (Shanower et al, 1993; Sétamou et al, 1993, 1995; Denké et al, 2000).

Biological Control (BC)

Classical BC: A project on biological control of cereal stemborers in subsistence agriculture in Africa was initiated by ICIPE in 1991. Thus far, the project has focused on the exotic pest *C. partellus* in East and Southern Africa. The larval parasitoid *Cotesia flavipes* Cameron was introduced from Asia into Kenya in 1993, and has become permanently established in the country, increasing maize yields by 10 per cent on average (Overholt et al, 1997; Zhou et al, 2001). In 1997–1998, following the success in Kenya, *C. flavipes* was released in eight more countries in the region. In 2000, a second Asian natural enemy of *C. partellus*, the ichneumonid pupal parasitoid *Xanthopimpla stemmator* Thunberg was introduced into the ICIPE labs. Suitability tests (which include *C. partellus*, *S. calamistis*, *E. saccharina* and *B. fusca*) showed that *X. stemmator* equally preferred all stemborers species offered and parasitization rates were similar. It is therefore planned to test *X. stemmator* in West Africa as a new association candidate against *S. calamistis* and *E. saccharina*.

Redistribution: The exchange (or 'redistribution') of natural enemies between regions of a continent as a solution to cereal stemborers has been proposed by several authors (Schulthess et al, 1997). The first success was the eulophid parasitoid *Pediobius furvus* Gahan, which was introduced from East Africa into Madagascar against *S. calamistis* (Appert and Ranaivosoa, 1971). Comparisons of species complexes and of behaviour of races yielded several promising candidates to be exchanged between eastern and western Africa: the tachinid parasitoid *Sturmiopsis parasitica* Curran has been introduced from West Africa into South Africa against *E. saccharina* on sugarcane (Conlong, 2001); a strain of *C. sesamiae* from the Kenyan coast was introduced into West Africa in the 1990s and became permanently established on *S. calamistis* in southern Benin (Schulthess et al, 1997); the scelionid egg parasitoid *Telenomus isis* Polaszek, an important natural enemy of noctuid stemborers in West Africa, was introduced into eastern Africa in 2003, where it does not exist (Schulthess et al, 2001).

CASE STUDY 2: BIOLOGICAL AND INTEGRATED CONTROL OF PESTS IN VEGETABLES

Background

Smallholder vegetable production has been expanding rapidly in sub-Saharan Africa in recent years and now contributes significantly to the income of many rural families. Vegetables are produced for home consumption, urban markets and increasingly for export. However, yields are still comparatively low. For example, tomato yields in 1999 were 10.2 t ha^{-1} in Kenya and 6.7 t ha^{-1} in Zimbabwe compared to an African average of 19.9 t ha^{-1} and a world average of 26.9 t ha^{-1} (FAO, 2001a). Since vegetables are a high value crop compared to food crops like maize, sorghum or cassava, and because of stringent quality and cosmetic requirements for export produce, pesticide use has been generally high.

Farmers in Kenya spray their French beans up to 12 times (Seif et al, 2001) and more than 90 per cent of tomato farmers in Kenya, Zimbabwe and Zambia use pesticides regularly (ICIPE, unpubl. data). Pesticide use in small-scale vegetable farming in Africa is associated with many problems. Farmers lack the appropriate training, and apart from contamination of farm workers due to inadequate safety equipment, use of the wrong pesticides, inadequate dosage (usually too low since farmers consider pesticides expensive) and inadequate application techniques are widespread (Sibanda et al, 2000). Biological and integrated control offer opportunities to reduce this dependency on pesticides. However, a much deeper insight into the biology and ecology of the target organisms than the traditional 'just spray' approach is required.

Biological Control (BC)

There is a need for taxonomic expertise. Parasitoids and predators are often very specific to their hosts, and without proper knowledge of the pest and its natural enemies, biological control efforts are bound to fail. This starts with the proper identification of the pest and the potential natural enemy. Two examples are given here to illustrate this point.

In about 1985, tomato farmers in southern Africa began to complain about a sudden upsurge of spider mites. The mites were usually identified as *Tetranychus urticae* Koch or *T. cinnabarinus* (Boisduval) by the relevant national authorities. These two species are the most common spider mites in the region and infest many different host plants, although they had not been known previously to be a major problem on tomato in southern Africa. The sudden upsurge in numbers was related to the eradication of natural enemies through increased use of broad-spectrum pesticides by tomato farmers. This phenomenon was well known from other parts of the world (van de Vrie et al, 1972). However, soon after the start of a project to address these questions, it was found that the species in question was *Tetranychus evansi* Baker & Pritchard, a mite of likely South American origin that was introduced to Africa (Knapp et al, 2003). As a result, most indigenous natural enemies do not feed on it. The project had to be re-designed completely and a major activity is now the search for natural enemies in Brazil that can be used in a classical biological control approach to control *T. evansi*.

The diamondback moth (DBM), *Plutella xylostella* L., is the most important pest of cabbage and its relatives in sub-Saharan Africa. A successful biological control programme for DBM using the parasitoid *Diadegma semiclausum* (Hellen) was executed in Asia (Ooi, 1992; Poelking, 1992; Talekar et al, 1992; Biever, 1996; Eusebio and Morallo-Rejesus, 1996; Iga, 1997). *Diadegma* species were also collected from DBM in Africa and were formerly identified as *D. semiclausum*. However, parasitation rates were much lower than in Asia (Löhr, pers. comm.). Later, a review of the genus *Diadegma* assigned all *Diadegma* from Africa to the species *Diadegma mollipla* Holmgren, which is mainly known as a parasitoid of the potato tuber moth, *Phthorimaea opercula* Zeller (Azidah et al, 2000). Further investigations using molecular techniques confirmed this and also allowed the discovery of a new *Diadegma* species from Ethiopia (Wagener et al, 2002). Consequently ICIPE started a programme to import *D. semiclausum* from Taiwan for releases in eastern Africa. To date, the parasitoid has been released in several areas in Kenya and parasitization rates have increased sharply.

These examples clearly show that proper identification of pest species and natural enemies is an indispensable prerequisite for successful biological control. Unfortunately, taxonomic expertise is often lacking, especially in Africa, and is on the decline in many other parts of the world. However, new molecular techniques are offering additional opportunities to classical morphology-based taxonomy.

Reduction of Pesticide Use Through Development of IPM Strategies

Apart from biological control, IPM programmes have the potential to reduce pesticide use by African smallholders. An IPM programme developed for French beans, the major export vegetable grown in Kenya allowed the reduction of the number of foliar pesticide applications from 12 to fewer than four per season, mainly through use of seed dressing. Further research would help in eliminating even this seed dressing or replacing it with biological or botanical products. Foliar pesticide application after pod formation is avoided completely because harvesting is done daily or every other day and there are no pesticides with such short pre-harvest intervals available (Seif et al, 2001).

In tomato, yields can be improved significantly with cropping techniques that improve the efficacy of acaricide treatments. Traditionally, most African farmers grow tomatoes without staking. This makes it difficult to reach spider mites, which prefer the lower sides of the leaves, with contact acaricides. This is the reason why farmers frequently complain about the inefficiency of pesticide applications, which they wrongly attribute to resistance of the mites to the pesticides used. ICIPE showed that staking and pruning of tomatoes not only significantly improved mite control, but also increased yield and revenue through other factors such as reduced disease incidence and higher numbers of larger tomatoes. The Zimbabwean farmers in the neighbourhood of these experiments who were highly sceptical at the start of the project, then quickly adopted staking and pruning in their tomato fields (ICIPE, unpubl. data).

CONCLUDING COMMENTS

Achieving a zero pesticide strategy in tropical agroecosystems may be easier than in temperate zones, as in many instances farmers have not yet begun the generalized use of pesticides. This gives special opportunities for scientists and farmers to work on a systems approach to minimize pest impact before agroecosystems have been disturbed. For those areas already heavily impacted by the use of pesticides, such as cash crops, horticultural crops and livestock, adjustment to pesticide-free systems management will take some adaptation. Time is needed to establish or re-establish conditions in the system that are conducive to increased natural control such as habitat management and agronomic practices, as well as to introduce farmers to new biological and physical control concepts and methods. In special cases where these will not provide adequate protection, the introduction of semiochemicals and botanicals may be called for. Farmers will need to undergo specific training in their use. For the optimal use of all pest management tactics, decision support tools will be needed.

The research presented in the first case study yielded various habitat management technologies, which showed the opportunity to reduce pest densities in maize and increase yields in farmers' fields. These technologies mainly aim at increasing plant biodiversity and at managing soils in order to increase the fitness of crop plants. In addition, new opportunities for classical biological control, new associations and an increase of the geographic range of indigenous parasitoid species and strains have been identified. In the second case study, we have seen how important are research findings in taxonomy, and also that there is still much more research needed on the discovery and understanding of natural enemies. Although these case studies are rather narrow in view of the large number of pest problems that affect tropical agricultural production, they point to some key issues for our goal of zero pesticide use. Reaching this stage, however, will demand more in terms of knowledge integration and community participation.

A unified regional level analysis and decision support system will be needed. A general modular ES (EcoSystem)/GIS system that can be used across different regions for different health and agricultural problems to serve as a basis for addressing new problems as they arise is already being developed. The modular form of the ES/GIS allows the addition of new components with ease, with outputs summarized as maps and economic production functions for regional economic and social analyses. This system would also provide important decision support tools for current and future research and implementation on tropical pests. Recommendations for the development of such systems have been made to the Consultative Group on International Agricultural Research (CGIAR) and to the United States Agency for International Development (USAID) for implementation on desktop computers in various regions of Africa (Gutierrez and Waibel, 2003; Gutierrez et al, 2003).

Considerable biological and ecological data is available on tropical crop pests and human and animal disease vectors, but these must be summarized as dynamic systems models for regional analysis. This would allow the ES/GIS to guide interventions in participatory adaptive management systems. An ES/GIS-based pest management strategy would provide a basis for low-cost analysis of complex agroecosystem problems in diverse regions worldwide for the benefit of local farmers and their national economies. However, it is equally important to ensure that all the elements at the lower level of integration, that is the individual and population levels, are well researched and information is available for the decision-making process. The knowledge at the lower levels will ensure that novel interventions that fit the 'no pesticide' label are developed, produced and marketed. Their availability will ensure the ultimate successful implementation of zero pesticide strategies.

Chapter 10

From Pesticides to People: Improving Ecosystem Health in the Northern Andes

Stephen Sherwood, Donald Cole, Charles Crissman and Myriam Paredes

INTRODUCTION

Since the early 1990s, a number of national and international organizations have been working with communities in Carchi, Ecuador's northernmost province, on projects to assess the role and effects of pesticide use in potato production and to reduce its adverse impacts. These are INIAP (National Institute of Agricultural Research from Ecuador), CIP (International Potato Center), Montana State University (US), McMaster University and University of Toronto (Canada), Wageningen University (the Netherlands), and the UN Food and Agriculture Organization's Global IPM Facility.

These projects have provided quantitative assessments of community-wide pesticide use and its adverse effects. Through system modelling and implementation of different alternatives, we have demonstrated the effectiveness of different methods to lessen pesticide dependency and thereby improve ecosystem health. Meanwhile, the principal approach to risk reduction of the national pesticide industry continues to be farmer education through 'Safe Use' campaigns, despite the safe use of highly toxic chemicals under the social and environmental conditions of developing countries being an unreachable ideal. These conflicting perspectives and the continued systematic poisoning of many rural people in Carchi have motivated a call for international action (Sherwood et al, 2002).

The project members have worked with interested stakeholders to inform the policy debate on pesticide use at both the provincial and national levels. Our position has evolved to include the reduction of pesticide exposure risk through a combination of hazard removal (in particular, the elimination of highly toxic pesticides from the market), the development of alternative practices and ecological education. The experience reported here has led us to conclude that more knowledge-based and socially oriented interventions are needed. These must be aimed at political changes for enabling new farmer learning and organizational capacity, differentiated markets and increased participation of the most affected parties in policy formulation and implementation. Such measures involve issues of power that must be squarely faced in order to foster continued transformation of potato production in the Andes towards sustainability.

POTATO FARMING IN CARCHI

The highland region of Carchi is part of a very productive agricultural region, the Andean highlands throughout Northern Ecuador, Colombia and Venezuela. Situated near the equator, the region receives adequate sunlight throughout the year which, coupled with evenly distributed rainfall, means that farmers can continuously cultivate their land. As a result, the province is one of Ecuador's most important producers of staple foods, with farmers producing nearly 40 per cent of the national potato crop on only 25 per cent of the area dedicated to potato (Herrera, 1999).

Carchi is a good example of the spread of industrialized agricultural technologies in the Americas during the Green Revolution that began in the 1960s. A combination of traditional sharecropping, land reform, market access and high value crops provided the basis for rural economic development (Barsky, 1984). Furthermore, as a result of new revenues from the oil boom of the 1970s, the Ecuadorian government improved transportation and communication infrastructure in Carchi, and the emerging agricultural products industry was quick to capitalize on the availability of new markets. A typical small farm in Carchi is owned by an individual farm household and consists of several separate, scattered plots with an average area of about six hectares (Barrera et al, 1998).

Not surprisingly, agricultural modernization underwent a local transformation. In Carchi, mechanized, agrochemical and market-oriented production technologies are mixed with traditional practices, such as sharecropping arrangements, payments in kind, or planting in *wachu rozado* (a pre-Colombian limited tillage system) (Paredes, 2001). Over the last half-century, farming in Carchi has evolved towards a market oriented potato-pasture system dependent on external inputs. Between 1954 and 1974 potato production increased by about 40 per cent and worker productivity by 33 per cent (Barsky, 1984). Until recently, the potato growing area in the province continued to increase, and yields have grown from about 12 t ha in 1974 to about 21 t ha today, a remarkable three times the national average (Crissman et al, 1998).

To confront high price variability in potato (by factors of five to 20 in recent years), farmers have applied a strategy of playing the 'lottery', which involves continual production while gambling for high prices at harvest to recover overall investment. Nevertheless, the dollarization of the Ecuadorian Sucre in 2000 led to triple digit inflation and over 200 per cent increase in agricultural labour and input costs over three years (World Development Index, 2003). Meanwhile, open trade with neighbouring Colombia and Peru has permitted the import of cheaper commodities. As a result of a trend towards increased input costs and lower potato prices, in 2003 Carchense farmers responded by decreasing the area planted in potato from about 15,000 ha in previous years to less than 7000 ha. It remains to be seen how farmers ultimately will compensate for the loss of competitiveness brought about by dollarization.

Carchi farmers of today rely on insecticides to control the tuber-boring larva of the Andean weevil (*Premnotrypes vorax*) and a variety of foliage damaging insects. They also rely on fungicides to control late blight (*Phytophtera infestans*). One economic study of pesticides in potato production in Carchi confirmed that farmers used the products efficiently (Crissman et al, 1994), and later attempts during the 1990s by an environmental NGO to produce pesticide-free potatoes in Carchi failed (Frolich et al, 2000). After 40 years inorganic fertilizers and pesticides appear to have become an essential part of the social and environmental fabric of the region (Paredes, 2001).

PESTICIDE USE AND RETURNS

Our 1990s study of pesticide use found that farmers applied 38 different commercial fungicide formulations (Crissman et al, 1998). Among the fungicides used, there were 24 active ingredients. The class of dithiocarbamate contact-type fungicides were the most popular among Carchi farmers, with mancozeb contributing more than 80 per cent by weight of all fungicide active ingredients used. The dithiocarbamate family of fungicides has recently been under scrutiny in the Northern Andes due to suspected reproductive (Restrepo et al, 1990) and mutagenic effects in human cells (Paz-y-Mino et al, 2002). Similar concerns have been raised in Europe and the US (USEPA, 1992; Lander et al, 2000).

Farmers use three of the four main groups of insecticides in 28 different commercial products. Although organochlorine insecticides can be found in Ecuador, farmers in Carchi did not use them. The carbamate group was represented only by carbofuran, but this was the single most heavily used insecticide – exclusively for control of the Andean weevil. Carbofuran was used in its liquid formulation, even though it is restricted in North America and Europe due to the ease of absorption of the liquid and the high acute toxicity of its active ingredient. Another 18 different active ingredients from the organophosphate and pyrethroid groups were employed to control foliage pests, although only four were used on more than 10 per cent of plots. Here the OP methamidophos, also restricted in North America due to its high acute toxicity, was the clear favourite. Carbofuran and methamidophos, both classified as highly toxic (Class 1I) insecticides by the World Health Organization (WHO), respectively made up 47 per cent and 43 per cent of all insecticides used (by weight of active ingredient applied). In sum, 90 per cent of the insecticides applied in Carchi were highly toxic. A later survey by Barrera et al (1998) found no significant shifts in the products used by farmers.

Most insecticides and fungicides come as liquids or wettable powders and are applied by mixing with water and using a backpack sprayer. Given the costs associated with spraying, farmers usually combine several products together in mixtures known locally as cocktails, applying all on a single pass through the field. On average, each parcel receives more than seven applications with 2.5 insecticides and/or fungicides in each application (Crissman et al, 1998). Some farmers reported as many as seven products in a single concoction. On many occasions different commercial products were mixed containing the same active ingredient or different active ingredients intended for the same type of control. Women and very young children typically did not apply pesticides: among the 2250 applications that we documented, women made only four.

Product and application costs together account for about a third of all production costs among the small and medium producers in the region. The benefit to yields (and revenues) from using pesticides exceeded the additional costs of using them (including only direct production costs such as inputs and labour but not the costs of externalities). Nevertheless, Crissman et al (1998) found that farmers lost money in four of ten harvests, largely due to potato price fluctuations and price increases in industrial technologies, particularly mechanized land preparation, fertilizers and pesticides, that combined can represent 60 per cent of overall production outlays. Unforeseen ecological consequences on natural pest control mechanisms, in particular parasitoids and predators in the case of insect pests and selective pressure on *Phytophtora infestans* in the case of disease, raises further questions about the real returns on pesticides (Frolich et al, 2000). As we shall see, long-term profitability of pesticide use is even more questionable when associated human health costs to applicators and their families are taken into account.

PESTICIDE EXPOSURE AND HEALTH EFFECTS

Based on survey, observational and interview data, the majority of pesticides are bought by commercial names. Only a small minority of farmers reported receiving information on pesticide hazards and safe practices from vendors (Espinosa et al, 2003). Pesticide storage is usually relatively brief (days to weeks) but occurs close to farmhouses because of fear of robbery. Farmers usually mix pesticides in large barrels without gloves, resulting in considerable dermal exposure (Merino and Cole, 2003). Farmers, and on larger farms day labourers, apply pesticides using backpack sprayers on hilly terrain. Few use personal protective equipment for a variety of reasons, including social pressure (e.g. masculinity has become tied to the ability to withstand pesticide intoxications), and the limited availability and high cost of equipment. As a result, pesticide exposure is high. During pesticide applications, most farmers wet their skin, in particular the back (73 per cent of respondents) and hands (87 per cent) (Espinosa et al, 2003). Field exposure trials using patch-monitoring techniques showed that considerable dermal deposition occurred on legs during foliage applications on mature crops (Cole et al, 1998a). Other studies have shown that additional field exposure occurs in the field during snack and meal breaks, when hand washing rarely occurs (Paredes, 2001).

Family members are also exposed to pesticides in their households and in their work through a multitude of contamination pathways. Excess mixed product may be applied to other tuber crops, thrown away with containers in the field, or applied around the house. Clothing worn during application is often stored and used repeatedly before washing. Contaminated clothing is usually washed in the same area as family clothing, although in a separate wash. The extent of personal washing varies but is usually insufficient to remove all active ingredients from both the hands of the applicator and the equipment. Separate locked storage facilities for application equipment and clothing are also uncommon. Swab methods have found pesticide residues on a variety of household surfaces and farm family clothing (Merino and Cole, 2003).

Pesticide poisonings in Carchi are among the highest recorded in developing countries (Cole et al, 2000). In active poisoning surveillance, although there were some suicides and accidental exposures, most reported poisonings were of applicators. While the extensive use of fungicides causes dermatitis, conjunctivitis and associated skin problems (Cole et al, 1997a), we focused our attention on neurobehavioural disorders caused by highly toxic methamidophos and carbofuran. The results were startling.

The health team applied the WHO recommended battery of tests to determine the effects on peripheral and central nervous system functions (Cole et al, 1997b, 1998a). The results showed high proportions of the at-risk population affected, both farmers and their family members. Average scores for farm members were a standard deviation below the control sample, the non-pesticide population from the town. Over 60 per cent of rural people were affected and women, although not commonly active in field agriculture, were nearly as affected as field workers. Alarmingly, both Mera-Orcés (2000) and Paredes (2001) found that poisonings and deaths among young children were common in rural communities.

Contamination resulted in considerable health impacts that ranged from sub-clinical neurotoxicity (Cole et al, 1997a, 1998a), poisonings with and without treatment (Crissman et al, 1994) to hospitalizations and deaths (Cole et al, 2000). In summary, human health effects included poisonings (at a rate of 171/100,000 rural population), dermatitis (48 per cent of applicators), pigmentation disorders (25 per cent of applicators), and neurotoxicity (peripheral nerve damage, abnormal deep tendon reflexes and coordination difficulties). Mortality due to pesticide poisoning is among the highest reported anywhere in the world (21/100,000 rural population). These health impacts were predominantly in peri-urban and rural settings. As was shown in Chapter 2, this high incidence of poisoning may not be because the situation is particularly bad in Carchi, but because researchers sought systematically to record and document it.

Acute pesticide poisonings led to significant financial burdens on individual families and the public health system (Cole et al, 2000). At the then current exchange rates, median costs associated with pesticide poisonings were estimated as follows: public health care direct costs of US$9.85/case; private health costs of $8.33/case; and lost time indirect costs for about six worker days of $8.33/agricultural worker. All of these were over five times the daily agricultural wage of about US$1.50 at the time (1992). Antle et al (1998a) showed that the use of some products adversely affects farmer decision-making capacity to a level that would justify worker disability payments in other countries. Neither group of researchers included financial valuation of the deaths associated with pesticide poisonings nor the effects of pesticides on quality of life, both of which would substantially increase the overall economic burden of illness estimates.

A MYTH – THE HIGHLY TOXICS CAN BE SAFELY USED

Following the research results, limitations in the pesticide industry's safe use of pesticides (SUP) campaign became apparent. In a letter to the research team, the Ecuadorian Association for the Protection of Crops and Animal Health (APCSA, now called CropLife Ecuador) noted that an important assumption of SUP was that exposure occurred because of *'a lack of awareness concerning the safe use and handling of [pesticide] products'*. Although our Carchi survey showed a low percentage of women in farm families had received any training on pesticides (14 per cent), most male farmers (86 per cent) had received some training on pesticide safety practices. Furthermore, labels are supposed to be an important part of the 'hazard communication process' of salesmen. Yet our work in Carchi indicated that farm members often could not decipher the complex warnings and instructions provided on most pesticide labels.

Although 87 per cent of the population in our project area was functionally literate, over 90 per cent could not explain the meaning of the coloured bands on pesticide containers indicating pesticide toxicity. Most believed that toxicity was best ascertained through the odour of products, potentially important for organophosphates with sulphur groups, but not generalizable to all products that are impregnated by formulators for marketing purposes. Hence even the universal, seemingly simple toxicity warning system of coloured bands on labels has not entered the local knowledge system. If industry is seriously concerned about informing farmers of the toxicity of its products, it should better match warning approaches to current perceptions of risk, such as considering using toxicity-related odour indicators.

In addition, the SUP campaign's focus on pesticides and personal protective equipment (PPE) is misguided. Farmers regard PPE as uncomfortable and 'suffocating' in humid warm weather, leading to the classic problem of compliance associated with individually oriented exposure reduction approaches (Murray and Taylor, 2000). Examination of the components of the classic industrial hygiene hierarchy of controls (Table 10.1) shows PPE to be among the least effective controls and suggests that the industry strategy of prioritizing PPE is similar to locking the stable after the horse has bolted. Our research has shown the ineffectiveness of product labelling (point 5). Isolation (point 4) is difficult in open environments such as field agriculture where farming infrastructure and housing are closely connected and some contamination of the household is virtually inevitable, particularly in poorer households. Priority should be given to other more effective strategies of exposure reduction, beginning with point 1: eliminating the most toxic products from the work and living environments. Likewise, this is the highest priority of the IPPM 2015 initiative.

Box 10.1 *Hierarchy of controls for reducing pesticide exposure*

Most effective
1   Eliminate more highly toxic products, e.g. carbofuran and methamidophos
2   Substitute less toxic, equally effective alternatives
3   Reduce use through improved equipment, e.g. low volume spray nozzles
4   Isolate people from the hazard, e.g. locked separate pesticide storage
5   Label products and train applicators in safe handling
6   Promote use of personal protective equipment
7   Institute administrative controls, e.g. rotating applicators
Least effective

Source: adapted from Plog et al, 1996

INIAP, the Ecuadorian agricultural research institute, is prepared to declare that alternative technologies exist for the Andean weevil and foliage pests and that highly toxic pesticides are not necessary for potato production and other highland crops in Ecuador (Gustavo Vera, INIAP Director General of Research, pers. comm.). Meanwhile, pesticide industry representatives have privately acknowledged that they understand that highly toxic pesticides eventually will need to be removed from the market. Nevertheless, the Ecuadorian Plant and Animal Health Service (SESA) and CropLife Ecuador have taken the position that they will continue to support the distribution and sale of WHO Class I products in Ecuador until the products are no longer profitable or that it is no longer politically viable to do so.

One seven-year study by Novartis (now Syngenta) found that SUP interventions in Latin America, Africa and Asia were expensive and largely ineffective, particularly with smallholders (Atkin and Leisinger, 2000). The authors argue that *'the economics of using pesticides appeared to be more important to [small farmers] than the possible health risks'* (p121). The most highly toxic products are the cheapest on the market in Carchi, largely because the patents on these early generation products have expired, permitting free access to chemical formulas and competition, and because farmers have come to accept the personal costs associated with poisonings.

POLICIES AND TRADEOFFS

Pesticide use in agricultural production conveys the benefit of reducing losses due to pests and disease. That same use, however, can cause adverse environmental and health impacts. Previously, we cited a study that showed that pesticide use by farmers was efficient from a narrow farm production perspective. Nevertheless, that study examined pesticide use solely from the perspective of reducing crop losses. If the adverse health and environmental effects were also included in the analysis, the results would be different. Integrated assessment is a method to solve this analytical problem. The Carchi research team devised an innovative approach to integrated assessment called the Tradeoff Analysis (TOA) method (Antle et al, 1998b; Stoorvogel et al, 2004).

The TOA method is an interactive process to define, analyse and interpret results relevant to policy analysis. At its heart is a set of linked economic, biophysical and health models inside a user shell called the TOA Model. Based on actual dynamic data sets from the field, we used simulations in the TOA method to examine policy options for reducing pesticide exposure in Carchi.

The policy options explored were a combination of taxes or subsidies on pesticides, price increases or declines in potatoes, technology changes with IPM, and the use of personal protective equipment. We examined the results in terms of farm income, leaching of pesticides to groundwater and health risks from pesticide exposure. Normally, policy and technology changes produce tradeoffs – as one factor improves, the other factor worsens. Our analysis of pesticide taxes and potato price changes produced such a result. As taxes decrease and potato prices increase, farmers plant more of their farm with potatoes and tend to use more pesticide per hectare. Thus a scenario of pesticide subsidies and potato price increases produce growth in income and increases in groundwater contamination and health risks from pesticide exposure.

With the addition of technology change to these price changes, the integrated analysis produced by the TOA Model showed that a combination of IPM and protective clothing could produce a win–win outcome throughout the range of price changes: neurobehavioural impairment and environmental contamination decreased while agricultural incomes increased or held steady (Antle et al, 1998c; Crissman et al, 2003).

TRANSFORMING AWARENESS AND PRACTICE: THE EXPERIENCE OF ECOSALUD

The unexpected severity of pesticide-related health problems and the potential to promote win–win solutions motivated the research team to search for ways to identify and break the pervasive cycle of exposure for the at-risk population in Carchi. The EcoSystem Approaches to Human Health Program of IDRC (www. idrc.ca/ecohealth) offered that opportunity through support to a project called EcoSalud (*salud* means health in Spanish). The EcoSystem Approaches to Human Health Program was established on the understanding that ecosystem management affects human health in multiple ways and that a holistic, gender-sensitive, participatory approach to identification and remediation of the problem is the most effective manner to achieve improvements (Forget and Lebel, 2001).

The EcoSalud project in Carchi was essentially an impact assessment project designed to contribute directly to ecosystem improvements through the agricultural research process. The aims were to improve the welfare of the direct beneficiaries through enhanced neurobehavioural function brought about by reduced pesticide exposure, and to improve the well being of indirect beneficiaries through farming innovation. The project design called for before-and-after measurements of a sample population that changed its behaviour as a result of the intervention. Consistent with IDRC's EcoSystem Health paradigm, the intervention was designed to be gender sensitive and increasingly farmer- and community-led.

EcoSalud started by informing members of three rural communities of past research results on pesticide exposure and health impacts. To illustrate pesticide exposure pathways, we used a non-toxic fluorescent powder that glowed under ultraviolet light as a tracer (Fenske et al, 1986). Working with volunteers in each community, we added the tracer powder to the liquid in backpack sprayers and asked farmers to apply as normally. At night, we returned with ultraviolet lights and video cameras to identify the exposure pathways. During video presentations, community members were astonished to see the tracer not only on the hands and face of applicators, but also on young children who played in the fields after pesticide applications. We also found traces on clothing and throughout the house, such as around wash areas, on beds and even on the kitchen table. Perhaps more than other activities, the participatory tracer study inspired people to take action themselves.

People, in particular mothers, began to speak out at community meetings. The terms *el remedio* (the treatment) and *el veneno* (the poison) were often used interchangeably when referring to pesticides. Spouses explained that the need to buy food and pay for their children's' education when work options were limited led to an acceptance of the seemingly less important risks of pesticides. They explained that applicators often prided themselves on their ability to withstand exposure to pesticides. As one young girl recounted (in Paredes, 2001):

> *One time, my sister Nancy came home very pale and said that she thought she had been poisoned. I remembered that the pesticide company agricultural engineers had spoken about this, so I washed her with lots of soap on her back, arms and face. She said she felt dizzy, so I helped her vomit. After this she became more resistant to pesticides and now she can even apply pesticides with our father.*

Despite stories such as this, many women became concerned about the health impacts of pesticides on their families. During one workshop, a women's group asked for disposable cameras to document pesticide abuse. Children were sent to spy on their fathers and brothers and take photos of them handling pesticides carelessly or washing sprayers in creeks. Their presentations led to lively discussions. The results of individual family studies showed that poisonings caused chronic ill-health for men and their spouses, and ultimately jeopardized household financial and social stability. Concern about the overall family vulnerability was apparent during community meetings, when women exchanged harsh words with their husbands over their agricultural practices that resulted in personal and household exposure to toxic chemicals. The men responded that they could not grow crops without pesticides and that the safer products were the most expensive. Communities called for help.

NIAP's researchers and extensionists in Carchi had gained considerable experience with farmer participatory methodologies for technology development,
including community-led varietal development of late blight disease resistant
potatoes. We know that such approaches can play an important role in enabling
farmers to acquire the new knowledge, skills and attitudes needed for improving
their agriculture. INIAP built on existing relationships with Carchi communities
to run farmer field schools (FFS), a methodology recently introduced to the
Andes. In part, farmer field schools attempt to strengthen the position of farmers
to counterbalance the messages from pesticide salespeople. As one FFS graduate
said (in Paredes, 2001):

> Prior to the field school coming here, we used to go to the pesticide shops to
> ask what we should apply for a problem. Then the shopkeepers wanted to sell
> us the pesticides that they could not sell to others, and they even changed
> the expiry date of the old products. Now we know what we need and we do
> not accept what the shopkeepers want to give us.

FFS have sought to challenge the most common of IPM paradigms that centres
on pesticide applications based on economical thresholds and transfer of single
element technologies within a framework of continuing pesticide use (Gallagher,
2000). In contrast, FFS programmes propose group environmental learning on
the principles of crop health and ecosystem management as an alternative to
reliance on curative measures to control pests. As a FFS graduate in Carchi noted
(in Paredes, 2001):

> When we talk about the insects [in the FFS] we learn that with the pesticides
> we kill everything, and I always make a joke about inviting all the good
> insects to come out of the field before we apply pesticides. Of course, it is a
> poison, and we kill everything. We destroy nature when we do not have
> another option for producing potatoes.

In practice, the FFS methodology has broadened the technical content beyond a
common understanding of IPM to a more holistic approach for improving plant
and soil health. The FFS methodology adapts to the diverse practical crop needs
of farmers, be they production, storage or commercialization. FFS ultimately
aspires to catalyse the innovative capacity of farmers, as exemplified by how a
graduate has improved cut foliage insect traps tested in his FFS (in Paredes, 2001):

> I always put out the traps for the Andean weevil, even if I plant 100 [bags of
> seed] because it decreases the number of adults. It is advantageous because
> we do not need to buy much of that poison Furadan. But I do use them
> differently. After ploughing, I transplant live potato plants from another
> field, then I do not need to change the dead plants every eight days.

In an iterative fashion, FFS participants conduct learning experiments on
comparative (conventional versus IPM) small plots (about 2500 m^2) to fill

knowledge gaps and to identify opportunities for reducing external inputs while improving production and overall productivity. After two seasons, initial evaluation results in three communities were impressive. Through the use of alternative technologies, such as Andean weevil traps, late blight resistant potato varieties, specific and low toxicity pesticides, and careful monitoring before spraying, farmers were able to decrease pesticide sprays from 12 in conventional plots to seven in IPM plots, while maintaining or increasing production (Barrera et al, 2001). The amount of active ingredient of fungicide applied for late blight decreased by 50 per cent, while insecticides used for the Andean weevil and leafminer fly (*Liriomysa quadrata*), that had commonly received the highly toxic carbofuran and methamidophos, decreased by 75 per cent and 40 per cent respectively.

Average yields for both conventional and IPM plots were unchanged at about 19 t ha but net returns increased as farmers were spending less on pesticides. FFS participants identified how to maintain the same level of potato production with half the outlays in pesticides and fertilizers, decreasing the production costs from about US$104 to $80 per tonne. Because of the number of farmers involved in FFS test plots, it was difficult to assess labour demands in the economic analysis. Nonetheless, farmers felt that the increased time for scouting and using certain alternative technologies, such as the insect traps, would be compensated by decreased pesticide application costs, not to mention decreased medical care visits.

In addition to the intensive six month FFS experience, EcoSalud staff visited individual households to discuss pesticide safety strategies such as improved storage of pesticides, PPE, use of low volume nozzles that achieve better coverage with less pesticide, and more consistent hygiene practices. Based on widespread disinterest in PPE, we were surprised when participants began to request help in finding high-quality personal protective equipment, that they said was unavailable at the dozens of local agrochemical vendors. EcoSalud staff found high-quality PPE (mask, gloves, overalls and pants) through health and safety companies in the capital city, costing US$34 per set, the equivalent of over a week's labour at the time. The project agreed to grant interest free, two-month credit towards the purchase price to those interested in buying the gear. Remarkably, 46 of the 66 participating families in three communities purchased complete packages of equipment. A number of farmers rented their equipment to others in the community in order to recuperate costs. Follow-up health studies are not complete, but anecdotal evidence is promising. As the wife of one FFS graduate who previously complained of severe headaches and tunnel vision due to extensive use of carbofuran and metamidophos said:

> *Carlos no longer has headaches after working in the fields. He used to return home [from applying pesticides] and could hardly keep his eyes open from the pain. After the field school and buying the protective equipment, he is a far easier person to live with* (pers. comm., farm family, Santa Martha de Cuba).

Complementary projects have supported follow-up activities in Northern Ecuador and elsewhere, including the production of FFS training materials (Pumisacho and Sherwood, 2000; Sherwood and Pumisacho, in press), the training of nearly 100 FFS facilitators in Carchi and nearby Imbabura, the transition of FFS to small-enterprise production groups and the establishment of farmer-to-farmer organization and capacity-building. Concurrently, over 250 facilitators have been trained nationwide and hundreds of FFS have been completed. Recently, Ecuador's Ministry of Agriculture decided to include FFS as an integral part of its burgeoning national Food Security Program. Furthermore, in part due to the successful experience in Carchi, FFS methodology has subsequently spread to Peru, Bolivia, and Colombia as well as El Salvador, Honduras and Nicaragua, where over 1500 FFS had been conducted by mid 2003 (LEISA, 2003).

ELIMINATION OF THE HIGHLY TOXICS: SEARCHING FOR CONSENSUS WITH THE PESTICIDE INDUSTRY

Tackling the broader context of pesticide use in agriculture requires the involvement of multiple stakeholders. To this end, EcoSalud was proactive in sharing research results on the harmful consequences of pesticide use and in advocating policies for improving the situation. A series of radio announcements and educational programmes were developed to broadcast throughout the province. EcoSalud staff also linked with strategic partners to advocate common interests. For example, the project nurse participated in provincial Health Council meetings and the project educator joined a local development consortium centred on one community. In addition, project staff lobbied interests with local, provincial and national political officials.

This initial work led to a province-wide stakeholders meeting entitled '*The impacts of pesticides on health, production, and the environment*' in October 1999, drawing 105 representatives from government, industry, development organizations, communities and the media. Presidents from the provincial councils of agriculture and health chaired sessions. Ministerial representatives from agriculture, health and education participated as well as the governor and mayors or representatives from each of the provincial municipalities. One outcome of the meeting involved the formation of a small committee composed of directors from INIAP, the Ministry of Education and the Ministry of Health, which drafted a 'Declaration for life, environment and production in Carchi'. The Declaration called for:

- assurance of greater control on the part of the Ecuadorian Agricultural Health Service (SESA) of the formulation, sale and use of agrochemicals, including the prohibition of highly toxic products (WHO Classes Ia and Ib);
- introduction of information concerning the impact of pesticides on health, the environment and farming productivity into the basic school curriculum;

- inclusion of IPM as part of degree requirements for university level agricultural technical training;
- commitment of further resources to research and training in integrated crop management with an orientation towards the reduction of pesticide use and safe use of pesticides;
- promotion of awareness raising in rural communities on the side-effects of agricultural practices and the use of more environmental and health friendly practices;
- the direct financial support of the agrochemical industry in the completion of these resolutions.

The first recommendation is in keeping with earlier cited research on SUP that concluded '*any pesticide manufacturer that cannot guarantee the safe handling and use of its toxicity Class Ia and Ib products should withdraw those products from the market*' (Atkin and Leisinger, 2000). The next points involve education and research initiatives by multiple stakeholders to gradually shift agricultural production to more sustainable practices. The last point raises the long-standing proposal for post-marketing surveillance, similar to that which is carried out on pharmaceutical drugs, and to be funded by agrochemical producers themselves (Loevinsohn, 1993).

The national and international pesticide industry (today called CropLife International and CropLife Ecuador, respectively) mobilized to pre-empt negative press coverage and to block potentially damaging measures to their financial interests. Industry representatives from the US, Central America, Colombia and the city of Guayaquil, where most Ecuadorian chemical companies are based, arrived at Quito days before a follow-up May 2001 conference to meet with the organizers and relevant government officials. Instead of requesting to learn more about the studies and recommendations for improving the situation, they seemed to lobby against the findings. Industry representatives expressed concern about the recommendation for eliminating WHO Class Ia and Ib insecticides and persuaded the Director of the Ministry of Agriculture's pesticide regulatory agency, SESA, and appointed President of the National Technical Committee not to support that measure. In fact, despite a central role in the meeting and his confirmation, the Director of SESA did not show up until after the conference was concluded.

Regardless, representatives from diverse FFS in Carchi travelled to the capital to attend the May 2001 meeting and made convincing presentations on their experience with IPM and their tested alternatives for substantially reducing dependency on the problematic products in question: carbofuran, methamidophos and mancozeb. They requested governmental attention to the Carchi declaration and National Pesticide Committee proposal. Officials from the Public Ministry of Health and the Pan American Health Organization indicated intentions to play a more active training, monitoring and advocacy role similar to other projects in Central America (Keifer et al, 1997). This event led to a television documentary on the pesticide crisis in Carchi that was shown throughout the country and subsequently was presented to select audiences in other parts of

Central and South America as well as in the US and Europe. Despite the receipt of multiple letters from farmer organizations, researchers and development professionals in Ecuador demanding government attention to the situation in Carchi, the Director of SESA never responded nor publicly expressed concern.

As a result of public alarm raised by the research findings, Bayer Corporation and CropLife International immediately implemented separate projects in Carchi to promote the safe use of pesticides. Initially, each agreed to work through INIAP's IPM programme, which centred on pesticide use reduction through FFS, but both projects ultimately refused to finance an impartial third party to design and run a pesticide reduction project. Instead, Bayer Corporation and CropLife hired its own people and limited activities to the promotion of the safe use of pesticides, despite clear knowledge of the findings in Carchi and elsewhere (for example, the aforementioned by Atkin and Leisinger, 2000) that the safe use of highly toxics was not realistic given the socio-economic exposure conditions in developing countries. In our opinion, this behaviour demonstrated that the industry was more interested in sales and profits than in the health of its patrons and rural people in general.

In July 2003, INIAP, CIP and the FAO launched its Spanish language book (Yanggen et al, 2003) that summarized the overall research to a forum of public officials, industry representatives and media and emphasized the recommendations presented previously. This was followed by in-depth radio programmes and newspaper articles that repeated the research findings. SESA officials and pesticide industry representatives responded with now familiar behaviour – seemingly disinterest over the alarming health impacts and disdain for calls for the market removal of the highly toxic insecticides. As per the findings of a BBC World Service radio programme that included interviews with government officials, pesticide salespeople, farmers and hospital personnel in Carchi, the official government position had become: 'We have established international standards of recommendation and force the pesticide industry to obey those rules' and 'We cannot be held responsible for farmers' misuse of pesticides'. Despite over a decade of research that clearly shows the hazards of pesticide use in Carchi as well as farmers' demonstrated means of economically reducing dependency on the highly toxic products, no significant change has occurred.

ADDRESSING ROOT CAUSES: SHIFTING THE EMPHASIS FROM PESTICIDES TO PEOPLE

Knowledge of how modern agriculture in Carchi became unsustainable is needed before we can hope to achieve more sustainable futures. In the 1950s, when the farmers of Carchi began to adopt agrochemicals, it was not necessarily their intention that future generations would become socially and ecologically dependent on them by the turn of the century. Nor were the health impacts experienced by farm families even comprehended at the time. Knowledge of the historical events that led to present-day pesticide dependency in Carchi can provide understanding of how best to move forward.

Economic, agronomic and biomedical research tends to hinge on deterministic explanations, such as the biological causes of insect pests or disease epidemics, with relative neglect of human social factors, in particular institutional cultural perspectives and power struggles that can shape human activity and lead to harmful environmental outcomes (Berkman and Kawachi, 2000; Röling, 2003). Social epidemiologists have found that issues of differential power and knowledge between different stakeholders will have to be confronted if societal and agroecosystem change are to be moved in more sustainable directions that promote human health in rural communities (Berkman and Kawachi, 2000; Watterson, 2000; London and Rother, 2000). Recent social sciences research on pesticide use in Carchi substantiate such perspectives (Mera-Orcés, 2000, 2001; Paredes, 2001; Sherwood et al, 2003).

Interventions that aim for pesticide use reduction in Carchi potato farming have tended to characterize resource poor farmers as a homogeneous group. But potato farming in the region is highly diverse, with the specific practices that farmers use in particular environments defined partly by ecology and partly by markets and technology. The set of practices can be analysed as farming styles. A farming style is a systematic and continuous attempt by farmers and their families to create a consistent set of practices within the biological and economic context within which they have to operate (van der Ploeg, 1994).

A characterization based on farming styles (culture, labour processes and decisions about technology and markets) in Carchi found that FFS were more attractive to certain social groups than others (Paredes, 2001). In particular, the most enthusiastic participants belonged to two groups: the *highly pragmatic and inquisitive* farmers and *landless labourers*. The first group was motivated by their interest in the FFS alternatives that allowed them a certain degree of independence from capital and input markets (credit and agrochemicals). The landless labourers, on the other hand, were primarily motivated by the unique opportunity to co-invest in production, which effectively afforded them access to land, as well as by more egalitarian treatment during training sessions. Meanwhile, field schools appeared to be of less interest to others. For example, the *high risk takers*, who commonly depended more on the capital and agrochemical input markets for potato production and who readily adopt (and abandon) technologies, as well as *intermediate farmers*, who tended to co-invest for production, were generally frustrated by the knowledge-intensive orientation of FFS methodology.

As farmers are not operating alone in farming domains, a farming style also represents a socio-technical network in which other actors, organizations and entities collectively define the apparent courses of action and development opportunities (van der Ploeg, 1994). Perceptions of risk, therefore, respond to a certain way of defining relevant problems and solutions within a socio-technical network. For instance, certain farming styles operate in networks in which pesticides came to be accepted as obligatory or unavoidable elements of good potato farming – despite the risk they represent for family's health. In other styles, less agrochemical use is desired to avoid risks, and, in this case, good farming also means safe farming. This can help us understand why interventions that aim towards behavioural change among farmers (for instance, safe use of pesticides) are limited when they view farmers as actors operating alone.

Thus, it is not only necessary to understand different farming styles but also to explore why such differences are possible within specific networks. When modernization schemes (e.g. privatization, open markets, agrochemical subsidies) promote certain styles of farming, they tend to inhibit other styles from developing. Initiatives to reduce pesticides need to take into account the complex social diversity found in Carchi not as a blueprint for intervention projects but as a way to find and understand farming styles that have developed in the same region. Programmes to decrease dependence on agrochemical markets must be based on a better understanding of farmers' social and technical knowledge that leads to such aims. This means that interventions will need to focus on learning first from the diversity of farmers' practices (social and technical) in a given region, including the agro-social networks of which they are part. The latter may include consumers, capital and product markets and the range of institutions that orient agriculture in particular ways at a given locality.

PUTTING MARKETS AND INSTITUTIONS TO WORK

Initial modelling of the potato-pasture ecosystem included consideration of disincentives to the use of highly toxic products, in particular a tax on carbofuran (Antle et al, 1998b). The use of taxes has been an important part of tobacco control strategies in North America. Such taxes could potentially shift the cost-based preference for cheaper, more toxic pesticides. Effective implementation of such market incentives would, however, face considerable enforcement obstacles, particularly with the current emphasis on deregulation and freer trade and the relatively porous border with Colombia.

It might also be possible to argue for the provision of adequate personal protective equipment through the same distribution channels as those existing for agrochemicals. The EcoSalud project team was impressed by the interest in PPE among farmer-applicators, as a short-term way of reducing their personal exposure. High quality PPE is routinely available from local agrochemical distributors in North America, partly as a result of the training requirements for pesticide applicators and partly due to the 'Responsible Care' ethic of companies that is more evident in some locations than others. In Ecuador, the EcoSalud project had to provide the market services to bring adequate, durable PPE to farmers.

Even these changes remain only short-term options that do not move the whole agroecosystem towards greater sustainability. In contrast, the steady growth of markets for organic produce in North America and Europe point to another important role for markets. When the EcoSalud team reflected on areas for future growth of alternatives, the development of 'ecological potatoes' (produced by means of cleaner production practices) was high on their list. In other parts of Ecuador, closer to the Quito urban market, organic production of vegetables, robust distribution systems and consumer demand are now part of the food economy.

Although completely pesticide-free production of potatoes in the Andes on a widespread basis may be difficult in the short term, a phase-in period without the use of highly toxic products and with considerably reduced amounts of pesticides may be of interest to both consumers and food processors, such as soup companies and potato chip producers. With increasing concern about how contaminants move through the food system and the need to stop them at source, food processors are increasingly interested in preferentially buying potatoes produced with fewer and less toxic pesticides. Some, however, may not wish to point out to consumers that any pesticides are used at all. In financial terms, the food processing industry is considerably larger and more powerful than the pesticide industry, and its vested interest in clean products represents an opportunity for change.

Increased agro-social learning and the development of differentiated commodity markets for cleaner production are examples of policy alternatives favourable to ecosystem health and agricultural sustainability. With financial support from international donors, research institutes may raise flags and NGOs may demonstrate alternatives with farmers in pilot locations, but ultimately widespread change requires the leadership of the most affected people (in this case, farmers and their communities) and governments committed to defending broader public interests.

Nevertheless, the current political system in Ecuador largely excludes farmers and rural communities from decision-making processes, while at the same time farmers have had limited experience with the degree of social organization and collective action needed to advocate interests in modern-day political forums. This situation is not a mere accident, but rather the result of political outcomes. Programmes of government 'modernization' in Ecuador and elsewhere, aimed at privatizing government services and decreasing government regulatory capacities, further complicate matters. Pressing challenges for Carchi include achieving greater government accountability to rural communities and the enabling of rural organizations to be capable of independently building consensus and acting on the collective interests of constituents.

As European experience has shown, a host of governmental and quasi-governmental support can have substantial impacts. Gerber and Hoffman (1998) have described the role of formal support (education system, information services, donors and state ministries), as well as more personal contacts among farmers and between farmers and consumers in the evolution of eco-farming in Germany. Pretty (1998) has summarized a wide range of policies that can work in the promotion of sustainable agriculture. The implementation of such policies requires efforts to be directed towards institutional change and reorganization. Perhaps they could be part of the current efforts to promote good governance, in ways that are beneficial to human and ecosystem health. Again similar to good governance efforts where power and control over resources are at stake, international support and pressure is likely to be required to overcome the influences of pesticide companies in developing countries.

CONCLUDING COMMENTS

Much conventional thinking in agricultural development places an emphasis on scientific understanding, technology transfer, farming practice transformation and market linkages as the means to better futures. Consequently, the focus of research and interventions tends to be on the crops, the bugs and the pesticides, rather than the people who design, chose and manage practices. Recent experiences of rural development and community health, however, argue for a different approach (see for example, Uphoff et al, 1998; Norgaard, 1994; Latour, 1998; Röling, 2000). Of course, technologies can play an important role in enabling change, but the root causes of the ecosystem crisis, such as that in Carchi, appear to be fundamentally conceptual and social in nature, that is, people sourced and dependent.

There is a general need for organizing agriculture around the development opportunities found in the field and in communities (van der Ploeg, 1994). Experience with people-centred and discovery-based approaches has shown promise at local levels, but ultimately such approaches do not address the structural power issues behind complex, multi-stakeholder, socio-environmental issues, such as pesticide sales, spread and use.

The search for innovative practice less dependent on agrochemical markets needs to focus on the diversity of farming and the socio-technical networks that enable more socially and ecologically viable alternatives. Progress in this area would require a new degree of political commitment from governments to support localized farming diversity and the change of preconceived, externally designed interventions towards more flexible, locally driven initiatives. In addition, local organizations representing the most affected people must aim to influence policy formulation and implementation.

Our modern explanations are ultimately embedded in subtle mechanisms of social control that can lead to destructive human activity. The social and ecosystem crises common to modernity, evident in the people–pest–pesticide crises in the Northern Andes, are not just a question of knowledge, technology, resource use and distribution, or access to markets. Experience in Carchi demonstrates that approaches to science, technology and society are value-laden and rooted in power relationships among the diverse actors – such as farmers, researchers, industry representatives and government officials – that can drive farming practice to be inconsistent with public interest and the integrity of ecosystems. Solutions will only be successful if they break with past thinking and more effectively empower communities and broader civil society to mobilize enlightened activity for more socially and environmentally acceptable outcomes.

Chapter 11

Breaking the Barriers to IPM in Africa: Evidence from Benin, Ethiopia, Ghana and Senegal

Stephanie Williamson

PESTICIDE USE IN AFRICA

Pesticide use in Africa is the lowest of all the continents, accounting for only 2 per cent of world sales, and averaging in the 1990s 1.23 kg ha^{-1} compared with 7.17 kg in Latin America and 3.12 in Asia (Repetto and Baliga, 1996; Agrow, 2001). This low use appears to suggest correspondingly low health and environmental hazards, and indeed that African agriculture may need to increase its pesticide use (Paarlberg, 2002). Regrettably, this assumption is wrong, as African farmers currently use many WHO Class Ia and Ib products, and few users take precautionary measures to prevent harm (Gerken et al, 2000; Hanshi, 2001; Matthews et al, 2003).

The largest requirements for pesticides are on cash crops, particularly cotton, cocoa, oil palm, coffee and vegetables, many of which are grown by smallholders. Many African countries also bear a legacy of pesticide provision by development assistance agencies, often supplied free and directly into the hands of untrained and non-literate farmers growing staple foods. In addition, control operations for locusts and malaria vectors have resulted in the stockpiling of large volumes of insecticides, which are often stored in unsafe conditions (FAO, 2001b; Vorgetts, 2001). In addition, recent years have seen widespread deregulation of input supply and consequently a rapid growth in informal pesticide trading, particularly of poor quality products (Mudimu et al, 1995; Macha et al, 2001; FAO/WHO, 2001).

Assessing use levels and trends from the limited data available in Africa is fraught with difficulties, and sales or import figures underestimate use since donations and informal trade are unrecorded (Winrock, 1994). Expenditure on pesticide imports increased steadily during the 1970s and 1980s, while the volume and value of imports into many countries declined by the early 1990s, associated with privatization and a sharp decrease in government subsidies. The leading users were those countries with a well-developed food export sector, notably Cameroon, Cote d'Ivoire, Kenya and South Africa. Annual pesticide imports fluctuated between US$486 and 580 million over the period 1995–2000, with import values estimated by the UN Food and Agriculture Organization (FAO) at US$503 million in 2000 (FAOSTAT, 2002). The latest industry figures show that the African market increased to 3 per cent of global share in 2002, with a value of US$753 million (Agrow, 2003).

There is evidence to suggest that more farmers are now using pesticides, particularly on food crops. In Benin, for example, some 60 per cent of cereal farmers were using pesticides in 1998, up from 40 per cent five years previously (Meikel et al, 1999), and in Uganda, 73 per cent of mixed cropping smallholders had increased their use of pesticides between 1995 and 1999, particularly on cowpea and groundnut (Erbaugh et al, 2001). Other countries, such as Rwanda, have seen a growing dependence on pesticides for both field and storage use (Youdeowei, 2000).

BACKGROUND TO STUDY SITES

Our research explored trends in pesticide practice by smallholder farmers in the context of liberalization, government restructuring and changing policies for food security and agricultural intensification. We examined cropping systems in four countries that were targeting both local and export markets. These systems were:

1 peri-urban vegetables – renowned for their profitability and the rapid growth in production as urban populations have risen, but also renowned for pesticide misuse;
2 cotton – promoted by several West African governments and donors as a successful livelihood strategy in the savannah zones, but with increasing concerns about pesticide use and misuse;
3 cowpea – a crop in transition from subsistence status to cash crop, with its increased cultivation being accompanied by the recent introduction of pesticides;
4 mixed cereal and legumes – a local food production system only recently moving into cash-oriented economies;
5 pineapple – an example of a lucrative export crop that smallholders have started growing in the last decade, and subject to compliance with stricter regulations on permissible pesticide residue levels in European markets.

The research was coordinated by PAN-UK with local partners in Benin (cotton, vegetables, pineapple), Ethiopia (mixed cereals, legumes), Ghana (cowpea, pineapple), and Senegal (cotton, vegetables). Farming community studies were conducted with over 400 smallholders and individual interviews with around 100 key informants from government agricultural research, extension, health and environmental agencies, input supply companies, growers' associations, NGOs, and development assistance agencies (Williamson, 2003). Table 11.1 provides data on the location, farming, climate and pressures in each site.

Table 11.1 *Description of case study sites and ecosystems*

Study site	Farming systems	Geography	Pressures
Benin – Kpako village, Banikoara district, Alibori Dept	Cotton; sorghum, millet, maize; livestock	Low potential lowland, Sudano-Sahelian savannah zone, erratic rainfall, v. limited access to irrigation	Deforestation, population increase, declining soil fertility, environmental degradation and food insecurity
Senegal – Diaobe, Sare Bounda, Linguewal and Nemataba villages, Kounkane district, Velingara Dept., Kolda Region	Cotton, millet, sorghum, maize, rice, cowpea, vegetables, livestock	Low potential low land, Sahelian savannah zone, erratic rainfall, v. limited access to irrigation	Deforestation, population increase, and degraded soils, food insecurity
Benin – Sekou village, Nr Cotonou, Atlantic Dept.	Peri-urban vegetables (okra, chilli, onion, tomato, leafy vegetables, cabbage, cucumber, courgette, aubergine, sweet pepper, carrot	High potential lowland on former swamps, ample rainfall and irrigation water	Urban encroachment, land tenure uncertainty
Senegal – Sangalkam, Gorom II, Tivaouane Peul and Wayembame villages, Sangalkam Rural Community, Les Niayes district	Peri-urban vegetable growing (tomato, green beans, cabbage, courgette, cucumber, aubergine, onion, sweet pepper, lettuce	Coastal savannah with high potential lowland and less fertile plateaux, unreliable rainfall	Population increase, water shortage, deforestation, land tenure uncertainty
Benin – Sekou village, Allada district	Pineapple, maize, sorghum, other fruits, livestock	Medium–high potential lowland, ample rainfall, some access to irrigation	Land concentration and tenure problems, food insecurity
Ghana – Fotobi and Samsam villages, Eastern Region	Pineapple, maize, cassava, vegetables, other fruits, livestock	coastal savannah ecological zone, medium-high potential hilly zone, regular rainfall, limited access to irrigation	Land tenure uncertainty, deforestation and erosion on hillsides
Ghana – Sakuba, Moglaa and Voggu villages, Tolon-Kumbungu District, Northern Region	Sorghum, millet, maize, yam, cassava, cowpea, groundnut and soya-bean, and livestock	Low–medium potential savannah lowland, very limited access to water	High fragmentation of farm lands, short or no fallows, declining soil fertility, high food insecurity
Ethiopia – Zenzelima, Yigoma and Fereswega villages, Bahir Dar Zuria district, Amhara Region	Maize, teff, sorghum, millet, legumes, vegetables, chat, eucalyptus, beekeeping and livestock	Medium potential highland zone, with regular rain failures but considerable access to irrigation water	Substantial population increase, landholding fragmentation, urban encroachment, grazing land and natural forest reduced, soil fertility declining

GROWING PESTICIDE USE

Several trends were apparent in all eight systems studied. More than half of all farmers reported an increased use of pesticides compared with the first half of the 1990s, despite rapidly rising costs in the last three to five years. In some cases, pesticides have only started being used recently. Ghanaian pineapple small-holders started to use agrochemicals when they began growing the Smooth Cayenne export variety, while those growing SugarLoaf for domestic markets very rarely use inputs. Herbicide use for weed control in pineapple is becoming more common in Benin among smallholders growing export varieties, while large-scale farmers have relied on herbicides for weed control for some time. In other cases, farmers indicated that they have become much more reliant on pesticides than in the past. Cotton farmers in Senegal remarked that if, for some reason, they could not get hold of pesticides, then they would be forced to abandon the crop.

An increase in application frequency was described in cotton, cereals and legumes in Ethiopia and most notably in vegetables. Vegetable farmers in Benin have increased the application frequency due to increased pest pressure, and now average 12–20 sprays per season on cabbage, up from three sprays in 1990, and 6–12 in 1995. However, only 44 per cent of Senegalese vegetable farmers considered that they applied higher volumes of pesticide now compared with ten years ago, mainly due to financial constraints. Yet the group consensus was that application frequency had jumped from every 15 days to every 3 days in some crops. Among Senegalese cotton farmers, 68 per cent said they had in-creased pesticide volume over the last decade, and those who had decreased were farmers who had abandoned cotton for groundnut. Ethiopian farmers only treat their more valuable crops (grasspea, chat, vegetables, teff and maize) but now spray three or more times a season, compared with once in the early 1990s.

Farmers reported widespread use of calendar spraying (except in cereals in Ethiopia and pineapple in Benin), and cocktail mixtures (although not in cotton in Benin and pineapple in Ghana). Deviation from recommended practice was widespread. Cotton farmers in Benin described how they spray 8–12 times per season, rather than the 5–6 applications recommended by the cotton companies and estimated that they applied 10.5 litres ha^{-1}, instead of 8 litres, the quota allocated. Many farmers also report increases in pest incidence in vegetables, particularly of mealybug and associated disease in pineapple in Benin, of cowpea in Ghana and in all the crops grown by Ethiopian farmers. Whether these perceptions are true or not, the result is further increases in pesticide use.

Pesticide expenditure thus now accounts for a large proportion of production costs. Table 11.2 provides an indication of this burden, estimated from partici-patory farm budgets conducted with farmer groups. Vegetable farmers in Senegal recalled how pesticide prices escalated over the last decade, first with the removal of government subsidies in 1990–1992, second by price rises follow-ing the devaluation of the West African franc in 1994, and more recently under market deregulation. Production costs have also risen with the introduction of hybrid seed varieties, which they said required more pesticides. Friedrich (2000)

Table 11.2 *The estimates of pesticide costs*

Crop and country	Pesticide costs per ha (US$)	Proportion of production costs (%)
Cotton, Senegal	58–65	22–31
Cucumber, Senegal	99	40
Pineapple, Ghana	185–220	16–20
Cowpea, Ghana	65–187	32–61
Mixed cropping, Ethiopia	22–38	10–54

suggests that as long as pest control costs remain below 10 per cent of total production costs, farmers would not have a problem with pest control. The figures from these studies suggest that many farmers are in serious difficulties regarding cost-effective pest management.

In Ethiopia, it is the richer farmers who apply more fertilizer to their fields, yet there is little difference between rich and poor farmers in terms of pesticide expenditure. In one village, the poorest category of farmers spent up to 54 per cent of production cash outlay on purchasing pesticides. Expenditure of US$20–25 per annum on pesticides is a significant burden on households whose estimated net cash income is well under the World Bank poverty indicator of US$1 per day. The riskiness of this high-cost production strategy was demonstrated in 2001, when cereal prices dropped by 40 per cent following good rainfall, and farmers were unable to cover their costs. Farmers incurred debts of US$35–117 and many had to sell off livestock assets.

Given farmers' high investment in pesticides, it is important to assess whether this investment is paying off. Results from these case studies show that this is not always the case. Many farmers complained that reliance on pesticides was increasing costs, decreasing profits and failing to achieve pest control. In Benin, national figures show that cotton income per hectare remained static during 2000–2001, while pesticide treatments costs rose by 80 per cent over the same period. Vegetable farmers were struggling to compete with cheap produce imported from neighbouring Togo and Nigeria, and so resorted to buying non-approved (and highly toxic) insecticides sourced from cotton and locust control supply channels, since these were available cheaply on the informal market.

This increasing application frequency combined with unregulated use may have increased pest problems. Whitefly and bollworm resistance to commonly used insecticides has caused major problems in the West African cotton sector since 1998, incurring large losses for farmers and the cotton companies. Ethiopian farmers catalogue a list of crops that they no longer cultivate, compared with ten years earlier, due partly to problems of declining soil fertility, but also to pest, disease or weed problems. For example, chickpea was no longer grown because of termite problems, and field pea due to persistent pest attack in the field and in storage.

Pineapple was the crop that continued to provide good returns on investment. However, women farmers in Ghana, who generally have less access to cash than men, found it harder to pay upfront for inputs as prices rose and some have had to abandon its cultivation. Women cowpea farmers also stressed that they found it hard to adopt improved varieties that require insecticides. Only the richest Senegalese vegetable growers grow bitter aubergine, one of the most lucrative crops for local markets, because it requires weekly sprays over a 12 month cycle.

A majority of key informants confirmed that pesticide use by smallholders is increasing. These informants included government agricultural research and extension staff, donors, NGOs and pesticide distributors. One government official indicated there was a trend towards the excessive use and misuse of pesticides in vegetable production in Senegal, mainly due to lack of training, and described how some farmers believe they need to apply continuously even though this is expensive and not always effective.

IMPACTS ON HUMAN HEALTH

The most commonly encountered pesticides in the eight cropping systems studied were endosulfan, dimethoate, cypermethrin, chlorpyrifos, fenitrothion, malathion, glyphosate, profenofos and deltamethrin. All are insecticides, except glyphosate, a broad-spectrum herbicide. Many farmers handle, apply, store and dispose of pesticides in ways that expose themselves, their families and some-times consumers to serious risks. At least one WHO Class Ia or Ib product or toxic fumigant was in use in all the cropping systems, the exception being pineapple in Ghana. Class Ia and Ib products were most commonly used on vegetables, and farmers often spray close to the harvest date, again putting consumers at risk. Most farmers do not use appropriate equipment or protective measures: in Senegal, for example, only 14 per cent of vegetable farmers use protective clothing. In Benin, 93 per cent of pineapple farmers store pesticides in their bedroom and 45 per cent of cotton farmers use empty pesticide containers to transport water for household purposes.

Farmers growing cotton and cowpea report that they regularly experience ill-health after spraying insecticides, including migraine, debilitation, stomach upset, skin and eye irritation, and sore throat and coughing. Sometimes they are hospitalized. In Ethiopia, farmers use toxic insecticides to treat headlice and bedbugs, and even apply them to cure open wounds, citing four recent fatalities in two villages. Regular sickness costs households money for medication, as well as time off work and lost productivity. One cowpea farmer in Ghana estimated she spent US$39 in treatment costs and was unable to work for a week following one particular poisoning incident. The organochlorine insecticide endosulfan was commonly cited by farmers in relation to poisonings. These acute toxicity incidents mirror similar experiences of cotton farmers in Benin and Senegal using endosulfan.

The OBEPAB (Beninois Organisation for Promotion of Organic Agriculture) has documented 619 cases of acute poisoning, 101 of which were fatal, in cotton growing regions over three seasons (1999–2002). These incidents were reported from 77 villages in 12 districts in two cotton growing regions of Borgou and Alibori, with endosulfan responsible for 88 per cent of fatalities in the 2000–2001 season (Tovignan et al, 2001). Translating OBEPAB's figures into annual poisoning incidence terms, reveals a figure of 21.3 poisonings per 100,000 population in 2000–2001 (the season with most cases) and 11.9 per 100,000 in 1999–2000 (the season with least). Fatality incidence per year ranged from 0.8 to 1.9 deaths per 100,000 people. OBEPAB's data also highlight the frequency of acute and fatal poisonings resulting from exposure to WHO Class II pesticides, indicating the problems associated with the use of these moderately hazardous products under conditions of poverty and poor education.

Incidence figures could not be calculated from the information supplied by Ethiopian farmers. However, their recall of six pesticide ingestion suicides and four fatalities from using undiluted malathion, with or without DDT, to cure open wounds or treat headlice in three villages with a population of 14,000, raises concerns about the level of poisoning. Yet official figures from this region indicate only 1.1 poisoning cases per 100,000 population, derived from those attending at clinics and hospitals. Such serious underestimation of poisoning incidence by official 'passive' vigilance systems is routine (Cole et al, 2000). Cotton and cowpea farmers in Ghana estimated that 490 people out of 1000 were adversely affected each season by pesticides, with a range of 333–600. Table 11.3 summarizes the health effects described in all eight cropping systems.

Table 11.3 *Synopsis of health effects reported by farmers in eight cropping systems*

Study site/crop	Occasional effects	Regular, acute effects	Very frequent acute effects	Hospitalization cases known	Fatalities known (accidental exposure)
Benin cotton		X	X		
Senegal cotton	X				
Benin vegetables		X			
Senegal vegetables	X				
Benin pineapple	X				
Ghana pineapple	X			X	
Ghana cowpea			X	X	X
Ethiopia cereals/ legumes		X		X	

In Benin, 81 per cent of pineapple farmers and 43 per cent of vegetable farmers reported that the effect of pesticides on their health was considerable. In Senegal, 24 per cent of cotton farmers and 20 per cent of vegetable farmers had witnessed or heard of cases of pesticide poisoning. Increased suicide, particularly of women and teenage girls, by ingestion of pesticides was mentioned by farmers in Ethiopia, Senegal and Benin as a growing worry in their village. Whilst there were no hospitalizations or fatalities in Kpako village in Benin, cotton farmers

Table 11.4 *Symptoms experienced after applying cotton insecticides by Kpako farmers in Benin (recall data over the previous five seasons, n = 45)*

Symptom	Proportion of farmers affected (%)
Migraine or headache	38
Skin irritation	34
Raised body temperature	32
Catarrh	21
Fatigue	12
Nausea	8
Dizziness	4
Fever	4
Itching	2
Aches	2
Cough, sore eyes, stomach upset, constipation	Each 1
No ill effects	1

reported a range of symptoms experienced after spraying (Table 11.4). Only 1 per cent of farmers noted no adverse effects on health.

Interviews in 2003 with cotton and cowpea farmers in Northern Region, Ghana revealed that insecticide-related ill health was widespread and considered by most to be a 'fact of farming life'. Farmers agreed that exposure during spraying made them so weak and sick that they had to stay in bed for 2–7 days afterwards to recover. Table 11.5 shows the number of days taken off sick after spraying insecticides per season, the routine preventative costs (mainly the purchase of milk drunk before or after spraying to mitigate poisoning), and the costs of more severe poisoning treatment at a local clinic or hospital (mainly administration of saline drips). Active ingredients in the products associated with these health effects by farmers were endosulfan, chlorpyrifos and lambda-cyhalothrin.

Table 11.5 *Estimated costs of days off work and treatment following insecticide spraying by Ghanaian farmers*

Average no. days off sick after spraying cotton	Cost in terms of average daily farm labour rate (10–20,000 cedis)	Average no. days off sick after spraying cowpea	Cost in terms of average daily farm labour rate (cedis)	Preventative treatment costs (cedis)	Medical treatment costs (cedis)
21.7	294,000	15.1	151,000	7800	452,000

Ill health is not confined to those directly applying toxic products. Women and children make up 30–50 per cent of recorded poisoning incidents, as confirmed by data from Amhara Regional Health Bureau in Ethiopia and from PAN partners' investigations in Senegal and Benin. The Ethiopian statistics for 2001 recorded 185 accidental poisonings reported at hospitals and health centres, of which 51 per cent were female. Children 5–14 years old made up 20 per cent of cases. The poisoning cases data in Table 11.6 were collected from farming communities during 2000–2001 by PAN Africa in the regions of Kolda, Kaolack and St Louis in Senegal and by OBEPAB in the departments of Borgou and Alibori in Benin. These studies used a pesticide incidents report form similar to that now recommended by the PIC Convention for documenting pesticides causing severe health or environmental effects in developing countries.

Table 11.6 *Documented poisoning cases by gender and age in two countries*

Group	Benin 1999–2000	Benin 2000–2001	Benin 2001–2002	Senegal 1999–2001
Male	86% (125)	75% (200)	61% (125)	67% (56)
Female	14% (23)	25% (65)	39% (81)	33% (28)
Total	**100% (148)**	**100% (265)**	**100% (206)**	**100% (84)**
Under 10	28% (41)	20% (54)	30% (60)	32% (27)
11–20	0% (0)	15% (41)	26% (55)	
Over 21	72% (107)	65% (170)	44% (91)	68% (57)
Fatal/total cases	7%	9%	32%	23%

Data from Benin show that food or drink contamination is an important exposure route (Table 11.7). In the 2000–2001 and 2001–2002 seasons, food contamination and re-using empty containers for food and drink accounted for 256 cases (57 per cent of all fatal and non-fatal poisoning cases) and 78 fatalities (86 per cent of all fatal poisonings). According to OBEPAB, pesticide poisoning may exacerbate existing ill health and malnourishment, and can trigger malaria attacks in those suffering chronic infection. Certain organophosphate (OP) pesticides have been shown to be immuno-suppressants by depleting T-cells and can therefore further suppress the immune system of people living with HIV/AIDS (HIV/AIDS Expert Group, 2001). Table 11.8 lists the pesticides responsible for poisoning cases, where these could be identified in Benin, drawing attention to the key role played by endosulfan in acute and fatal cases.

IMPACTS ON LIVESTOCK, BENEFICIAL ORGANISMS AND WILDLIFE

Several livestock poisoning incidents were recalled in Ethiopia, and among Ghanaian cowpea and Beninois cotton farmers. Farmers revealed that farm animals could be poisoned by feeding on pesticide-treated foliage or grain, as a result of baiting for rodents, or other animals, and via direct application of insecticides misused as treatments for animal parasites. Farmers reported that

Table 11.7 *Poisoning route in Benin and Senegal (actual number in brackets)*

Poisoning route	Benin 1999–2000	Benin 2000–2001	Benin 2001–2002	Senegal 1999–2001
Field application	53% (78)	17% (45)	13.5% (28)	33% (28)
In store	–	9% (24)	–	
Contaminated food	17% (25)	68% (180)	68% (140)	–
Involuntary ingestion	–	3.5% (9)	3.5% (7)	–
Suicide/attempt	21% (31)	2.5% (7)	8% (17)	14% (12)
Re-use of empty container	–	–	7% (14)	17% (14)
Seed treatment	2% (3)	–	–	–
Tick/headlice treatment	0.5% (1)	–	–	5% (4)
Inhalation in room	0.5% (1)	–	–	–
Murder	1% (2)	–	–	–
Playing	–	–	–	2% (2)
Confusion	–	–	–	10% (8)
Unknown	5% (7)	–	–	19% (16)

Table 11.8 *Products responsible for poisoning (actual numbers in brackets)*

Active ingredient(s)	Benin 1999–2000	Benin 2000–2001	Benin 2001–2002	Senegal 1999–2001
Endosulfan	60% (89)	83% (219)	53% (109)	12% (10)
cypermethrin + dimethoate (Sherpa, **Cystoate**)	13% (20)	–	4.5% (9)	1% (1)
cypermethrin + chlorpyrifos (Nurelle)	6% (9)	–	–	–
chlorpyrifos (**Dursban**)	–	2% (6)	10% (21)	–
lambda-cyhalothrin + profenofos or cypermethrin (Cotalm)	4% (6)	10% (26)	16.5% (34)	–
carbofuran + thiram + benomyl (Granox)	–	–	–	6% (5)
Other named products	17% (25)	1.5% (5)	4.5% (9)	8% (7)
Undetermined products	–	3.5% (9)	11.5% (24)	73% (61)

Products or active ingredients in **bold** were responsible for at least one death amongst this data.

livestock poisonings had increased considerably in the Benin cotton zone since the introduction of endosulfan insecticide in 1999. Ethiopian farmers in one village recalled 13 of their cattle dying as a result of insecticides applied to treat skin parasites. Livestock are a highly valued asset for many farming households and their loss is a severe financial blow. One Senegalese farmer who lost five goats from poisoning estimated their value at US$97, a major loss when most farmers in his village earn only US$70 net income from their entire cotton crop.

Ethiopian farmers report bee kills after pesticide treatment of grasspea at flowering stage, a favourite nectar source for bees. They estimate that their annual honey production had declined from an average 10 kg to 7 kg in traditional beehives over the last ten years, which they attribute partly to pesticide kills and partly to the decline of natural woodlands. Repeated poisonings of wild animals and beneficial organisms have been observed by cotton and cowpea farmers (Table 11.9). Beninois farmers described how the first pesticide application for cotton on newly cleared land kills large earthworms. After two to three years' cultivation, any remaining earthworms that are killed are smaller, thinner and less numerous. Pesticide treatments have also caused the disruption of termite and ant colonies, and the deaths of snakes, toads and birds.

Table 11.9 *Accidental animal poisonings observed by cotton farmers in Benin (n = 45)*

Type of animal killed	Proportion of farmers who have observed dead animals after spraying cotton fields (%)
Earthworms	69
Non-target insects	65
Snakes	63
Toads	47
Bees	36
Rats and other rodents	13
Small birds	9
Ants	8
Termites	6
Domestic animals	4
Lizards	1

PROMOTION OF IMPROVED VARIETIES

Government promotion of improved, higher yielding crop varieties is closely associated with the introduction of and growing use of insecticides in cowpea in Ghana and maize in Ethiopia. These varieties are susceptible during storage to weevil damage. High-yielding cowpea varieties, such as Bengpala, Valenga and Milo grown in Ghana, are also more susceptible to field pests, due to their softer seed coat (Adipala et al, 2000; Dr A. Salifu, pers.comm.). Cowpea farmers indicated that four of the improved varieties require 'light spraying' (2–3 insecticide applications per season), and one 'heavy spraying' (4–5 applications), while local varieties, such as Apagbala, Wasaie and Zonfabihi, did not need to be sprayed at all. Ethiopian farmers describe how the yield benefits from improved maize may sometimes be cancelled by high losses during storage and that most granaries have become continuously infested with weevils since their introduction.

Improved varieties certainly offer many yield advantages, particularly when conditions are good. But farmers have not always benefited. With the recent transition to maize-based cropping systems, farmers in the Amhara Region of Ethiopia now rely on selling maize to pay off credit incurred on seed and fertilizer inputs. But cereal price crashes in 2001 due to good harvests left farmers with serious debts as their income no longer covered the cost of crops produced using externally purchased inputs. Many farmers explained that because of the debts incurred on maize, they would not be able to afford any fertilizer in 2002. Senegalese vegetable farmers reported how their production costs had risen with the use of hybrid varieties, while in Benin, vegetable farmers explained that the need for pesticides was much higher on non-native temperate crops such as cabbage, carrot and cucumber than on the indigenous vegetables amaranth, *Vernonia* and *Solanum*.

THE LEGACY OF GOVERNMENT AND DONOR SUPPORT FOR PESTICIDES

In all four case study countries, agrochemical provision for smallholders was formerly facilitated by the state, either through commodity boards or via extension services. Pesticides were provided on a subsidized basis to farmers growing cash crops of importance to the national economies, such as coffee, cotton and cocoa. These were also often provided for cereal or legume staples or for control of outbreak pests such as locusts, armyworm and quelea birds. Village Pest Control brigades were set up from the late 1980s in several Sahelian countries, including Senegal, to organize resource-poor farmers to apply pesticides provided by the government or donors. Some of these still exist, although there has been much criticism of their effectiveness, safety and of the unequal distribution of pesticide access and benefits amongst villagers (de Groot, 1995).

Recent cuts in government funding have not only reduced pesticide distribution but severely weakened extension advice and training services. Women growing vegetables in one cotton village had been provided with a 'white powder' free of charge from the extension service by their village chief (probably fenitrothion). PAN Africa's investigations into pesticide poisoning in Kolda Region in 2000 showed that some Village Pest Control Committee members took pesticides for their own use (PAN Africa, 2000), while farmers and key informants in 2001 implicated government staff in the resale of donated pesticides.

The four case countries have reduced or eliminated direct subsidy of pesticides since the mid 1990s, yet the practice of supporting access to pesticides continues. In Ghana, price subsidies on pesticides were removed in 1996 but herbicides, fungicides and growth regulators are exempt from import duty and VAT. Insecticides, rodenticides and fumigants should attract an import duty of 10 per cent but importers almost always apply for a waiver on duties (Gerken et al, 2000). Government subsidies and a monopoly over pesticide distribution were removed in Ethiopia from 1995, but the government continues to provide pesticides free of charge to farmers for the control of outbreak pests, such as armyworm and quelea. In 2003, Plant Health Clinic staff in Tigray Region distributed over 2500 kg and 1490 l of insecticides (fenitrothion, malathion, carbaryl and chlorpyrifos) to 127 peasant associations in 16 districts for control of an armyworm outbreak. They did not, however, provide any protective equipment.

State subsidies were completely removed on pesticides in Benin in 1991. Subsidies were phased out in the mid 1990s in Senegal, yet in 2001, the government subsidized the new and expensive insecticide Prempt (fenpropathrin plus pyriproxyfen), which is used for whitefly on cotton, so that farmers could overcome critical problems of whitefly resistance to older generation products. The remaining form of state subsidy in Ghana is the sale by the government of Japanese pesticide donations, estimated to have a value of US$1 million in 1999. Clearly, pesticide donations and direct subsidies are not solely responsible for excessive pesticide use, although there is concern that donated pesticides are not used for their intended purpose. These programmes also send out a strong message about the indispensability of pesticides. Sissoko (1994) highlighted how the indirect consequence of large-scale locust control operations in Mali was that peasant farmers, extension staff and decision-makers all came to consider pesticides as an essential component of agriculture and that pest control was a responsibility of government and so should be free to farmers.

LIBERALIZATION AND PUBLIC–PRIVATE SECTOR CHANGES

In all four countries studied, there has been a proliferation of informal pesticide trading following liberalization during the past decade. Vegetable farmers in Senegal now purchase their pesticides from retail outlets of national distribution companies, small-scale informal traders operating in local shops, itinerant peddlers, and open markets in larger towns. The last three sources frequently repackage and re-label products, the contents of which may have been diluted or mixed, and they do not always correspond to labels. Farmers' lack of cash has encouraged the development of village-level trading of pineapple inputs in very small volumes in Benin, by the glassful or one eighth of a litre, compared with authorized outlets that mainly sell one or five litre containers. Pineapple farmers admitted to often using products without knowing their identity, name or characteristics. Farmers explained that as agrochemical prices have increased, they look to obtain them via cheaper, informal sources. The advantage of the informal channel is that it is quick, readily-accessible and the cash outlay for small volumes is within their means. However, they run the risk of being sold adulterated or fraudulent products.

Unapproved and sometimes illicit supplies may also be obtained via unauthorized cross-border trade. In Ghana, such trade is common from Côte d'Ivoire and Togo, as evidenced by the widespread sale of pesticides labelled in French, violating one of the key labelling requirements in the FAO Code of Conduct (FAO, 2002). Cross-border trade is further encouraged by wide price and exchange rate differentials. In 2001, some registered cotton insecticides used in northern Ghana cost ten times more than their counterpart products in Côte d'Ivoire.

Until 1991, the state cotton company SONAPRA was the sole supplier of insecticides to Beninois cotton farmers. By 2001, there were 23 private sector distribution companies operating in the country. More than half of Beninois pineapple farmers source their pesticides exclusively from informal outlets. Senegalese vegetable farmers describe how poorer farmers were obliged by price to source only from informal outlets, despite the risk of being sold adulterated or fraudulent products. Ethiopian farmers reported a similar trend, including easy access to DDT, which is no longer approved for agricultural use. Governments are acutely conscious of the problems with the informal sector and some, such as Ghana and Ethiopia, are gradually attempting to register and train dealers. However, these channels pose a major challenge to regulatory authorities with few resources.

INADEQUATE RESEARCH AND EXTENSION FOR IPM

Ministries in Senegal, Ghana and Ethiopia, and to a lesser extent in Benin, continue to play an active role in pesticide promotion through the distribution or sale of pesticide donations. Extension staff may work as pesticide dealers – an activity that sends out mixed messages to farmers and the public about apparent government commitments to reducing reliance on pesticides. Integrated pest management (IPM) remains poorly understood in the four countries. Some officials had very little idea about what IPM involves, but many were broadly aware of the concept and supportive in theory about changing to IPM strategies in the future. Yet many expressed the view that 'we cannot expect farmers to reduce pesticide use until alternatives are available'. Some conceived of IPM as a matter of replacing specific pesticides with biological pesticides and other external inputs that are not readily accessible in African markets. Some Ethiopian stakeholders saw IPM as an imposition from intensive commercial systems in industrialized countries and not directly applicable to mixed smallholder cropping in the tropics.

Even where there is enthusiasm for IPM, very few government staff in research or extension have practical experience in IPM implementation and so lack the confidence to champion its cause. Successful pilot projects have not been scaled up nor their achievements and potential publicized and lobbied for at the highest political levels. The farmer field school (FFS) pilot IPM training programme in Ghana has shown great promise, reducing pesticide use in vegetables, rice and plantain by 95 per cent, while increasing yields and farm income (Fianu and Ohene-Konadu, 2000). Yet the training approach is not integrated fully into national extension programmes and at district level, FFS activities have not been given priority in work plans. FFS training has been enthusiastically supported by the Amhara Regional Bureau of Agriculture but it is proving difficult to gather political support for IPM and FFS methods at federal level. The Ministry's extension services remain wedded to the demonstration-package method focusing on external inputs.

OPPORTUNITIES FOR TRANSFORMING PEST MANAGEMENT IN AFRICA

Major new initiatives to invest in African smallholder agriculture by donors are welcome reversals of a long-term decline in farming support. It is not yet clear, however, whether this investment will encourage or reduce the use of harmful pesticides, even though there is widespread recognition amongst most donors of the problems associated with agrochemical dependency. A recent report advocated the urgent need to develop alternative approaches to soil nutrient management and viable alternatives to costly pesticides, through pro-active, IPM-based, farmer-centred pest management (Dixon and Gulliver, 2001). Ghana's new Agricultural Services Subsector Investment Program (World Bank, 2000) includes IPM as its chosen strategy for export and local crop protection, yet it mentions the low use of pesticides in Ghana and equates this with low risk. The first IPM-related activities in 2001 were for registering and training field pesticide inspectors and pesticide dealers, not for farmer IPM training.

A second challenge is to encourage governments, donors and agribusinesses to make a serious political and institutional commitment to IPM and pesticide use reduction at policy and programme levels. Growing requirements in European markets on pesticide residue levels and broader aspects of pesticide practice have concentrated minds on pesticide use and compliance in recent years (Dolan et al, 1999; COLEACP, 2003), and prompt a shift in the culture of pest management in agro-export production. Whether these trends will exert a positive influence on pesticide practice in domestic markets, or contribute to increasingly double standards for food safety and worker protection between local and export systems remains to be seen.

Demand for organic and fair-trade produce in export markets continues to grow (IFOAM, 2003). Middle Eastern importers are interested in organic pulses from Ethiopia, while organic pineapple is now being sourced from Ghana for fresh fruit salads by UK supermarkets. Organic cotton projects in Benin, Senegal, Uganda, Tanzania and Zimbabwe recognize the need to achieve economies of scale to meet European demand (Ton, 2002). Concern about pesticide residues is beginning to spread among African consumers too, as evidenced by the growth of 'conservation grade' vegetables in Kenya (Pretty and Hine, 2001) and community marketing of organic and pesticide-reduced produce in Senegal (Ferrigno, 2003). These initiatives are often limited by a lack of relatively modest amounts of capital investment and marketing support, rather than technical constraints on production.

Regional harmonization of pesticide registration policy is underway in West Africa and southern and eastern Africa (FAO/ECOWAS, 2002; COLEACP, 2003) but it is too early to predict whether these processes will raise or lower existing standards of pesticide approvals and controls. Promotion of small- and medium-sized enterprises for alternative control products is yet to take off, although commercial neem seed extract production is supported in Kenya (Förster, 2000) and Ghana's Cocoa Research Institute is now sourcing neem seed collected by communities for pilot scale production (Buffin et al, 2002). Facilities for the production of *Bacillus thuringiensis* (*Bt*) biopesticide are under development in at least four African countries and regional discussion on facilitating the registration of *Bt* and other microbial products has been taking place since 2000 (Langewald and Cherry, 2000).

Concluding Comments

This research on eight cropping systems in four countries revealed increasing interest in IPM training, but it remains uncertain whether sufficient demand can be exerted on research agendas to transform conventional crop protection paradigms. IPM and agroecological concepts need to be mainstreamed into agricultural college and school curricula, with practical educational materials adapted for African cropping systems. The experience of many successful farmer participatory IPM programmes elsewhere is that sufficient proven techniques already exist to justify immediate investment in training programmes.

Safe and sustainable pest management without reliance on harmful pesticides should not be impossible to achieve. Dozens of successful small-scale initiatives already exist in Africa, using a wide range of biological and cultural methods (Pretty, 2001; Pretty et al, 2003b). The best of these draw on both indigenous and formal knowledge, and employ participatory learning and experimentation methods. The problem is that most of these initiatives remain small and relatively uncoordinated, with only limited potential for dissemination to other farmers, since they require investment in human and social capital to expand and evolve. As Pretty et al (2003b) note, such initiatives have generally succeeded despite policy support at national level rather than because of it. Persuading more decision-makers and other important stakeholders to accept the IPM concept and its practical implementation is a vital priority in transforming African farming systems for the benefit of rural communities and their consumers.

Chapter 12

Towards Safe Cocoa Pest Management in West Africa

Janny G.M. Vos and Sam L.J. Page

THE COCOA CROP

Cocoa, *Theobroma cacao* L., originated in South America and is now cultivated in West Africa (mainly Côte D'Ivoire, Ghana, Nigeria, Cameroon), South America (including Brazil and Ecuador) and Asia (mainly Indonesia and Malaysia). Up to 90 per cent of the world's cocoa is produced by smallholder farmers, each cultivating less than three hectares, with the remainder grown on large farms, mainly in Asia and Brazil. Some smallholder farmers cultivate their cocoa trees within the natural forest, but more often in agro-forestry zones where selected forest trees provide shade. In other systems, specially planted tall trees and herbaceous food crops, such as plantain and banana, are used as shade. In Côte D'Ivoire, cocoa is grown without shade and, although this provides increased yields in the short-term, it requires higher levels of inputs because of increased weed and pest problems (Anon, 2001).

Cocoa beans are the seeds of the cocoa tree, which are formed inside pods that grow on the tree trunk. After harvest, the seeds are fermented to change the seed colour and cause the chocolate flavour to develop. The seeds are then dried and sent to processors as the raw material for the production of cocoa mass, cocoa powder and cocoa butter. These, in turn, are used by confectionery and other industries as base material for chocolate production. The cocoa market is notoriously volatile and world prices are currently low, leading to low revenues for smallholder cocoa farmers.

As cocoa is an exotic plant in West Africa, it has contracted a number of serious 'new encounter diseases' (NED), which originate from the indigenous flora but to which exotics have not co-evolved defence mechanisms. It has been suggested that when cocoa is in its natural habitat, in the upper reaches of the Amazon rainforest, it is protected from infection by a range of endophytes (Holmes and Flood, 2002). An example of a cocoa NED in West Africa is cocoa swollen shoot virus (CSSV), which originated from *Sterculiaceae* and *Bombacaceae*. It is believed that the aggressive *Phytophthora megakarya* is also an NED. Currently, cocoa growers in West Africa need to deal with a wide range of pests, diseases and weeds (Table 12.1). In addition, there is a looming threat of the introduction of devastating, but as yet exotic, diseases from South America, such as frosty pod and witches broom, through the global movement of plant material.

In Ghana, cocoa is a fundamental component of the rural livelihood system, with many farmers committed to this major cash crop, and Ghanaian cocoa is of renowned quality. The general emphasis of recent development focuses on encouraging the intensification of existing cocoa production, rather than clearing rainforest for fields. There are several different options for this process of intensification, and the key question is – can cocoa continue to be productive through interventions that do not cause harm to the environment or to human health?

Table 12.1 *Priority pest problems in West Africa*

Priority	Pest problem	Causal agent(s)
1	Black pod	*Phytophthora palmivora* and *P. megakarya*
2	Mirids	*Sahlbergella singularis* and *Distantiella theobroma*
3	Stem borer	*Eulophonotus myrmeleon*
4	Mistletoe	*Tapinanthus* sp. (mostly *bangwensis*)
5	Termites	*Nasutitermes* sp. & *Microtermes* sp.
6	Weeds	*Chromolaena odorata* and others
7	Swollen shoot	cocoa swollen shoot virus (CSSV)

Source: Vos and Neuenschwander, 2002

PESTICIDE USE IN COCOA PRODUCTION

A range of products are currently registered for use on cocoa in Ghana (Table 12.2). However, pesticide hazard information is not easily available and often confusing, as some agencies refer to products whilst others to active ingredients. Data derived from recent surveys by the Sustainable Tree Crops Programme (STCP) in West Africa show that only about half (49 per cent) of all cocoa farmers in Ghana use pesticides. After harvesting, during on-farm fermentation, drying and transportation, no pesticides are used (Vos, 1999).

Few of the products used for cocoa pest management in the field are entirely safe. However, the WHO Class I products could be phased out without major agronomic problems as there are Class II and III products that can replace them for black pod and mirid control. According to Padi (pers. comm.), stem borer control using Gastoxin (aluminium phosphide) is still in the experimental stage as there are insufficient stocks to provide this pesticide on a large scale. Alternatives for stem borer control are currently practised by farmers, such as inserting skewers into borer holes to mechanically kill the borers (Vos, 1999). With regard to storage treatments, there are already low-toxic alternatives available to control the cocoa storage pests *Ephestia cautella* (moth) and *Araecerus fasiculatus* (beetle). To prevent these pests from multiplying during shipment, cocoa used to be

Table 12.2 *Officially registered products for use in cocoa production in Ghana and their characteristics*

Pest, disease or weed	Active ingredient(s)	WHO Class	Side-effects
Black pod	Cupric hydroxide	III	Copper based fungicides
	Cuprous oxide	II	are highly persistent and
	Metalaxyl + Cuprous oxide	II	impact soil micro-organisms
	Cupric hydroxide	III	
	Cuprous oxide	II	
	Cuprous oxide	II	
Mirids	Imidacloprid	II	Imidacloprid is highly toxic to bees
	Pirimiphosmethyl + Bifenthrin	II	Pirimiphosmethyl is harmful to parasitic wasps
	Carbamate	I	
Stem borer	Aluminium phosphide*, precursor to phosphine fumigant	I	Phosphine is highly toxic to humans, all insects, mites and rodents, but method of recommended application reduces risk
Mistletoe	–		
Termites	Chlorpyrifos	II	Broad spectrum insecticide and acaricide, toxic to fish and persistent in soil
Weeds	Paraquat	II	Paraquat is toxic to bees, but method of recommended application reduces risk
	Glyphosate	U	Glysophate is slightly harmful to lacewings and ladybird beetles
Swollen shoot	No pesticides are recommended for vector control		

U = Unlikely to present acute hazard in normal use
*Undergoing screening; not officially recommended yet
Sources: Manu and Tetteh, 1987; CRIG, pers. comm.; WHO Recommended Classification of Pesticides by Hazard and Guidelines to Classification 2000–2002; IPCS INCHEM pesticide data sheets; IOBC data on side effects of pesticides on beneficial organisms

routinely fumigated with either methyl bromide or phosphine. These products are both highly hazardous, with methyl bromide being a known atmospheric ozone depleter. Alternative technologies, such as controlled atmospheres containing low oxygen concentrations are now replacing the use of these toxic fumigants.

It is clearly more desirable on environmental and human health grounds for farmers to be using as few Class I and II products as possible. However, in practice, replacement is not always simple. The statement by Bateman (2001) that it is appropriate and desirable for Class I products to be withdrawn from general use, and that there would be immediate complications if Class II products were withdrawn at the same time, is consistent with our findings. To illustrate the banning process of a Class II product, Box 12.1 summarizes the phase-out process of lindane in Europe.

Although, in principle, successful in terms of eliminating a toxic product from West African cocoa production, the ban lindane campaign has, through its heavy focus on food residue issues, by-passed the cocoa growers in West Africa. Cocoa farmers in Ghana are now advised to substitute the far more expensive imidacloprid and pirimiphos methyl/bifenthrin mix for mirid control. Both imidacloprid and bifenthrin are WHO Class II products. Due to the cost implication of the replacement products, a cheap option for pest management has been phased-out for cash-strapped smallholders who have already been seriously affected by the low cocoa market prices. An approach to reduce costs in application would be to do timely spot applications only on congregated mirid populations in openings in the crop canopy. Although still using a WHO Class II product, this practice would be consistent with reducing the amount of toxic products used in agriculture.

INTEGRATED PEST MANAGEMENT FOR COCOA

In Ghana, the Cocoa Research Institute Ghana (CRIG) recommends a range of IPM practices to manage major cocoa pests (Table 12.3) and is emphasizing the generation of more environmentally sound and low-toxic alternatives. Despite considerable attention to the production of alternatives, it has to be noted, however, that current adoption rates by farmers of the recommended practices are low, mainly due to the high costs of inputs (Padi et al., 2000). Vos (1999) confirmed that synthetic pesticide use by cocoa farmers in Ghana is indeed lower than would be expected in light of pest problems and the farmers' general desire to apply chemicals. This shows the need for more support to fine-tune and implement cost-effective alternatives to the currently registered products for cocoa black pod and mirids control so that the former could effectively be phased-in while the latter are de-registered.

Black pod is the most important crop health problem in the region. Control is normally based on regular phytosanitary procedures integrated with the use of fungicides applied using backpack sprayers, which are effective against black pod caused by *P. palmivora*. Unfortunately neither of these methods are giving effective management of the currently spreading and more aggressive form caused by *P. megakarya*, which can cause complete pod loss. Most recommended fungicides for cocoa are copper-based, but they do have moderate to slight hazards in addition to their adverse effects on soil micro-organisms.

Sucking insects such as mirid bugs and capsids (the second most important pest problem), can also cause severe damage to cocoa in West Africa, particularly in the absence of shade. Annual losses are estimated at 25 per cent, but far more serious yield losses are experienced in the absence of control measures

---

Box 12.1 *The long process of lindane phase-out in Europe*

Lindane is an organochlorine product that has been in relatively widespread use in both industrialized and developing countries. In developing countries it is used for the control of a broad spectrum of plant-eating, soil-dwelling, public health pests (including children's headlice), and animal parasites because it is relatively cheap, its patents having expired. Lindane persists in the environment and produces residues that accumulate in the adipose tissue of mammals and birds. Acute exposure mainly affects the central nervous system causing symptoms such as vomiting and diarrhoea, leading to convulsions. Lindane is also an endocrine disrupter and a suspected carcinogen (IARC, 1979). Since lindane was first banned in 1979 by the Dutch government, it has taken more than 20 years to extend this to a European-wide ban. Following the European ban in 2002, Ghana de-registered lindane for use on cocoa.

The time-scale of the banning procedure for lindane is as follows:

1979:	lindane was first banned in the Netherlands in 1979 because of *'the persistence of impurities and its high bio-concentration in the food chain'*.
1985:	lindane was placed on the Pesticide Action Network (PAN) Dirty Dozen list
1997:	lindane was subjected to the UN FAO and UNEP's Prior Informed Consent (PIC) procedure
1998:	PAN Europe co-signed a declaration calling for a ban on the use and production of lindane, and shortly afterwards the Austrian authorities recommended that lindane should be de-registered
July 2000:	the EU Standing Committee on Plant Health voted for a ban on lindane for all agricultural and amateur gardening uses
December 2000:	the European Commission ratified this decision
December 2001:	the ban came into effect and the UN sponsored Codex Alimentarius committee reduced the maximum permitted limit of lindane residues in food to zero
June 2002:	lindane was completely phased out by all member states

---

(Padi et al, 2002a). Until the 2001–2002 cocoa season, recommended applications of the organochlorine lindane and the carbamate propoxur using motorized mist blowers were most effective to control these pests. Both are classified as WHO Class II insecticides and, following the banning of lindane in Europe, the use of this product is now being phased-out of cocoa production in West Africa.

In contrast to the lengthy and externally imposed implementation of the lindane phase-out process, the current challenge is to find ways to phase-out hazardous products whilst simultaneously phasing-in locally validated alternatives, that are cost-effective and less hazardous. We therefore summarize the availability and further development needs of low-toxic alternatives for the management of the two major cocoa pest and disease problems, black pod and mirids.

Table 12.3 *Recommended IPM practices and development of new IPM methods for management of priority cocoa pests in Ghana*

Pest problem	Recommended IPM practice	Potential future options
Black pod	Regular fungicide application from onset of rainy season and cultural practices including sanitation and shade management	Trunk injection using phosphonic acid and others; biocontrol using parasitic or antagonistic fungi; new black pod resistant varieties
Mirids	2–4 insecticide applications over August–December and cultural practices including shade management	Use of sex pheromones and mycopesticides; use of indigenous mirid natural enemies; botanical pesticides such as neem; new mirid resistant/tolerant varieties
Stem borer	None available as yet	Rational pesticide use applying to infected trees only; sanitation
Mistletoe	Pruning	–
Termites	Insecticide soil drench application in holes before and after transplanting; no mulching in termite endemic areas	Botanical pesticides to replace the current chemical pesticides in termite control
Weeds	Hand weeding or 4 herbicide applications per year	–
Swollen shoot	Eradication of CSSV infected trees and their contacts and replacement with CSSV resistant varieties	Biocontrol of CSSV vectors (mealybugs); new CSSV resistant varieties

Source: Padi et al, 2002a

NEW OPTIONS FOR LESS HAZARDOUS BLACK POD MANAGEMENT

Black pod, caused by various *Phytophthora* species, is the main constraint to cocoa production in West Africa and can cause complete yield loss in some years (Holmes and Flood, 2002). In Ghana, control of black pod by the use of regular phytosanitary procedures is only successful in areas where cocoa is infected with *P. palmivora*. The more aggressive pathogen *P. megakarya* is spreading and is now threatening neighbouring Côte d'Ivoire.

Table 12.4 indicates that there is a bleak outlook for non-chemical management of *P. megakarya*, which is not effectively controlled even when sanitation is combined with regular copper-based fungicide applications (Holmes and Flood, 2002). However, in the future an alternative could be available, using chemical elicitors of induced systemic resistance. Previous studies in Papua New Guinea have shown that *P. palmivora* can be managed effectively through the application of phosphonic acid as one trunk injection per year (Guest et al, 1994). Research in Ghana by CRIG, in collaboration with CABI and Natural Resources

International (NRI), has shown that black pod losses due to *P. palmivora* and *P. megakarya* are reduced when phosphonic acid is applied as a trunk injection (Opoku et al, 1998).

Phosphonic acid is a systemic product, used as a foliar fertilizer, with some direct effect on pathogens. However, its main action is considered to be through the activation of the host plant's own defence mechanisms. It induces a cascade of responses, which results in an induced systemic resistance. Its compliance with the principles of safe pest management will need to be further investigated, but available information indicates its use as a stem injection is a more environmentally sustainable method of disease management than the present use of copper-based sprays. The application system results in a low level of environmental and operator contamination.

A variety of cultural methods are available for black pod management, which accords with the current CRIG IPM recommendation for regular fungicide applications from the onset of the rainy season. Farmers need to weigh up the advantages of cultural practices to manage black pod with their impact on other pests, such as mirids, and the economics of application. Labour is known to be a limiting factor for Ghanaian cocoa farmers, both due to the cost and unavailability in cocoa producing areas.

Table 12.4 *Efficacy of non-chemical control of cocoa black pod in West Africa*

Causal agent	Method	Limitation
P. palmivora	Frequent removal of infected pods	Effective, but labour intensive
	Manipulate shade and pruning	Too little shade will increase mirids
	Plant resistant varieties	Highly resistant clones have been identified, but varieties are not yet available to farmers
	Apply antagonists, e.g. *Clonostachys* and *Trichoderma*	Under investigation in Cameroon, Côte d'Ivoire and Ghana, products expected in the short-term. Some success shown in South America
	Use endophytes	Research has been initiated in Cameroon, products expected in the long term
P. megakarya	Frequent removal of infected pods	Ineffective
	Manipulate shade and pruning	Too little shade will increase mirids
	Apply classical biocontrol	Sourcing of potential agents in progress
	Plant resistant varieties	Some tolerant cultivars have been identified in Ghana
	Apply antagonists, e.g. *Clonostachys* and *Trichoderma*	Under investigation in Cameroon
	Use endophytes	Basic research has been initiated in Cameroon

Source: Vos and Neuenschwander, 2002

Various aspects of biological control are being investigated (Holmes and Flood, 2002), with expectations that either fungal mycoparasites (fungi that directly parasitize the target pathogen) or antagonists (fungi that inhibit the target pathogen through chemical means) can be found that could be used as mycopesticides. A number of research institutes in West Africa are investigating the use of locally isolated biocontrol agents as a means to control black pod, primarily the Institute of Agricultural Research for Development (IRAD) in Cameroon and the National Centre for Agricultural Research (CNRA) in Côte d'Ivoire. Research on endophytes (organisms with all or part of their lifecycle inside the host plant's tissues, where they may act to reduce disease incidence through induction of the host plant's own resistance mechanisms, competitive exclusion or direct antagonism) is being investigated by CABI for management of the fungal diseases frosty pod rot and witches broom in Latin America. Studies have been initiated by IRAD in Cameroon to identify local endophytes from cocoa for the management of black pod.

Under a global germplasm utilization and conservation programme, local selections with resistance to black pod have been identified. Classical breeding should, in time, offer additional pathways to reduce the use of chemical pesticides. Chocolate manufacturers are sponsoring research for enhanced black pod resistance breeding through the use of genetic markers in conventional breeding programmes. To date, genetic modification is unlikely to be accepted by consumers and has not yet been accepted by the cocoa industry.

NEW OPTIONS FOR LESS HAZARDOUS MIRID MANAGEMENT

An integrated package for mirid management involving cultural practices, tolerant or resistant varieties, biological control and judicial use of neem shows the most promise. Unfortunately, biological control techniques are still in the testing stage and breeding for resistance could only offer a solution in the long run (Table 12.5).

Light shading (at 10 large or 15 medium trees per hectare) can help reduce mirid damage. Shade trees, particularly forest trees left after clearing the land have a very important role to play in the conservation of forest and associated fauna. Where all the indigenous trees have been removed, fast-growing trees such as *Gliricidia sepium, Terminalia ivoirensis, Ricinodendron leuclotii* and *Spathodea campanulata* may be planted. In young plantations, food crops such as banana and plantain can provide temporary shade.

In a collaborative project between CRIG and NRI, investigations are focusing on the use of sex pheromones to manage mirids on cocoa. Pheromones have been synthesized for the two major mirid species in Ghana, with laboratory and field tests on the synthetic blends of pheromones initiated (Padi et al, 2002a). The pheromone (sex attractant) chemicals have been found to be effective in attracting male mirids to traps. An efficient trap has now been identified (Padi et al, 2002b) and small-scale on-farm trials are being conducted to test the pheromone blends for optimizing mirid trapping for monitoring and control.

Table 12.5 *Efficacy of non-chemical control of cocoa mirids in West Africa*

Causal agent	Method	Limitation
S. singularis & D. theobroma	Manipulate shade Use pheromone traps	Too much shade will increase black pod Pheromones have been identified, but not yet available to farmers
	Apply mycopesticides e.g. *Beauveria bassiana* Spray botanical 'neem'	Field testing will begin shortly, not yet available to farmers 74–80% effective, impact on natural enemies questioned
	Plant tolerant varieties	Recommended
S. singularis	Encourage natural enemies e.g. *Euphorus sahlbergella*	*E. sahlbergella* only 30% effective

Source: Vos and Neuenschwander, 2002

Five isolates of *Beauveria bassiana*, including one that was isolated in Ghana, are being screened in the laboratory against cocoa mirids. This work is being done by scientists from Ghana's CRIG in collaboration with CABI Bioscience (Padi et al, 2002a). Results indicate that its growth and viability, together with its ability to sporulate profusely, make it a good candidate and field testing is being planned. Further work will be needed to test its specificity, and optimize formulation and application techniques.

Neem is a broad spectrum botanical pesticide derived from neem trees (*Azadirachata indica*). These are widespread in West Africa and farmers can prepare neem-seed extract easily. In addition, neem products are available commercially. Locally funded research in Ghana has compared the commercial preparation, Neemazal, and crude aqueous neem seed extract with synthetic insecticides. The neem killed 80 per cent of target pests, compared with 95 per cent with synthetic pesticides (Padi et al, 2002a). Current work in Ghana involves the application of 20 per cent crude aqueous seed extract and 5 per cent Neemazal. Similar work in Côte D'Ivoire compared a neem spray with diazinon (OP: WHO Class II) at recommended rates. The neem seed extract killed 74 per cent of cocoa mirids while the conventional insecticide gave a 99 per cent kill (Kebe et al, 2002). Both groups of workers observed that the neem also had a repellent effect and in Ghana a lowering of the mirid reproductive rates was noted.

ORGANIC AND IPM COCOA PRODUCTION

Organic production of cocoa has been successfully implemented in Mexico and Central America (Belize, Nicaragua, Costa Rica, Panama) as well as Ecuador. Smallholder cocoa growers in Cameroon, Papua New Guinea, Peru, Honduras, Dominican Republic and Ecuador are said to be interested in converting to this type of production (Kennedy, Green & Blacks, pers. comm.). Until recently, organic cocoa growing in Ghana could only be described as 'organic by neglect' as there is no technically supported training available to farmers. A recent Conservation International initiative to train farmers in ecological cocoa management in the periphery of Kakum National Park endeavoured to change this situation.

In Central America, however, there are already active plant nutrition and pest management programmes, with soil microbial health being addressed and farmers being trained in ecology and crop management, including wildlife conservation (Krauss, CABI Bioscience, pers. comm.). Pest management is achieved through the promotion of biological diversity and is said to be working well in indigenous forest areas. However, very little technical information is available on natural pest management for cocoa in West Africa. It is clear that in order to produce cocoa organically far from its centre of origin, there is a need for appropriate farmer training and research. It is also questionable whether it will be possible to grow cocoa organically in *P. megakarya* hotspots once copper fungicides are no longer accepted in organic agriculture.

IPM strategies cannot be packaged in the same way as pesticide recommendations because they will have to be adapted by farmers to local agroecosystems. Therefore, the IPM strategies must be fine-tuned at grass-roots level with farmer participation being crucial to their success. Such an approach requires a paradigm shift on behalf of research and extension, from a top-down approach to one in which the farmers are respected as valuable research partners. Smallholder farmers need technical support and innovative training methods to understand the concepts of sustainable methods to manage pests and for improving agricultural practices, including soil fertility management and on-farm fermentation of shelled cocoa beans.

One example is the Sustainable Tree Crops Programme (STCP). This was created under a public–private partnership and provides a framework for stakeholders to collaborate towards the improvement of the economic and social well being of smallholders and the environmental sustainability of tree crop farms. One of the aims of the STCP is to train farmers in pilot project areas in Cameroon, Côte d'Ivoire, Ghana and Nigeria, with a long-term strategic objective to institutionalize farmer participatory training into farmer organizations and national extension services.

The programme links in various projects funded by public and private sponsors and supports the development of participatory extension approaches for addressing the major production constraints, black pod and mirids. Local and researcher knowledge are being developed and shared through farmer participatory experimentation. The approach will place the farmers and their contextual environment at the centre of technology dissemination and adoption processes. The attention will focus on increasing productivity, lowering dependence on costly agrochemicals, and improving the quality of cocoa.

TOWARDS SUSTAINABLE PEST MANAGEMENT

Sustainable pest management inherently means taking an approach that moves away from providing the most cost-effective way of controlling crop pests in relation to the farm-gate value of the crop to protecting producers and their environments. The success of any phase-out, phase-in programme will, however, depend on the availability and use of compatible and cheap alternative technologies in pest management. WHO Class I products can be phased out from cocoa production in Ghana as Class II and III products are available for the major pests. However, more time will be needed to replace WHO Class II products, as less toxic alternatives are currently only under development.

A Danish study on the consequences of partial or total phase-out of pesticides (Bichel Committee, 1999) proposed a different approach consisting of a three-pronged strategy for reducing pesticide use, including a general reduction of pesticide use, a reduction of the exposure of biotopes, and increased organic restructuring. The reduced use of pesticides would be achieved by only applying pesticides for the most harmful pests (called 'optimization'). The Bichel Committee expected a timeframe of 5–10 years for this to happen, without significant operating and socio-economic losses. This first step, of reducing the use of agrochemicals through optimization in addition to rational pesticide use, should be considered for cocoa in West Africa with a focus on the most harmful pests, black pod and mirids. At the same time, farmer participatory training in the implementation of good agricultural practice, such as sanitation and known cultural practices that disadvantage pests, could be instituted immediately as the basis for further IPM programme activities. A continued focus on research and the implementation of less toxic alternatives will be needed to feed into and enhance such farmer participatory training programmes.

At the time of writing, there is no information available on less hazardous and environmentally sounder pesticides that could be used for cocoa mirid or black pod control. An active dialogue with agrochemical companies will be needed on the provision of information on less hazardous (WHO Class III) products. These could become registered for use on cocoa in West Africa, so as to reduce dependence on the currently registered more hazardous products. The lessons learned from the banning of lindane experience and its impact on cocoa smallholders in Ghana suggest that phase-out can occur provided that cost-effective methods are generated and effectively phased-in. Clarity will be needed on which methods (chemical and non-chemical) will prove to be safe, and there will need to be a shift of investments into farmer training to assure adoption of alternatives.

A key component will be the organization of farmers into formal groups that can participate in the political and scientific dialogue. Liaison with campaign groups and cooperatives, such as Kuapa Kokoo in Ghana, can enhance their call for economically viable alternatives to currently registered pesticides. Organized farmers' groups can also explore ways of marketing sustainably produced cocoa as an alternative to organics and fair trade niche markets. This is one of the reasons why the current Ghana STCP pilot project works directly with Kuapa Kokoo to implement cocoa IPM farmer field schools.

National farmer support systems, including extension and research, should be mandated and supported to provide farmers with appropriate IPM practices and raise awareness on pesticide toxicity as well as changes in pesticide legislation. Joint ventures with the private sector could emerge for large-scale production and dissemination of alternative methods, such as improved seed varieties and biocontrol agents. Extension needs to focus on filling the gaps in farmers' knowledge rather than the delivery of blanket recommendations, whereas research (both national and international) needs to become more focused on meeting farmers' special needs and demands.

At the other end of the food chain, the food industry in the cocoa importing countries, comprising cocoa bean buyers and chocolate manufacturers, are dealing with a luxury market that is vulnerable to negative publicity, especially when there are suggestions about adverse effects on farmers' health. The campaign to ban lindane illustrates consumer power in enforcing a ban on hazardous pesticides. It also shows, however, the urgent need to increase consumer awareness regarding the impact of banning pesticides on smallholder livelihoods in developing countries. The current support given through the chocolate industry for farmers to develop and implement low toxic methods for pest management needs encouragement and expansion.

CONCLUDING COMMENTS

Low toxicity alternatives to pest management in cocoa production in West Africa are being developed. With continued support for further development, sustainable cocoa production in West Africa will be realistic, provided that strong support is made available for farmer training. Smallholder cocoa farmers will need to be able to access information and knowledge transfer channels to become better informed managers of their farms, whereas other stakeholders in the IPM network will need to re-focus current strategies. A long-term process of re-education and re-organization of farmer support systems should be considered as a way forward for smallholder farmer groups to ultimately benefit from alternative pest management as well as to compete more successfully in the global economy.

Chapter 13

Agroecological Approaches to Pest Management in the US

Carol Shennan, Tara Pisani Gareau and J. Robert Sirrine

INTRODUCTION

An agroecological approach to agriculture involves the application of ecological knowledge to the design and management of production systems so that ecological processes are optimized to reduce or eliminate the need for external inputs. Nowhere is this more apparent than in the management of agricultural pests. There are a wide range of potential approaches to deal with crop pests (arthropods, nematodes, pathogens and weeds) in different types of cropping systems. Any single ecological approach will not provide a 'silver bullet' to eliminate a pest problem, rather successful management requires a suite of approaches that together create an agroecosystem where pest populations are maintained within acceptable levels (Liebman and Davis, 2000; SAN, 2000; van Bruggen and Termorshuizen, 2003). Thus agroecological management requires the careful integration of multiple techniques appropriate for each situation.

Ecological pest management (EPM) seeks to weaken pest populations while at the same time strengthening the crop system, thus creating production systems that are resistant and/or resilient to pest outbreaks. Combinations of cultural and biological approaches are used to achieve these goals. In this chapter, we discuss the ecological processes that inform EPM decisions, the suite of practices farmers may use for EPM, and some specific examples of EPM practices currently in use in the US. Table 13.1 illustrates the range of techniques available and places them within a framework based on the type of effect produced and the scale of application.

APPLICATION OF ECOLOGICAL KNOWLEDGE TO PEST MANAGEMENT

Agroecological approaches are based on an understanding of the ecology of pests and crops, and larger scale ecosystem processes that define the context in which pest–crop interactions take place. As yet this ideal is rarely attained, but innovative farmers and researchers are making progress toward this goal. Most pest

Table 13.1 *A framework for ecological pest management*

Goal	Mechanism	Types of practices	Method	Scale	Practices
Weaken pests	Mating disruption	Biological control	Pheromone release	*Field*	
	Increased levels of predation, parasitism or disease	Biological control	Release introduced natural enemies	*Field to landscape*	
			Augment indigenous natural enemies	*Field*	
			Increase vegetation diversity	*Field/Farm*	In-field insectaries, field margin vegetation, habitat patches and corridors
				Landscape	Distribution, size and connectivity of habitats, habitat and farm field mosaics
		Cultural practices	Diversify cropping systems	*Field*	Intercrop
				Farm	Crop rotation and fallows
Weaken pests and strengthen crops	Shift competitive balance in favour of crop	Biological control	Apply antagonistic organisms or their extracts	*Plant/field*	Leaf sprays, seed treatments
		Cultural practices	Use allelopathic crops and pest resistant varieties	*Field*	Intercrop, surface mulches, and varietal selection
			Enhance antagonistic organisms	*Field*	Soil amendments, fertility management
			Use competitive crops	*Field/Farm*	Crop rotation
			Modify crop canopy	*Field*	Plant spacing/density, intercrop, nutrient management
Strengthen crop	Optimizing crop environment and resource availability	Cultural practices	Enhance soil quality	*Field*	Fertility management, soil amendments, cover crops, and tillage
			Mechanical pest removal	*Field*	Tillage, bug vacs, flame weeders, and trash removal
			Micro-climate adjustment	*Field*	Plant spacing, time of planting, and irrigation management
				Farm	Windbreaks and shade trees

management efforts focus on directly managing pest–crop and pest–natural enemy interactions, but increasing attention is being given to habitat manipulation and landscape level processes. There are three main areas of ecology that provide insights for pest management: population and community dynamics, plant–environment and pest–environment interactions, and pattern and scale in ecosystems.

Population and Community Ecology

Concepts developed in population and community ecology are central to EPM and focus on interactions amongst populations (individuals of a single species), communities (an assemblage of populations), and the physiochemical environment. Useful examples include reducing crop palatability (nutrient management), making crop plants less 'apparent' (intercropping, rotations), reducing pest reproductive success (mating disruption), and increasing predation or parasitism of herbivores (biological control).

Natural enemies play important roles in agroecosystems by regulating pest abundance, sustaining the fitness of pest populations, and also providing selective forces in pest evolution (Price, 1997). Effects of predation are now thought to extend to lower trophic levels (such as plants) in some situations – a phenomenon referred to as a trophic cascade. Effects of top predators on lower trophic levels have been demonstrated in aquatic systems (Paine, 1974; Power, 1990) and among arthropods and plants in terrestrial systems (Letourneau and Dyer, 1997). As Letourneau (1998) suggests, enhancing natural enemy populations that reduce crop pests, resulting in a 'top-down' effect on crop productivity, is a fundamental tenet of biological control. However, plants may also have 'bottom-up' effects. For example, plant chemistry can affect the behaviour of natural enemies directly through chemical cues for finding prey or indirectly through the herbivore pest. Moreover, plants can provide habitat, beneficial microclimate, or alternative food sources to enhance natural enemy survival (Barbosa and Benrey, 1998). Both top-down and bottom-up processes should be considered in biological control.

Plant structure and diversity also affect herbivores, and growing crops in large monocultures subverts mechanisms that regulate their populations in natural systems. These mechanisms revolve around 'apparency' and concentration of host plants, and the levels of natural enemy populations. Apparent species, such as trees, generally have chemical defences that become more toxic over time so reducing leaf digestibility to many herbivores (Feeny, 1976; Rhoades and Gates, 1976). Conversely, less conspicuous annual plants rely on escape in space and time as their main defence (Price, 1997). Planting large monocultures and breeding out defensive toxins has subverted both escape and any chemical defences in annual crops. Increased herbivory in monoculture may also relate to how concentrated the food source is, leading to greater ease of finding host plants, higher pest tenure time, and higher herbivore feeding and reproductive rates. In addition, the lack of plant diversity decreases the diversity and abundance of predators and parasitoids (Root, 1973; Gurr, et al, 1998).

Plant–Pest–Environment Interactions

Understanding plant–environment and pest–environment interactions is critical for effective agroecological management, since a central principle is to create field conditions that both maximize crop health and minimize conditions conducive to pests. Temperature tolerances and growth-reproductive responses to microclimate, water availability, and nutrient availability can all affect the timing and severity of pest damage to crops. Utilizing these interactions forms the basis of cultural pest management methods such as crop rotation, time of planting, planting arrangement, fertility and water management. Models for predicting pest and disease incidence based on climate records are increasingly used to help make pest management decisions. Manipulating field conditions through management can also shift the competitive edge toward the crop rather than the weeds, or toward antagonists versus pest organisms.

Pattern and Scale in Agroecosystems

Ecosystem ecologists are increasingly concerned with spatial patterns and issues of scale and hierarchy in ecology (O'Neill et al, 1986; Peterson and Parker, 1998). These concepts have been applied to the ecology of land management and plant disease suppression (King et al, 1985; van Bruggen and Grunwald, 1994). Scale has been explicitly considered in EPM by efforts to apply the theory of island biogeography (MacArthur and Wilson, 1963, 1967) to agricultural settings (Price, 1976; Price and Waldbauer, 1994). By treating crop fields as islands, the theory predicts that the number of species inhabiting a field is a result of immigration and extinction rates, and that over time an equilibrium of species results. Larger islands are expected to support more species and have lower extinction rates and islands closer to sources of colonists to have higher immigration rates. Island biogeography theory can be a useful guideline for EPM, where the goal is to reduce the island size and increase the distance from source pools for pests, while increasing the island size and decreasing the distance from colonizer sources for natural enemies. In essence, this can be achieved by increasing vegetational diversity in the field, farm and landscape. Manipulating colonization rates requires regional land management coordination as the process occurs at larger scales than the field. Differences among colonizing species and species-trophic structure relationships must also be taken into account (Letourneau, 1998).

Price and Waldbauer (1994) argue that successful biological control requires a broad approach across multiple scales. Landscape features affect species interactions and weather patterns, and can have notable effects on pests by changing habitat patterns (Colunga-G et al, 1998). Agricultural fields and non-agricultural areas form a structural mosaic of habitats with insects and other mobile organisms moving between them. Field borders, cover crops and non-agricultural areas can function as refuges for both pests and their natural enemies. Landis and Menalled (1998) also point out that disturbance takes place at multiple spatial and temporal scales. Crop choice and farm characteristics determine tillage regimes, pest susceptibility, resources for natural enemies, and the level of chemical intervention; perhaps the greatest disturbance to biological control processes. These effects are magnified when multiplied over many farms across a region. For example, over 30 species of aphids act as vectors of the soybean mosaic potyvirus (SMV), and virus transmission is determined by frequency and rate of aphid dispersal across the landscape rather than aphid colonization of individual soybean fields. This is because infected aphids transmit the virus each time they sample a plant irrespective of whether or not they remain in that field (Irwin, 1999).

Cultural Practices for Ecological Pest Management

Cultural practices focus on manipulations at the crop, field and farm level both to strengthen crops and to weaken the competitive advantage of pests. Practices encompass direct physical removal of pests from the system (flame weeding, tillage, vacuums and pest habitat removal) to more complex manipulations of field conditions that prevent pest population build-ups (nutrient and water management, crop rotations, intercropping and plant density adjustments).

Mechanical Removal of Pests

The most common type of mechanical pest removal is soil tillage to control weeds, and this is particularly important for organic farmers who do not use pesticides. An excellent resource on tillage techniques and equipment is available from the Sustainable Agriculture Network (SAN) (Bowman, 1998). Depth and type of tillage, residue management, and timing all affect weed populations, and if improperly managed can lead to increased weed seed banks or increased spread of perennial weed roots and stolons (Renner, 2000). Tillage is not a panacea for weed management and needs to be part of an integrated weed management system utilizing multiple techniques (Bond and Grundy, 2001; Liebman and Davis, 2000), particularly given the increased popularity of conservation tillage systems for erosion control, organic matter retention and reduced fuel consumption. Ironically, this has also greatly increased herbicide use for weed control, although efforts are underway to develop non-chemical weed control options for conservation tillage systems, using interplanted cover crops to suppress weeds.

Low mobility arthropod pests may also be controlled by tillage. For example, using a combination of shredding and tillage to bury cotton plant residue effectively kills pink bollworm larvae (*Pectinophora gossypiella*, Gelechiidae) (Daly et al, 1998). However, it is critical that tillage be practised in this way across a region, to prevent the pest re-establishing from adjacent fields. Further, reduced and no-tillage can enhance natural enemy abundance as shown for corn-soybean systems (Hammond and Stinner, 1987), but the effect varies with previous cropping history and insecticide application.

A less common removal technique is the use of tractor-mounted vacuums to suck up arthropod pests, although beneficial insects are also removed (Kuepper and Thomas, 2002). These Bug Vacs are effective in removing lygus bug in strawberries, and also show promise in potato fields for controlling the Colorado potato beetle (Grossman, 1991; Zalom et al, 2001). Yet cost, increased labour requirements, potential for disease spread, and general ineffectiveness have been cited as reasons for their limited success (Kuepper and Thomas, 2002).

Field sanitation is important for the removal of diseased plant material or pest habitat. Yet maintaining a completely 'clean' field through cultivation and habitat removal may not be desirable from a whole system perspective. Non-crop vegetation can fulfil many beneficial ecological functions, and even weeds have been shown to support beneficial insects, such as knotweed *Polygonum aviculare* (Bugg et al, 1987) and species of the Brassicaceae family (Nentwig, 1998; Gliessman, 2000).

Nutrient Management

Many studies have shown that plant quality can influence herbivory, and that in turn nutrient and water availability affect plant quality (Awmack and Leather, 2002). It follows, therefore, that fertility and water management can affect pest damage and predator–prey relationships as shown for mites in apples (Walde, 1995) and leafminers in bean (Kaneshiro and Jones, 1996). Yet, despite many studies, no clear principles relating nutrient levels and herbivory have emerged. In 60 per cent of studies, herbivore populations increased with nitrogen fertilizer addition, yet no effect or negative responses to nitrogen were observed in the other 40 per cent. However, recent work shows that ratios of nutrients can have stronger effects on herbivores than individual nutrient levels (Busch and Phelan, 1999; Beanland et al, 2003).

High plant and soil nitrogen levels can also increase susceptibility to pathogens as shown in organic and conventional tomato production systems in California (Workneh et al, 1993). Lower disease incidence in organic farms was apparently related to microbial interactions, soil nitrate and/or ammonium levels and plant tissue N content. By increasing crop growth and canopy closure, N additions also led to higher powdery mildew damage in no-till wheat (Tompkins et al, 1992). Crop-weed competitive balance can be greatly affected by nitrogen availability depending on the responsiveness of each species to increased N supply, with highly N responsive crops typically showing lower weed pressure with increased N availability (Evans et al, 2003; Liebman and Davis, 2000).

Water Management

Manipulation of soil moisture levels can also be a strategy in pest management. For example, in irrigated vegetable production, the use of drip lines restricts moisture to the crop root zone, limiting weed emergence between the crop rows. Growers also pre-irrigate to stimulate weed germination prior to planting followed by shallow cultivation to remove the weeds (Sullivan, 2001). In-row weeds may be controlled by creating a dry 'dust mulch' around the crop plants to reduce weed germination (Sullivan, 2001). Furthermore, while excess soil moisture encourages the development of many diseases, in some situations periods of flooding can also be used as a pest control technique. The flood period serves both as a fallow, removing the pest organism's food source, and creates conditions of low oxygen that are detrimental to pest survival (Katan, 2000). Periodic flooding can be particularly effective at controlling plant parasitic nematodes (Sotomayor et al, 1999; Shennan and Bode, 2002), and manipulation of flooding depth and timing can be used to manage weeds in California rice production systems (Caton et al, 1999).

Soil moisture, humidity within the crop canopy, and temperature are also major determinants of disease incidence, as reflected in the continuing development of disease forecasting models for major crop pathogens (Bhatia and Munkwold, 2002; Gent and Schwartz, 2003). Forecasting models help growers predict the onset of conditions conducive to disease development, enabling them to not only limit applications of fungicides and organically-approved materials, but also to select appropriate planting times, seeding rates and planting arrangements. Soil salinity is strongly linked to irrigation management, and can also increase disease susceptibility in some crops (Snapp and Shennan, 1994).

Spatial Arrangement, Plant Density and Intercropping

Changing planting density and pattern (row spacing, distance between plants, broadcast versus drilled seeding) affects the relative competitiveness of crops and weeds, with denser plantings reducing weed growth. However, other interactions must be considered. For example, in no-till wheat, powdery mildew severity is related to complex interactions between nitrogen application, crop phenological stage, row spacing and seeding rate. Narrow row spacings restricts early season disease spread by reducing air movement along the rows, whereas high seeding rates increases later season disease severity by increasing canopy density and humidity (Tompkins et al, 1992).

Ecological studies suggest that intercropping, the practice of planting more than one crop together, should reduce pest incidence and enhance biological control (Andow, 1991). Indeed evidence shows that diversified systems tend to support greater species richness and abundance of natural enemies (Andow, 1991), but pest damage is less predictable. For example, in a North Carolina vegetable cropping study, intercropping increased the abundance of a variety of natural enemies, but arthropod damage levels also tended to be higher (Hummel et al, 2002a, 2002b). Further, foliar diseases may increase due to a more conducive (humid) microclimate in the intercrop canopy. There are a number of examples of successful intercrops used in the US, most commonly intercropping cereals with forage legumes, although less work has been done with horticultural crops (Carruthers et al, 1999; Liebman and Davis, 2000; Hummel et al, 2002b).

Mulches and Allelopathy

Surface mulches can be used to suppress weeds and soil pathogens by a variety of mechanisms. The main types of mulches include plastic, dead plant residue or living plants. Clear plastic is applied to bare soil during hot parts of the year to kill a wide range of soil pathogens by raising the temperature of the soil above tolerance limits (Pinkerton et al, 2000; Stevens et al, 2003). Conversely, black or dark green plastic suppresses weeds by blocking light. Use of plastic mulch may have fewer environmental risks than pesticides, but still represents an external input with disposal issues. When practical, the use of plant residue or living mulch is preferable for EPM. Residue mulches can control weeds by reducing light transmission, as found in cherry orchards where a suppressive mulch layer was shown to inhibit weed growth and increase yields by 20 per cent over conventional herbicide tree row management (Landis et al, 2002). Clover planted into winter wheat also effectively controls common ragweed (Mutch et al, 2003).

In addition to light reduction, weed suppression can also be enhanced by using residues from plants with allelopathic properties. Allelopathy is defined as '*the effect(s) of one plant (including micro-organisms) on another plant(s) through the release of a chemical compound(s) into the environment*' (Bhowmik and Iderjit, 2003). The challenge with using allelopathy is to achieve good weed suppression without stunting crop growth. While the use of allelopathic residues as surface mulch is most common, the options of breeding allelopathic crops, or extracting allelopathic compounds to use as 'natural herbicides' are being considered (Bhowmik and Iderjit, 2003). Finally the use of selected species as 'living mulches' has also been studied (Hartwig and Ammon, 2002), but here the challenge is how to kill or remove the living mulch at the right time to prevent competition with the crop.

Crop Variety, Rotation and Cover Crops

Selecting appropriate crops or crop varieties is fundamental to a preventative pest management system. The use of resistant varieties together with rotations to non-susceptible crops can limit pest build-up within a field. While breeding for disease and nematode pest resistance is well known, much less effort has been focused on breeding crops for greater weed suppressiveness. The potential for improved weed control was illustrated in a recent study where highly competitive rice varieties significantly reduced weed biomass without any corresponding loss in yield under non-weedy conditions (Gibson et al, 2003).

Crop rotation is a powerful and commonly used tool for preventing severe pest outbreaks, and works by removing the pest's food source, breaking pest lifecycles, and growing crops that suppress pests either by allelopathy or by being highly competitive against weeds. For example, Michigan farmers have experimented successfully with a maize, maize, soybean, and wheat rotation with cover crops sown into each crop. The wheat benefits from nitrogen fixed by the soybean, and then suppresses perennial weeds for the following corn crop

(Landis et al, 2002). Also, in Ohio, introducing wheat into soybean rotations suppressed potato leaf hopper populations (Miklasiewicz and Hamond, 2001).

Strawberry production in California is an interesting example, where the use of methyl bromide as a soil fumigant has resulted in continuous strawberry production. Growers have generally specialized in either producing strawberries or vegetables, but rarely both. With the projected loss of methyl bromide as a fumigant, the development of economic pest suppressive rotations is critical to both conventional and organic producers, but is leading to new growers' collaborations to alternate management during rotation cycles. Researchers at UC Santa Cruz are working with local farmers to develop strawberry-cover crop–vegetable rotations that include disease suppressive cover crops (mustards) and vegetables (broccoli) (Muramoto et al, 2003). The potential for anaerobic decomposition of crop residues as a disease control strategy is also being investigated.

The term cover crop refers to crops that are grown primarily for soil improvement, nutrient retention, erosion control or weed control, rather than for harvest. The cover crop itself may act directly on the pest by mechanisms such as allelopathy or enhanced competition against weeds, or indirectly via their impacts on soil quality or by providing food or shelter resource for beneficial arthropods. The benefits and management issues involved with using cover crops have been extensively studied and cover crops are now used in a number of annual and perennial production systems across North America. The choice of cover crop species or mix of species is very important (Abawi and Widmer, 2000; SAN, 2000), and how the crop residue is managed. For example, a cowpea summer cover crop enhances weed suppression in lettuce fields, but yields are greatest when the residue is incorporated prior to lettuce planting versus being left as a surface mulch (Ngouajioa et al, 2003).

Soil Amendments

It is well known that soils can become suppressive to particular pathogens and considerable effort has gone into understanding what makes a soil suppressive and whether management can enhance this effect (Weller et al, 2002). Organically managed soils have been found to be more suppressive of some diseases than conventionally managed soils (Workneh et al, 1993), and the potential for organic soil amendments to suppress pathogens has received considerable attention. Compost additions and cover crop residues have been found to reduce fungal, bacterial and nematode pathogens in a number of systems, although the effect can be highly variable depending on the specific crop–pathogen–amendment combination (Abawi and Widmer, 2000). For example, compost reduces certain fruit diseases of tomato but not others, increases foliar disease levels, and has differential effects depending on the tomato cultivar and whether the plants were grown organically or not (Fennimore and Jackson, 2003).

BIOLOGICAL APPROACHES FOR EPM

Biological control focuses on weakening pests through interventionist or preventative strategies that take advantage of ecological and evolutionary processes between pests and their natural enemies. Biological control methods are based upon increased resource competition, mating disruption, or increased predation, parasitism or disease of the pest organism.

Increased Crop Competitiveness or 'Shifting the Competitive Balance'

Cultural practices (tillage, plant spacing) that shift the competitive balance in favour of crops over weeds were discussed above. Biological control methods can also reduce pest competitiveness and are proving particularly interesting as non-pesticide options for control of foliar diseases. This field has developed rapidly in recent years, but is still in its infancy compared with the biological control of arthropod pests. The strategies being developed include increasing competition for nutrients on leaf surfaces by enhancing saprophytic fungal, bacterial and/or yeast populations. This approach is already showing promise for controlling grey mould, *Botrytis cinerea*, on grapes, tomato and potted plants (Wilson, 1997), but is limited to pathogens that require nutrients to grow and infect the plant. For others that penetrate the leaf rapidly and do not require nutrients, enhancing rates of mycoparasitism is more effective and has successfully controlled powdery mildew on grapes in coastal California (Wilson, 1997). The bacterial disease fireblight on apple and pear, caused by *Erwinia amylovora*, is also controlled by increased populations of *Pseudomonas flourescens* on the leaf surface when applied as a spray or disseminated by honey bees prior to bloom (Wilson, 1997; Landis et al, 2002).

One effect of crop rotation or intercropping may be to alter the composition of soil microbial communities and affect the competitive balance between beneficial and pathogenic micro-organisms. This effect was demonstrated when annual rye grass or various wheat cultivars were planted prior to establishing an apple orchard. Selected wheat cultivars reduced the incidence of root infection of apple seedlings, apparently by increasing the populations of pseudomonads in the soil (Gu and Mazzola, 2003). Clearly different genotypes as well as species affect soil ecology in different ways, which opens up the possibilities of screening crop cultivars to enhance soil suppressiveness to pathogens through increased competition.

Mating Disruption

The main approach to mating disruption makes use of the signals transmitted between males and females by some arthropods. Females emit a pheromone when ready to mate and the males then follow the scent towards the female. Large amounts of pheromone applied across a broad area confuse males and prevent them from locating mates. Successful applications of this strategy include for control of Oriental Fruit moth (*Grapholita molesta*) in orchards (Rice and Kirsch, 1990), grape berry moth in vineyards (Dennehey et al, 1990; Trimble, 1993) and Pink Bollworm (*Pectinophora gossypiella*) in cotton (Daly et al, 1998). The greatest success with mating disruption has been in perennial systems with pheromone products used on an estimated 37 per cent of fruit and nut acreage in the US including over 9700 hectares of apple and pear orchards for codling moth control (Office of Technology Assessment, 1995). This compares with pheromone use on only 7 per cent of vegetable crops.

But mating disruption is not a panacea for the eradication of insect problems because not all pests respond to pheromones, each pheromone is specific to a single pest, and the timing of pheromone release is crucial for effective control. Use of pheromone traps for monitoring pest abundance as part of EPM programmes, however, has considerable potential in a variety of agroecosystems. Release of sterile insects has also been proposed as another form of mating disruption, but its use is controversial and adoption in agricultural settings has been limited (Krafsur, 1998).

Increasing Predation, Parasitism or Disease Levels on Pests

Biocontrol via increased predation, parasitism or disease involves either artificially releasing natural enemies or pathogens (classical biocontrol), or transforming the agroecosystem to create favourable conditions for natural enemies (conservation biocontrol) (Dent, 1991). In cases where exotic pests have colonized a region, classical biocontrol typically involves the importing of a natural enemy from the area of pest origin. One early example involved the Cottony Cushion Scale of citrus in California during the late 1800s. The Scale was not native and lacked natural enemies, resulting in population outbreaks that threatened the citrus industry until a predaceous ladybird beetle was introduced from Australia that eradicated the Scale within a year. Since that time, numerous biocontrol efforts have been attempted with varying degrees of success. An estimated 2300 parasitoids and predators were released worldwide from 1890 to 1960 with 34 per cent establishment and some control in 60 per cent of cases (Hall and Ehler, 1979; Hall et al, 1980). However, exotic natural enemies may compete with native natural enemies for alternate prey, thereby interfering with biological control processes or otherwise cause unexpected effects.

Where natural enemies are present, augmenting local populations with mass releases can result in rapid pest suppression or greater suppression early in the season. This approach is used with some success in fruit, nut and vegetable systems. According to the Office of Technology Assessment (OTA) (1995), augmentative releases are used on an estimated 19 per cent of fruit and nut acreage, and 3 per cent of vegetable acreage. Different species of predatory mites are released on 50–70 per cent of California's strawberry acreage to control the two-spotted spider mite, *Tetranychus urticae* (Hoffman et al, 1998; Parella et al, 1992). By the early 1990s, the sale of natural enemies reached US\$9–10 million annually with 86 US companies offering hundreds of species (OTA, 1995).

Conservation biocontrol on the other hand, does not rely on external sources of beneficials, but aims to sustain or increase indigenous populations by careful habitat manipulation, timing of pesticide applications and crop rotations. This approach is preferred over mass releases from an agroecological and sustainability perspective since it relies on inherent agroecosystem properties rather than repeated external inputs. Over time, pest populations may also stabilize, with the system becoming more resilient to pest outbreaks.

Manipulation of Community Dynamics by Conservation Biological Control

To attract and maintain natural enemy populations in the agroecosystem, habitats should provide food resources (host prey, pollen and nectar, alternate prey), and shelter for overwintering. Vegetational diversity varies in terms of species composition, structure, spatial arrangement and temporal overlap of the plants (Andow, 1991). Habitat management can encompass a gradient of diversification from within-field to field margin to regional vegetation management strategies, often referred to as farmscaping. The degree to which farmscaping is effective at enhancing natural enemy populations will depend on the scale and combination of techniques applied. Inter-planting insectary mixes are in-field management strategies, whereas wildflower borders, grassy buffer strips, windbreaks and hedgerows are examples of field margin diversification techniques. The distribution and connectivity of landscape features such as hedgerows, habitat fragments and riparian vegetation are now being considered as having both pest management and biodiversity conservation benefits. In the following sections, we use Californian examples to illustrate some of the successes and limitations of farmscaping.

In-field Insectary Plantings

Inter-planting crops with flowering herbaceous plants is a common farmscaping technique, since pollen and nectar are essential to the fecundity and longevity of several natural enemy species (Jervis et al, 1993; Idris and Grafius, 1995). In a study that evaluated natural enemy to pest ratios for different insectary plantings, Chaney (1998) found that sweet Alyssum, *Lobularia maritima*, had consistently higher natural enemy to pest ratios than other plants tested. Natural enemy densities were high and aphid populations low within 11 m of the insectary (Chaney, 1998) suggesting that alyssum planted every 20th bed would maintain effective biological control in lettuce fields. Californian farmers now often incorporate alyssum in lettuce systems, although they tend to interplant at a smaller interval.

Incorporating perennial crops or vegetation in an agroecosystem can provide stable sources of food and refuge sites for overwintering natural enemies while fields are tilled, fallowed or crops are young. Blackberry and prune trees serve important functions in vineyard systems by providing habitat for alternative hosts of the parasitic wasp, *Anagros epos*, which later preys upon the vineyard leafhopper pest, *Erythroneura elegantula* (Doutt and Nakata, 1973; Murphy et al, 1998). The maintenance of varied successional stages of perennial crops may also be important. For example in California, alfalfa that was strip harvested retained *Lygus hesperus* populations in the alfalfa field where it is not a pest, whereas completely harvested alfalfa fields created an unfavourable microclimate for *L. hesperus*, causing survivors to migrate into other crop fields where they did become pests (Stern et al, 1964).

Field Margin Habitat Management

California is known as the vegetable and fruit basket of the US, but mild climate, high land prices, and short-term leasing structures compel farmers to produce high-value horticultural crops throughout most of the year. While Californian agro-landscapes tend to be more diverse than other parts of the country, the agroecosystems have high levels of disturbance and large areas in sequential monocultures that limit the development of multi-trophic arthropod communities. High input costs and a desire to improve the health of the farm have caused a growing number of Californian producers to 'perennialize' their farming systems by transforming unmanaged or clean cultivated field margins into habitat for beneficial insects. Diversification of field margins with hedgerows and linear features comprised of flowering perennial shrubs, grasses and forbs, is becoming more common (Bugg et al, 1998).

Unlike European hedgerows, which are often relic hedges connected to a network of hedgerows or woodlots, Californian hedgerows are generally recently planted assemblages unconnected to larger areas of natural habitat. John and Marsha Anderson, at Hedgerow Farms, were some of the first to start experimenting with hedgerows in the 1970s in California's Central Valley. Their hedgerows now attract beneficial insects, suppress weeds and provide habitat for wildlife (Imhoff, 2003).

Although hedgerows show promise for serving multiple agroecological functions, many knowledge gaps about their function in biological control still exist. The ability of natural enemy populations in perennial habitats to decrease crop pests in adjacent agricultural fields is not well understood (Letourneau, 1998). Biological control may not be enhanced by hedgerows if the availability of pollen and nectar is so high within hedgerows that natural enemies do not disperse into adjacent agricultural fields to feed on crop pests (Bugg et al, 1987). Alternatively, non-crop vegetation may attract new pests, non-pest prey that natural enemies prefer over the crop pest; or top predators that prey on the natural enemies of interest (Pollard, 1971; Altieri and Letourneau, 1984; Rosenheim et al, 1995; Bugg and Pickett, 1998). The extent to which pests migrate from hedgerows into crop fields, or that hedgerows serve to maintain predator populations in the area, is relatively unknown.

Ultimately the goal of farmscaping for conservation biological control is to create heterogeneous crop and non-crop mosaics that provide stable and productive habitat for natural enemies and facilitate their dispersion into crop areas to regulate pest populations. Nicholls et al (2001) demonstrated the importance of connecting border plantings to in-field floral corridors to encourage natural enemy movement and biological control in vineyards. Natural enemy dispersal ranges, which can vary from a few metres to over a kilometre for some parasitoid species (Corbett, 1998), will determine the effectiveness of various habitat patterns at enhancing biological control. Successful conservation biological control is thus contingent upon matching vegetational scale to the movement range of desired natural enemies in relation to their primary food sources. The varying scales at which natural enemies disperse suggest a need to expand the focus of habitat management from the farm and field level to incorporate larger landscape patterns and processes, a relatively unexplored area.

Effects of Fragments of Native Vegetation in Agricultural Landscapes

The development of multi-trophic arthropod communities depends on both spatial processes (dispersal and foraging), which occur at larger scales than the farm, and temporal processes (overwintering and reproduction). Habitat fragmentation, caused by farming or urban development, can disrupt both types of processes. By reducing the area of habitat and increasing the distance between habitat patches, fragmentation isolates small natural enemy populations from one another, increasing the likelihood of local extinction (Kalkhoven, 1993). Thus, the quality, quantity and connectivity of habitat patches in the agricultural landscape are all significant for sustaining diverse communities of natural enemies and their hosts.

Ricketts (2001) also highlights the importance of land between habitat patches. Many species that live in habitat patches also utilize resources outside the habitat patch, a desirable attribute for biological control since we want natural enemies to migrate into agricultural fields. Therefore, it is important to consider both the characteristics of habitat patches/corridors and cultural practices in crop fields, and coordinate activities to minimize disturbances to natural enemies while they are in the agricultural matrix. Many of the practices we have discussed can accomplish this. Unfortunately there is no simple recipe for designing an agricultural mosaic that supports all important natural enemies, although structurally complex landscapes can lead to higher levels of parasitism and lower crop damage (Marino and Landis, 1996; Thies and Tscharntke, 1999).

Habitat restoration work has also tended to focus on habitat enhancement for species of preservation concern, such as developing habitat 'stepping stones' and corridors. Yet little value or focus has been placed on conservation of agriculturally important species, despite the over 400 million kg of pesticides applied to US agricultural lands each year (EPA, 1999). It is our hope that increasing collaborations among agricultural and ecological researchers and land managers will improve this situation.

One such example involves an innovative approach to conservation that ties together economic activity (farming) with biodiversity conservation in a National Wildlife Refuge in Northern California (Shennan and Bode, 2002). The project tested a strategy for improving wetland habitat diversity and sustaining economically viable agriculture in the same landscape. It involved rotating parcels of land between wetland and cropland over various time scales. The concept is that the cropping phase serves as a disturbance to remove overly dense perennial marsh vegetation with limited wildlife benefit, decompose the large amounts of residue, and recreate open areas that could then be managed to establish early successional habitats currently absent from the refuge.

Conversely, the wetland phase benefits crop production by eliminating a number of key soil-borne pests (notably plant parasitic nematodes and fungal pathogens) as well as improving soil fertility, organic matter and physical properties. The project demonstrated that productive seasonal wetlands could be re-established in areas cropped for more than 40 years within one to three years of wetland management. Furthermore, plant parasitic nematodes were eliminated after one year of seasonal flooding, and when these new wetlands were returned to crop production after three years, crop yields exceeded county averages even with reduced fertilizer additions. If adopted, this system would create a mosaic of seasonal and permanent wetlands in various successional stages, interspersed with areas in crop production, greatly increasing habitat diversity in the region. Historically, lease revenues from cropped land have exceeded the budget of the refuge, creating the possibility of a self-sustaining refuge with strong connections to local communities. Unfortunately, while this approach has great potential, wider issues such as water allocation policy, crop prices and markets, and institutional challenges are most likely to shape future refuge management.

Putting It All Together

As a farmer, the challenge of EPM is to find approaches that also work in regard to other facets of farm management, such as water quality protection, erosion control, economic feasibility and labour demands. Deciding to shift to a preventive pest management strategy rather than a largely reactive one requires sustained commitment and can involve considerable risk. Further, it requires farmers to examine tradeoffs between different goals that are rarely encompassed in research studies. For example, recommended species for field margin plantings to reduce run-off or soil loss may conflict with recommendations for biocontrol, or choosing a cover crop for weed suppression may mean forgoing the nitrogen benefits of a less competitive legume. Most farmers, therefore, make gradual changes as they discover new methods that work and integrate them into their system.

One example of a study that tested the use of multiple EPM strategies together is an apple orchard experiment described by Prokopy (2003). An ecologically based system was designed that used disease resistant rootstock, a suite of cultural and behavioural pest control strategies, landscape vegetation management and minimal pesticide use. This was compared with both unmanaged apple trees and standard commercial IPM systems. Over 20 years of pest suppression in the ecological orchard was comparable to the commercial systems. Such 'whole systems' studies are becoming more common and, while they have limitations, they are important complements to more narrowly focused studies and serve as demonstrations of EPM systems.

THE SOCIO-POLITICAL CLIMATE FOR EPM

Our discussion has focused on ecological and agronomic aspects of EPM but, ultimately, success hinges on the extent to which farmers will move away from pesticide-based systems towards an agroecological vision of farming. Important elements that can help or hinder this process include public policy, market characteristics, public demand and how effectively the necessary knowledge is developed and communicated to the farming community. These topics merit a chapter to themselves; however we will briefly touch upon key issues.

EPM Adoption

It is clear that interest in agroecology and sustainable agriculture continues to increase in the US, but little information is available to assess how widely agro-ecological approaches are being used. Some data are available on the extent of IPM adoption, but the definition of what constitutes 'IPM adoption' can vary from simply adjusting pesticide applications based on pest monitoring to true eco-logical pest management where multiple cultural and biological strategies are used. Data are available on the amount of certified organic acreage, but these numbers not only underestimate the use of EPM by missing non-certified growers, but again can include a spectrum of approaches from simple input substitution to more ecological pest management systems (Guthman, 2000). Nonetheless, we can say that IPM is used to some degree on a substantial portion of agricultural land in the US, and that the bulk of tactics used operate at the field level, whereas landscape or regional approaches are rare (Irwin, 1999; Kogan, 1998).

Various studies have tried to examine why some farmers adopt ecological practices while others do not, and a variety of influences have emerged. In a study of Iowa farmers, Lighthall (1995) found that risk and scale were key factors that influenced farmers decisions, and small (32–160 ha) to mid-size (190–320 ha) operations tended to use sustainable agriculture practices more than large operations (3656–870 ha) due to issues of timing production and market win-dows, and more complex management requirements (Lighthall, 1995). In con-trast, the highest use of IPM was found in small and very large operations in a study of Californian fruit and vegetable growers (Shennan et al, 2001). The latter study also identified the importance of Pest Control Advisors (PCAs) to Cali-fornian growers and that the intensity of IPM use (based upon number of tactics employed) was greatest when an independent PCA was consulted rather than one affiliated with an agrochemical company. Nonetheless, a significant number of growers who used multiple IPM tactics used PCAs affiliated with chemical companies and reported that their PCA had encouraged them to use cultural and biological control strategies. In addition, Guthman (2000) and Lohr and Park

(2002) identified regional norms and the availability of institutional support as determining factors for the use of EPM among organic farmers.

Agricultural Policy

Public policy undoubtedly plays a critical role in shaping agricultural systems and, for many years, US policy has favoured the large-scale industrial model of agriculture that is heavily dependent on external inputs. Indeed, some have argued that the structure of price supports and commodity programmes in the US has effectively discouraged the use of environmentally beneficial management strategies by distorting the economics of production (Faeth et al, 1991). This situation improved somewhat in subsequent years, with the 1996 Farm Bill being hailed as a significant advance for environment and conservation (Baker, 1996). For example, this bill included US$200 million for an Environmental Quality Incentives Program (EQIP) that assisted farmers and ranchers with the implementation of conservation practices through incentive payments and technical support. Unfortunately, while many conservation, environment and rural community programmes were still included in the 2002 Farm Bill, they were eventually weakened and disproportionately favoured large agribusiness (Mital, 2002). Recently, the bulk of funding for these programmes has been eliminated, while commodity support programmes remain essentially untouched, which does not bode well for the future.

The situation facing Tart Cherry producers in Michigan can be used to illustrate how the complex mix of ecological, socio-economic, cultural, market and policy issues define farmers' production decisions and adoption of agroecological practices. Michigan is the nation's primary producer of tart cherries (*Prunus cerasus* L.) representing 75 per cent of US production and an important component of Michigan's economy with an annual value surpassing US$100 million. Michigan growers are challenged by natural constraints, environmental concerns, weakening economic status, drastic market fluctuations, political pressures and social pressures. Farmers' attempts to overcome these pressures have often conflicted with environmental considerations. Further, federal policy established in the 1940s forces tart cherry growers into pesticide use by imposing a 'zero-tolerance' national standard for tart cherries. This prohibits processors from accepting cherries with any cherry fruit fly and from purchasing fruit from the entire orchard that season. Thus, while the majority of growers use some form of IPM, almost all growers spray insecticides on several occasions per season to avoid total crop forfeiture (Curtis, 1998). As one large grower commented, growers would have more flexibility to reduce pesticide use if this zero tolerance policy was changed.

Conservation easements (voluntary legal agreements between farmers and conservation or land trust agencies) may help to stimulate conservation practices on farms. The farmer gives up certain property and development rights to a piece of ecologically sensitive farmland, while retaining ownership of the land and gaining tax benefits.

Federal policy programmes, such as the Conservation Reserve Program (CRP), which pay farmers to set agricultural land aside for conservation, may also facilitate EPM by serving to restore natural biological control processes. A criticism of CRP, however, is that the voluntary contracts between government and farmers are short-term, and renewal is contingent upon federal budget allocations and farmers' current needs and interests. For this reason, conservation easements are attracting increasing attention. Overall, the use of both conservation easements and policy tools to encourage agroecological farming represents a largely untapped opportunity (in part due to the infancy of US agri-environmental policy) that should be pursued more vigorously.

The Role of Research and Extension

There is relatively little support for research into ecological agricultural systems. However, two mechanisms have increased the effectiveness of research. As sustainable and organic farmers have found limited information in mainstream research and extension systems (Lipson, 1997), informal information exchange networks have become especially important to farmers pursuing alternative approaches (Lohr and Park, 2002). Apprenticeships on organic farms are also used as a means of learning organic farming techniques. Second, the use of farmer–researcher partnerships has been promoted as a way to improve information exchange, raise the visibility of alternative approaches, and help to disseminate information through farmer networks. These teams involve groups of farmers working closely with research and extension personnel to define research needs, design and carry out predominantly on-farm research, and evaluate results as they are generated.

Also, there are an increasing number of excellent publications targeted at farmers that take an ecological approach to agriculture. These are becoming available through many university extension programmes, non-profit organizations and federal programmes (see Cavigelli et al, 2000; and the Sustainable Agriculture Research and Education (SARE) programme, Sustainable Agriculture Network (SAN), Appropriate Technology Transfer for Rural Areas (ATTRA), University of California (UC) Integrated Pest Management (IPM) Online, the Wild Farm Alliance).

Consumer Influence

Finally, the role of consumers in creating demand for food and fibre grown in environmentally sound and/or socially-just production systems cannot be underestimated. In many contexts, labels are being tested as a means of distinguishing products based on various criteria to create a market advantage for growers. The most successful label to date is certified organic, but fair trade, various eco-labels and labels based on labour standards are also being used. Consumer perceptions about food production, their willingness to pay more for pesticide-free, organic, local produce or to support family farms have been studied in many locations across the US (Goldman and Clancy, 1991; Williams and Hammitt, 2000; Loureiro et al, 2002). It is difficult to generalize, but many consumers are concerned about perceived health and environmental issues, and may be willing to pay more for food if it were produced more sustainably. Yet, consumers are also often confused about the meanings of labels, like organic (Organic Alliance, 2001). There is clearly a need for more effective consumer education about food production issues and what terms like organic, sustainable, eco-friendly and fair trade really mean.

THE FUTURE?

In summary, despite the evolution of US agriculture toward intensive, large-scale monocultures maintained by high-cost, off-farm inputs, farmers do have an increasing variety of cultural and biological management tools available that can maintain low levels of pest damage with little use of external inputs. We have shown how EPM works by either weakening the pest and/or strengthening the crop system, and illustrated different methods and approaches that are being used in farming systems across the US. While some general management principles can be broadly applied, it is also apparent that location and system specific combinations of practices are needed to achieve the best results.

At the same time it is clear that we have a long way to go. Knowledge gaps still exist, and these are important constraints to the widespread use of EPM. While innovative approaches to pest management are being explored by researchers in many disciplines within agricultural and ecological science, there is still limited collaboration across these disciplines. Too often, it is left to farmers and extension agents to work out how to integrate different kinds of practices. Further, managing an agroecological farming system requires more detailed knowledge and familiarity with multiple techniques than a high-input based system. Many farmers will need economic and other incentives to move away from chemical pest control and take the risks involved in transitioning to a more ecological vision. We believe that the potential is there, but for agroecological approaches to become more mainstream, growers, research and extension agents, policy makers, economists, environmental groups and consumers all need to be involved to make this happen.

Chapter 14

Towards Safe Pest Management in Industrialized Agricultural Systems

Stephanie Williamson and David Buffin

INTRODUCTION

This chapter explores how consumer concerns about pesticides are driving change in a growing number of food chain businesses and how these market forces are leading to innovations for pesticide use reduction and integrated pest management (IPM) in Europe and North America. We describe five case studies in different cropping systems and analyse some examples of government-led programmes to reduce the risks and use of pesticides. The chapter draws together some common themes and success factors in these programmes and concludes with an assessment of current opportunities and constraints for safer and more sustainable pest management.

PESTICIDE USE AND INTEGRATED FARMING IN INDUSTRIALIZED SYSTEMS

Although industrialized countries are the largest consumers of pesticides, the trend in very recent years has been towards using less active ingredient. Official figures from the European Union (EU) recorded an increase in use from 200,000 tonnes of active ingredient in 1992 to almost 250,000 tonnes in 1998, followed by a small decrease (Eurostat, 2002). Agrow's analysis for the period 1998–2001 shows a decrease in European sales from US$9054 to $5902 million and from US$9860 to $7951 million for North America (Agrow, 2003a). Sales in some countries are up, such as by 6 per cent in Denmark, and by 9 per cent in Finland, whereas they fell by 19 per cent in Australia.

However, decreases in volumes of pesticides do not necessarily translate into decreased toxicity or reduced environmental or health impacts. In UK arable systems, there was a marked decrease in the weight of pesticides applied between 1970 and 1995, but an increase in herbicide applications per field (Ewald and Aebischer, 2000). Similarly, there was a 40 per cent decrease in volume between 1983 and 1998 in Ontario, Canada, but an increase of 20 per cent in the compounds with certain risk categories, including for aquatic organisms, surface water and acute human health (Brethour and Weersink, 2001).

Over the last decade or so, integrated pest and crop management has become increasingly common in North America (Reis et al, 1999; LaMondia et al, 2002), Europe (IOBC, 1999; de Jong and de Snoo, 2002) and Australasia (Mensah, 2002). Supporting research to reduce pesticide use and risks in specific cropping systems has become more common (Heydel et al, 1999). Despite the efforts of many governments, however, integrated crop management (ICM) is practised on only 3 per cent of EU farmland (Agra CEAS, 2002). In addition, there are many interpretations of what constitutes IPM, ICM, integrated production or even integrated farming, which makes it harder to assess progress.

Some reasons for limited uptake of integrated approaches may include poor farmer understanding or lack of incentives to change practice. In the UK, a survey of nine different arable and horticultural sectors revealed that many of the 1163 respondents had only a vague idea of the meaning of the terms IPM and ICM: only 40 per cent of arable farmers had heard of IPM, while 30 per cent of those growing field vegetables had not heard of ICM (Bradshaw et al, 1996). Over 50 per cent wanted more information on non-pesticide methods and the results also suggested that if markets acted as drivers for IPM/ICM, farmers would adopt these methods.

Organic agriculture, on the other hand, has witnessed extraordinary growth in the past decade, with markets growing at a rate of 30 per cent per annum since 1998 in Europe and at 20 per cent for over a decade in the US (Vetterli et al, 2002). Eight European countries now have more than 5 per cent of land area under organic management, with Liechtenstein in the lead with 17 per cent, followed by 11.3 per cent in Austria, 9.7 per cent in Switzerland, and 5–8 per cent in each of Czech Republic, Denmark, Finland, Italy and Sweden (Yussefi and Willer, 2003). In North America more than 1.5 million hectares are under organic production on 45,000 farms, although this amounts to only 0.58 per cent of the area in Canada and 0.23 per cent in the US. Organic farmers may apply certain mineral or botanically derived pesticides approved under national or international organic standards, which include copper and sulphur-based fungicides, rotenone and pyrethrum, horticultural oils and microbial biopesticides, all of which may also used by conventional growers. Nevertheless, organic farming rules require any pesticides to be used minimally, only as a last resort and at lower rates than in conventional farming (Baker et al, 2002).

Some environmental organizations and researchers note concerns about continued use of pesticides in organic farming (Vetterli et al, 2002), particularly in relation to aquatic and soil organisms. They also argue that organic systems do not necessarily bring about positive environmental benefits, or achieve better environmental and sustainability outcomes than the best of integrated systems (Rigby et al, 2001). Other studies point to improvements in pollution reduction, biodiversity conservation, soil fertility maintenance and other indicators for organics, in comparison to conventional and integrated systems (Fliessbach et al, 2000; Reganold et al, 2001). Much depends on the interpretation of what is meant by integrated production and how far a particular farming system differs from conventional, intensive systems.

A recent Consumers Union analysis of US pesticide residue data found that organic produce was much less likely to contain pesticide residues than conventional or IPM grown produce and also less likely to have multiple residues (Baker et al, 2002). Nevertheless, synthetic pesticide residues were detected in 23 per cent of organic samples, 40 per cent of which were banned organochlorines most likely derived from contaminated soil. Studies on residues in Swiss organic wine gave clear evidence that organic wine from grapes grown in areas with high spray intensity contained more residues than organic wine from lower intensity areas. Major sources of cross-contamination were spray drift, uptake from soil and contaminated processing, revealing as much about the contamination of non-target habitats by conventional agriculture as the problem of striving for pesticide-free produce in organic systems (Tamm, 2001).

CONSUMER AND RETAILER CONCERNS ABOUT PESTICIDE RESIDUES

The European Commission (EC) publishes annual reports on Member States' residue monitoring programmes and its own coordinated monitoring of 36 pesticides in five commodities (apples, tomatoes, lettuce, strawberries and table grapes). Latest data show that 51 per cent of samples were without detectable residues, and maximum residue levels (MRL) were exceeded most often in lettuce (3.9 per cent), followed by strawberries (3.3 per cent), and that the maneb group of fungicides were most frequently encountered (EC, 2002). The EC concluded that, since 1997, there has been only a slight increase in the percentage of samples with residues at or below MRLs, but samples containing multiple residues have increased significantly. Residues in British lettuce, for example, exceeded MRLs in 9 per cent and 17 per cent of samples in 2000, while around 40 per cent of the 36 samples contained multiple residues (PRC, 2000). For several years, non-approved pesticides have been found on lettuce sold in the UK, and in 2002, two samples contained the OP dimethoate, and another had oxadixyl and pyrimethanil, none of which are approved for use.

Current risk assessment methodology does not quantify the public health implications of residues in the diet but consensus does now exist, at least in the US, that dietary pesticide residues are a significant public health concern, especially for young children (Baker et al, 2002). A World Health Organization (WHO) study with the European Environment Agency (EEA) made a similar conclusion, stressing that dietary and environmental exposure and risks related to age and sensitivity are not addressed when establishing average daily intakes or MRLs (WHO/EEA, 2002). Together with increasing progress by government agencies in improving pesticide use regulation, civil society organizations are increasingly putting pressure on private companies in the food chain, especially supermarkets, to take action to reduce residue levels. Friends of the Earth publishes an annual ranking of UK supermarkets using the level of produce containing residues, and the Dutch Foundation for Nature and Environment published research on residues in grapes from Italy, Spain and Turkey sold by the retailers Lidl and Aldi in the Netherlands, finding that 75 per cent of the grapes exceeded Dutch residue limits (PAN Europe, 2003). The European Consumers' Organization, BEUC, also conducts independent residue analysis, with grapes again topping the residue charts with 92 per cent of samples containing residues in Italy, 67 per cent in the UK and 47 per cent in Belgium (Kettlitz, 2003).

The UK Food Standards Agency (FSA) has recognized that whilst levels of pesticide residues typically found in food are not normally a food safety concern, consumer preference was for food that did not contain residues. As a result, the FSA board has agreed that the FSA will develop a pesticide residue minimization action plan that will look at reducing pesticide use in five crops: cereals, apples, pears, potatoes and tomatoes (FSA, 2003). Interviews with over 1000 British consumers in a recent survey showed that they were deeply concerned about the effects of pesticides (Co-op, 2001). When prompted with a series of questions, the following percentage of those questioned were concerned that pesticides are harmful to wildlife (76 per cent); leave residues in food (60 per cent); pollute water courses (60 per cent); are harmful to growing children (50 per cent); are harmful to 'me' (49 per cent); and damage the health of farm workers (48 per cent) (Kevin Barker, Co-op, pers. comm.). Some nine out of ten believed that retailers have a responsibility to inform shoppers about pesticides used in the production of the food they sell.

The Co-operative Group retailer, which is also one of Britain's largest farming enterprises, took a policy decision in 2001 to insist that its suppliers stop using pesticides with specific health and environmental concerns (Co-op, 2001). Its quality assurance department, in liaison with pesticide health experts and environmental organizations, drew up a list of suspect pesticides and decided to prohibit the use of 24 of these, for which alternatives existed, in its fresh and frozen produce. It restricted a further 30 in 2001, which now require prior approval by the Co-operative Group before they can be used (Buffin, 2001a). The decisions were made not only on the basis of acute toxicity, but also carcino-genicity, mutagenic, reproductive, endocrine disrupting, persistent and bio-accumulative effects. Marks and Spencer supermarket then announced that it was prioritizing action on 79 pesticides and would prohibit their use on produce it purchased via a phased approach starting in 2002, including 12 active ingredi-ents approved by UK regulators (Buffin, 2001b). The company currently has 60 prohibited pesticides worldwide and a further 19 on a restricted basis (Marks and Spencer, 2003).

The result of these actions is that both retailers have taken stricter action than required under the law. Their prohibitions and restrictions apply to produce from any source, not only from UK or other European growers. Both retailers publish the results from their own residue testing on their websites and cite consumer demands and their companies' desire to achieve a competitive edge in food quality and safety as the key drivers influencing the company decisions for pesticide reduction. Both are working closely with their suppliers to enable them to comply with the new requirements, via advice for growers, resources and research opportunities for safer alternatives.

RETAIL AND GROWER INITIATIVES ON PESTICIDE REDUCTION AND IPM

The limited progress of conventional policies to reduce pesticide risk or impacts has been noted by some analysts (Falconer, 1998; Archer and Shogren, 2001). Traditional, economic solutions of price incentives to correct market failure with regard to externality problems have not worked either (Zilberman and Millock, 1997). However, over the past decade, certain farmers and other actors in the food supply chain have started a series of separate initiatives to respond to consumer concerns and demand, as well as to promote the economic and ecological survival of their enterprises. The growing influence of retail and food companies in the fields of sustainable agriculture and pesticide reduction is notable and reflects the vertical integration taking place in the global food chain and concentration of food sourcing, processing and sale by multinational companies and supermarkets (Reardon and Barrett, 2000; Thrupp, 2002).

We describe five case studies of integrated pest and crop management production for a range of field crops in Europe and North America. Cases were selected which would emphasize different elements of collaborative working, pesticide reduction approaches, research and systems development, training and advice for farmers, marketing, and leadership by farmers, companies or partnerships.

IPM Production and Marketing of Apples and Pears in Belgium

The Wallonia Group of Fruitgrowers (GAWI) was created in 1988 and now covers 45 growers cultivating 800 hectares of apples and pears in the Walloon region of Belgium (Denis, 2003). GAWI trains members in integrated production (IP), and helps to develop and validate environmental protection methods under their FRUITNET guidelines. The FRUITNET certification mark was established in 1996 and currently involves around 90 growers throughout Belgium, who annually produce 50,000 tonnes of apples and pears, some 10 per cent of Belgian production. Around 20 producers in northern France have become associated with GAWI, which also collaborates with IP fruit growers in Switzerland.

GAWI has promoted and adapted IP methods based largely on the work by the International Organisation for Biological Control. In 1996, Belgium was the first European country to set an official protocol for apple and pear integrated production and regulation. But FRUITNET protocols exceed these requirements, mainly with additional environmental measures. GAWI first tested IP methods on pilot orchards in the early 1990s, before gradually introducing these to members and incorporating them into the current protocols. These include obligatory ecological methods to diversify orchard flora and fauna, including the installation of nest boxes and perches for insectivorous birds, and one metre wide herbaceous borders along at least one side of each plot. Optional methods include hibernation cages for beneficial insects such as lacewings and earwigs, and the cultivation of indigenous flower species. Growers must also establish multi-species hedges as a refuge for beneficial insects and other wildlife.

Pest management focuses on eliminating broad-spectrum insecticides, especially pyrethroids, carbamates and organophosphates, which kill important beneficial insects. Scouting, sampling and pheromone trapping ensure careful timing of pesticide applications. However, the elimination of some pesticides poses problems for certain secondary pests that were previously controlled under conventional systems and GAWI has developed a colour-coded list of approved and restricted pesticides.

Green list products are selective and can be used when justified, while yellow list ones may only be used under strictly defined conditions. Orange list products may only be applied with permission from the certification body. For weed control, use of simazine and diuron herbicides is banned for FRUITNET growers, while other herbicides must be confined to within 75 cm of tree trunks with a maximum of three annual applications. Mechanical and thermal weed control is encouraged as far as possible. Reducing fungicide use remains a major challenge, particularly for apple scab disease and during the post-harvest period. GAWI encourages members to use regular disease warning bulletins and on-farm observation to ensure applications are closely related to infection risk. Orchard density may not exceed 3000 trees per hectare, in order to avoid the use of growth regulators, prevent pest development and favour the use of non-synthetic chemical weed control techniques.

GAWI promotes the marketing of FRUITNET produce as a means of adding value for growers and a guarantee of quality and origin for consumers. In 1996 Delhaize supermarket chain agreed to buy only FRUITNET produce for its Belgian apples and pears, and more recently the FRUITNET protocol was recognized by the EUREP-GAP retail initiative at European level. Traditional varieties of apple and pears, which are very difficult to grow profitably under current market requirements, are sold via direct farm sales and via FRUITNET-certified juices. GAWI produces information, educational and recipe leaflets for the public and children (GAWI, 1998).

Pesticide-Free Production (PFP) in Canadian Arable Systems

Increasingly simplified farming systems on the Canadian prairies have resulted in a growing dependence on pesticides, especially herbicides. While farmers are keen to reduce pesticide costs, current systems are not sufficiently robust to cope with pesticide elimination without high yield penalties and increased pest pressure. The PFP initiative was set up in 1999 by a group of farmers, researchers and extension workers in Manitoba province because of low commodity prices, rising input costs and a desire to reduce pesticide use (Nazarko et al, 2002). The aim was to produce a crop without the use of pesticides, one year at a time, in order to avoid some of the adoption problems suffered by other reduced-use projects. The 50 PFP farmers agreed to avoid seed treatments and all in-crop pesticide use within a season, but they were permitted to use a pre-seeding burn-off with a non-residual herbicide, such as glyphosate.

The system is flexible – if for any reason it becomes unavoidable to apply a pesticide, the crop can be marketed conventionally and the farmer can try for PFP status again the following season. Since 2002, the PFP Farmers Co-operative has begun to market PFP crops, following market research showing that consumers would be willing to pay up to 10 per cent more for food products grown without pesticides. Those crops currently grown under PFP principles include rye, barley, oats, flax, wheat and sunflower and other oilseeds. Crop rotation is extremely important for reduced-pesticide systems, especially to reduce weed pressure. For example, PFP canola (oilseed rape) grown in a canola–oat–wheat rotation yielded the same as canola sprayed under a conventional regime and rotated only with wheat. Many of the most successful farmers in the PFP programme now grow a wider range of crops than other farmers and commonly run livestock operations too. Better weed control is also achieved by banding rather than broadcasting fertilizer, high seeding rates and shallow seeding. Use of resistant cultivars is the best option for reducing fungal attack (Macfarlane, 2003).

The main motivation for growers to adopt PFP was to reduce the reliance on pesticides and to network with like-minded farmers. PFP farmers tend to be younger and better educated than the average and with an interest in alternative farming. A third of those participating in 2000–2001 were in conversion to organic and around 20 per cent were zero-tillage farms. Most PFP farmers found that they gained financially by growing PFP, mainly due to savings on input costs. Yield penalties were minor, with average yields for PFP crops recorded as 93 per cent of the ten-year average for Manitoba. The challenge now is to build on the economic and ecological success of PFP and persuade more Canadian farmers of the benefits of changing their systems.

Healthy Grown Potatoes, Wisconsin, US

This initiative started in 1996 as a collaboration between World Wide Fund for Nature (WWF), the Wisconsin Potato and Vegetable Growers Association (WPVGA) and the University of Wisconsin (UW) to promote the development and industry-wide adoption of bio-intensive IPM practices and to respond to and increase consumer demand for environmentally responsible produce (Protected Harvest, 2001). In the first three years, the collaboration focused on industry-wide change by setting one, three and five year goals for pesticide risk reduction and five and ten year goals for 'bioIPM' adoption. In 2000, the collaboration began work on a set of eco-potato standards and, in 2001, a not-for-profit organization, Protected Harvest, was established to certify growers and approve standards. By 2003, some 3400 ha were certified as meeting Protected Harvest standards for reduced pesticide risk and improved IPM (WPGVA, 2003).

From the outset, the importance of credible methods to monitor growers' progress in meeting pesticide reduction and IPM adoptions targets was emphasized. This focus resulted in the development of a multi-attribute pesticide risk measure, a 'toxicity index', that calculated 'toxicity units' for all pesticides used in potato production in Wisconsin. The rating compares active ingredients on a weight basis, taking into account acute and chronic mammalian toxicity, ecotoxicity to fish, birds and aquatic invertebrates, the impact on beneficial organisms, and resistance management. One success of the programme is reflected in the reduced use of toxic products by Wisconsin potato producers by some 250,000 kg between 1997 and 2000 (BNI, 2001). To qualify for certification, growers must achieve a minimum number of points under production and toxicity standards. Certain practices are obligatory while others are designated as bonus points, usually those practices that are cutting edge and not used by all farmers.

The eco-label standards are set so that growers minimize the amount of high-risk pesticides applied in a given year. To qualify, growers must eliminate the use of 12 specific pesticides, and cannot exceed 800 toxicity units for short-season and 1200 units for long-season potatoes. Restrictions are placed on the use of certain high-risk pesticides that are needed for resistance management and/or because no lower-risk alternatives are available. Toxicity ratings are adjusted for late blight disease pressure, the most serious challenge for pesticide use reduction. Protected Harvest recognizes the economic threat posed by late blight and has specific provisions for controlled fungicide use only in cases of infestation.

WPVGA created the brand Healthy Grown to market its Protected Harvest certified potatoes, which are sold in over 70 stores within Wisconsin, and in 14 other states. Protected Harvest/Healthy Grown potato bags also carry the WWF panda logo, in an effort to give market place recognition to a credible, third party certified eco-label. Marketing is backed up by a range of information for store managers and consumers. Protected Harvest differs from many other eco-labels in the transparent process by which standards are developed, enforced and modified in cooperation with growers, scientists and environmentalists. Healthy Grown potatoes are priced between conventional and organic retail levels, to give growers a fair return for high quality produce in a healthy environment. Researchers are now focusing on alternatives to toxic soil fumigants, with the goal of a 50 per cent reduction by 2007, and are collaborating with the International Crane Foundation to develop farm management plans for conserving important prairie, wetland and forest habitats.

Vining Peas in England

The Unilever agrifood company began its Sustainable Agriculture Programme in 1998 to define and adopt sustainable practices in its supply chain. The company's motivation was to secure continued supply of its raw materials, and to gain a competitive advantage for its products by responding to consumer demand for sustainable production practices. The programme initially selected five of its crops (peas, spinach, tea, palm oil and tomatoes) and developed sustainability indicators to measure changes in soil health, soil loss, nutrients, pest management, biodiversity, product value, energy, water, social and human capital, and the local economy (Unilever, 2003a).

Unilever annually produces 35,000 tonnes of peas for frozen produce sold under the Birds Eye brand. Vining peas are grown in eastern England and the programme began by working with nine farms and its own research farm to develop and implement integrated farm management, including IPM practices (Unilever, 2003b). The quantity of pesticides applied per hectare had already decreased by a half over the previous 20 years, during which time many specific products had been eliminated from use. Unilever developed its own pesticide profiling system to draw up a list of preferred pesticides, based on efficacy, human and environmental hazards, residue risk and consumer perception. Vining pea IPM now includes the use of disease-resistant varieties, routine cultural and physical control, field life history, action thresholds and monitoring for key pests, regular field observation, use of broad-spectrum insecticides only when no alternative is available, and seed treatments to avoid the need for foliar fungicides. Weed control is achieved via careful crop rotation, good seed-bed preparation, manual removal of contaminant weeds, reduced rate herbicide application and careful choice of product.

The permitted list of pesticides is reviewed annually and all 500 of Unilever's growers must keep accurate records of their pesticide use. Company agronomists and farm managers are expected to be able to justify the use of each pesticide and all applications must be made by suitably qualified operators, with spray equipment properly maintained and calibrated. Detailed control recommenda-tions are now available for all key invertebrate pests, disease and weeds and in many cases, the recommendation is not to treat the problem at all. Unilever promotes research on improved pea varieties to reduce the need for pesticide inputs, biocontrol agents for all key pests and insect pheromones for pea midge control, weed mapping for a more efficient targeting of herbicides, reduced rate pesticide application and options for mechanical weeding. Unilever works closely with the International Biocontrol Manufacturers Association to look for alternatives to synthetic pesticides to use in peas. It also studies pesticide leaching at catchment level to better understand their impact. Loss of biodivers-ity is a key concern for Unilever and it collaborates with a range of wildlife organizations to monitor and improve biodiversity in and around its pea fields.

Unilever supports its growers to group together to share information and take an active part in research and development, with meetings, training events and close contact with company agronomists. In 2003, it initiated the Farmers' Forum for Sustainable Farming, a farmer-led organization with representation from all growing areas and farm types as well as external stakeholder representation from Forum for the Future, a leading NGO promoting sustainability. The Forum aims to stimulate more grower decision-making in the pea business and works to improve relationships between farmers, consumers, local communities and policy makers. The Forum has been active in helping to develop and communi-cate a set of Sustainable Agriculture Standards for peas, including standards on pest management and biodiversity, which is a contractual requirement for all Unilever pea growers from 2004.

Arable Crops and Field Vegetables in the UK

The UK Co-operative Group works with experts worldwide to develop crop and pest advisory sheets with information and practical guidance on IPM and preventative measures, concentrating on biological, cultural and physical controls as the first choice, and also providing growers with more information on the hazards of particular pesticides so that they can make better informed decisions. The Co-operative Group's farming business, *farmcare*, is a wholly owned subsidiary that farms Co-operative Group land and contract manages farms on behalf of other landowners. *Farmcare* points out that many farmers are unaware of the issues surrounding the synthetic pesticides that public interest groups would like to see banned and supports the Co-op's designation of 'red' and 'amber' products as a simple way to alert farmers to any dangers (Gardner, 2002). Advisory sheets are available for carrots, potato, cauliflower, mushrooms, and for avocado and pineapple from overseas suppliers.

The advisory sheets give growers information on first preventing a particular problem from occurring, managing it via cultural, biological or mechanical methods as second choice, and finally, synthetic chemical control as a third choice. The sheets also give basic information on environmental and human health hazards and persistence, and other factors to consider in decision-making.

The Co-operative Group has supported research into integrated farm management practices since 1993 on one of the *farmcare* arable farms. A ten-year assessment of its Probe programme (Profit, Biodiversity and the Environment) found that integrated farm management methods are comparable to conventional methods in terms of profitability (Jordan et al, 2000). Crop protection costs were 30 per cent lower than under conventional practice and the volume of active ingredients almost halved (*farmcare*, 2003). In 2002, wheat was grown successfully without any use of foliar insecticides, slug pellets or plant growth regulators. The significant reduction in pesticide use over the ten years was achieved mainly by crop rotations, resistant varieties, thresholds and diagnostics for improved decision-making and some tolerance of certain pests, use of stale seedbed technique and careful targeting of nitrogen to reduce disease pressure. However, more management time and in-field observation are needed to allow input reduction to work.

GOVERNMENT PROGRAMMES FOR PESTICIDE REDUCTION

In Europe, state commitment to pesticide reduction programmes has been strongest in the Scandinavian countries, with a high political priority given to environmental protection and health and consumer safety. Switzerland, and to a lesser extent the Netherlands, have also been active in seeking to decrease pesticide dependency. But there is controversy over terminology. The term 'pesticide use reduction' is advocated by only certain governments and interest groups, as it explicitly means decreasing the volume of pesticides used and therefore also sold by agrochemical companies. The agrochemical industry, many farmer associations, governments and others argue that what is needed is a reduction in the risks of causing adverse effects on human health and the environment.

Denmark is one of the few countries with a proactive programme for pesticide use reduction. It was successful in reducing use by over 50 per cent by 1997, from a baseline of average use in 1981–1985. However, this was achieved mainly by the adoption of newer pesticides, notably pyrethroid insecticides and sulfonylurea herbicides (Jensen and Petersen, 2001). The problem was that a reduction in volume was not accompanied by a reduction in application frequency, nor necessarily of risks from particular hazardous products. Consequently, the Danish Minister for the Environment appointed a group of experts in 1997, known as the Bichel Committee, to assess scenarios for reducing and eliminating pesticide consumption over a ten-year period. The committee recommended that the treatment frequency index (TFI) could be cut from 2.5 in 1997 to between 1.4 and 1.7 over a decade without any serious loss to farmers or the economy. Reduction in TFI formed the basis for the subsequent Pesticide Action Plan, aiming to reduce TFI to 2.00 by 2002.

The Danish Plan II included a range of activities such as the establishment of demonstration farms and information groups, more use of decision support and warning systems for pests and diseases, extra training for farmers and advisers, obligatory record-keeping by farmers, buffer zones along water courses, and increased support for organic production (Nielsen, 2002). A pesticide tax since 1996 set at 54 per cent of wholesale price for insecticides and 33 per cent for herbicides and fungicides has sent economic signals to farmers, yet the guideline figures from the Bichel report and Plan II were instrumental in convincing many farmers that they were over-using pesticides, including the fact that some farmers already used far less than the norm without any reduction in yield. The positive engagement of farmer and pesticide distributors has been key to the success of the plan, showing that reduction only starts once farmers are themselves motivated.

By 2000, Plan II TFI targets had already been achieved and the latest plan, developed in 2003, aims to reduce TFI from the 2002 level of 2.04 to 1.7 by 2009. The Danish Ecological Council is urging this target to be achieved by 2005, and to 1.4 in 2008, based on studies which show that TFI could be reduced to 1.4 without changes in crops and without special costs (H Nielsen, pers.comm.). This progress has been made not with a specific IPM programme, but by using crop-specific guidelines and clear targets. Farmers agree individual action plans with their advisers and can calculate their TFI via internet-based programmes. Exchange of experiences in Experience Groups has been an effective way of sharing information on using reduced dosage. One project had led to the formation of 95 of these groups, each with five to eight farmers. The groups meet in the field several times each season to discuss topics such as herbicide selection and dosage and mechanical control options (Jensen and Petersen, 2001).

Norway's two Action Programmes in the 1990s did not have specific targets for pesticide use reduction but aimed for as much as justifiable through a suite of measures similar to those in Denmark. The new plan for 1998–2002 sought to reduce risks for negative health impacts and environmental contamination by 25 per cent over the period (Sæthre et al, 1999). New IPM initiatives include setting national IPM guidelines for each crop, mandatory education in IPM in order to obtain a pesticide operator certificate and IPM training for extension staff and others. All farmers must now keep pesticide application records. In

Sweden, consistent regulation over 15 years has proven that government policies can reduce usage as well as risks. Based on a set of indicators, the environmental risks were reduced by 63 per cent in 2000 and health risks by 77 per cent, compared with the reference period 1981–1985 (Ekström and Bergkvist, 2001).

The Dutch government, under strong pressure from environmental organizations, has been working for many years to tackle its high intensity of pesticide use, with a particular emphasis on water contamination. It has banned several problematic pesticides still used by its European neighbours. The Dutch have devised a practical 'yardstick' for assessing the environmental impacts of specific active ingredients, which serves as a decision support tool for farmers and crop advisers, as well as a monitoring tool to evaluate progress (Reus and Leendertse, 2000). The latest government agreement with the farming sector and NGOs is on measures to reduce harm to the environment caused by pesticides by at least 95 per cent by 2010 from the 1998 baseline (Jehae, 2003). Key objectives will be alternative ways to combat potato blight disease, which accounts for 20 per cent of all pesticides used, and to persuade the 20 per cent of farmers who make up 80 per cent of national pesticide consumption to adopt good agricultural practice as a minimum requirement.

One particular regulatory barrier to pesticide reduction is the cost and difficulty of registering non-chemical pest control products, such as microbial biopesticides, insect pheromones and botanically based products, including neem seed and garlic sprays. The EU's Pesticides Authorisation Directive 91/414 severely limits the range of non-chemical pest control methods available to European farmers by insisting on the same extensive registration data requirements as those for approving pesticides. In contrast, the US and Canada adopt a more pragmatic approach for the registration and fast-tracking of less-toxic options. This means that a North American company can register a product in 6–9 months at a cost of around £27–34,000, compared with over £300,000 in Europe (Chandler, 2003). In the UK, for example, company attempts to register garlic granules and spray for the control of cabbage root fly have been stalled for two years, with the Pesticides Safety Directorate requesting more efficacy data. There is now a real lack of control options for this pest as most of the organophosphates have been phased out and approval for chlorfenvinphos, the remaining active ingredient, expires in 2003. With over 320 active ingredients being withdrawn from the European market in 2003, the challenge is likely to grow.

WAYS FORWARD

The cases in this chapter point to some common themes and principles for engaging with farmers to reduce pesticide use, and build more sustainable pest management systems. The Belgian fruit growers project is the most notably farmer-led, and is also strong in farmer participatory research, combined with innovative marketing of labelled fruit via supermarkets and direct sales, with considerable effort put into consumer awareness raising. Canada's Pesticide Free Production is probably the most conventional in that it is public sector-led, but in a short time it has achieved considerable interest from farmers, designed new production systems, tested and adapted with farmers, and is branching out into specialized marketing. Healthy Grown potatoes are the most advanced in terms of own branding and eco-labels, backed by a strong partnership between growers, environmentalists and formal research. The Protected Harvest concept sets growers specific impact reduction targets, as well as listing prohibited and restricted active ingredients. In the UK, two company initiatives concentrate more on lists and production protocols, with Unilever using risk reduction goals and sustainability indicators and the Co-op utilizing hazard levels and use reduction . Both companies' staff work closely with their growers to encourage and support proactive changes in management practice, rather than merely policing compliance.

Key players in the agrifood and retailing business have recently launched joint initiatives to boost consumer confidence in the safety of food products, as well as to maintain their competitive edge. EUREPGAP, the initiative on Good Agricultural Practice started in 1997 by a consortium of Europe's leading retailers, has been hailed by its supermarket and supplier members as one of the greatest recent success stories in the retail industry (Möller, 2000). EUREPGAP encourages ICM schemes for fresh produce within Europe and worldwide by benchmarking schemes, standards and traceability, and setting up a single recognized framework for independent verification. By 2003, over 200 companies and organizations had signed up, with more than 10,000 growers certified EUREPGAP compliant in 32 countries (EUREPGAP, 2003a). EUREPGAP protocols act as normative documents for the standards to which growers must adhere. Pesticide selection and justification, handling, application, storage and disposal are all covered, as well as the goals of minimizing pesticide use by employing preventative non-chemical means as far as possible (EUREPGAP, 2003b).

Debate around the effectiveness of having specific eco-labels and price premiums for food produced under reduced pesticide, crop assurance and other schemes will continue. US consumer surveys have shown that a willingness to pay 10–25 per cent more for IPM produce is closely related to consumer knowledge about IPM, and that more knowledge tends to be associated with higher income and female consumers, while public education at pesticide problems and alternative choice needs to be targeted at men, low income households and those raising children (Govindasamy et al, 1998). GAWI fruit, Protected Harvest potato and PFP combinable crops all employ eco-labels or marketing methods referring to their reduced use of pesticides. The Co-op doubts that IPM labeling is a viable selling point in the UK, due to poor consumer awareness, although some of its sister organizations elsewhere in Europe do use eco-labels. What is clear is that consumer confidence in brands and eco-labels relies on clarity and transparency.

Another factor which could play an increasingly important role in food production and pest management choices is the rapidly expanding sector of local food systems, which link consumers directly with farmers through farmers' markets, box schemes and community-supported agriculture programmes. Many of these produce organic or IPM produce. Amongst our cases studies, GAWI fruit has done most to link directly with consumers, although it also sells through one of Belgium's biggest supermarkets. The PFP Farmers' Cooperative is expanding in Manitoba and Alberta provinces in Canada, while Unilever places an emphasis on social capital under its sustainability programme, via new dialogue between pea growers and local communities in East Anglia.

To what extent might local food system and farmer association schemes be better at delivering a radical shift from pesticide reliance than the corporate agrifood sector? Can local level, bottom-up approaches ever hope to have as much impact on agricultural production on a large scale as recent moves by the multinationals, such as EUREPGAP and the Sustainable Agriculture Initiative (SAI) set up in 2002 by Unilever, Nestlé and Danone and now expanded to ten major food companies (SAI, 2003)? Like EUREPGAP, the SAI is supported by key agrochemical companies and it is hard to envisage how these programmes could result in a substantial challenge to the prevailing crop protection paradigm. The Co-operative Group in the UK decided not to join EUREPGAP and questions whether such assurance schemes can deliver more value for farmers, rather than merely 'policing' production (Barker, 2003).

Purchasing decisions and produce selection by supermarkets can act as a powerful brake on promoting pesticide reduction, as can EU decision-makers who set quality and appearance standards for food regulation. A significant amount of pesticide use takes place to guarantee the cosmetic appearance of fresh produce and farmers are often unwilling to run the risk of having their high cost produce rejected (Gardner, 2002). A study of apple production systems in Germany found little pesticide reduction in IPM, compared with conventional methods, notably in fungicide use for scab disease. One local variety with high scab resistance, a good yield, taste and appearance was not available in super-markets as it did not fit the image of red, shiny skin. In such cases, consumer and supermarket choices must change considerably to attain pesticide reduction goals (PAN Germany, 2001).

In reality, a flexible combination of better regulatory controls and public incentives for reducing pesticide dependency, along with significant investment in independent advice and training for farmers and participatory research, supported by progressive agricultural and food production policies and pro-grammes in public and private sectors, will advance the cause of sustainable pest management in industrialized agricultural systems. More active decision-making and participation opportunities by farmer groups and consumers in demanding and developing profitable, equitable and ecologically sound farming strategies is just as necessary, and regional governments and communities are beginning to support such action, such as by funding nature-friendly food production. The case studies in this chapter offer lessons on different ways to change pesticide practice in a market context. How replicable these programmes are will best be judged by whether they expand and evolve successfully over the next decade.

Chapter 15

Policies and Trends

Harry van der Wulp and Jules Pretty

INTRODUCTION

This book has shown that problems associated with pesticide use are still widespread and continue to carry a high cost to users and society. In many situations, a reliance on harmful pesticides as the primary mode for crop protection is still high, while available alternatives remain under-used. This book has also shown that there are several important developments that are converging towards more sustainable crop production and a clean-up of agricultural pest management. Many governments, both in industrialized and developing countries, are tightening policies relevant to pest and pesticide management.

There is a growing consumer demand for safe and wholesome food. The food sector is responding with more critical attention being paid to the crop production practices of their suppliers. There is also a growing awareness of the social and economic consequences of unsustainable food production. In many parts of the world, farmers are learning that a heavy reliance on pesticides is not necessary to maintain or increase production, and that each unnecessary pesticide application represents a health risk and a loss of income. There is a broadening availability and acceptance of alternative pest management approaches and products.

The many examples presented in this book show that pesticide use can be reduced in many different farming systems. Where pesticide use remains justified, a better selection of plant protection products can often reduce the adverse effects on human health and the environment. There is enormous potential to further detoxify agriculture. In this final chapter, we review the policy and market trends that are driving the process of change.

POLICY RESPONSES TO PESTICIDE ISSUES

In many farming systems, pesticides play a role in supporting current levels of production. At the same time, they are often over-used or otherwise misused, and have many undesirable side-effects. The role of government is to find a responsible balance between enabling judicious pesticide use where such use is necessary to achieve desirable crop production levels, and reducing the adverse health, environmental and agronomic risks. To determine such a balance, it is important to know when the use of a pesticide is necessary and what risks are involved. This balance has often been tilted in favour of pesticide use due to overestimation of the benefits of pesticide use and underestimation of the risks. Ignorance about crop-pest ecology and an absence of good information about alternative pest management approaches and products have been key factors contributing to overestimations of the benefits of pesticide use. This has been reinforced by pesticide industry marketing and their promotion of pesticide use. Under-estimation of the risks and negative impacts of pesticide use, particularly in developing countries, has often been a reflection of limited or biased research into this area. The pesticide industry has contributed to such underestimation through 'safe use' campaigns, which promote the perception that training will reduce health problems associated with the use of highly hazardous pesticides. Research described in Chapters 4 and 10 has shown that this perception is often incorrect and misleading.

Governments have a range of policy instruments to influence this balance. Pesticide legislation offers possibilities for regulating the availability and use of pesticides. The use of dangerous products can be banned or restricted to certain crops, users or circumstances. Important additional factors include the choices governments make for budget allocations on the enforcement of pesticide legislation, for monitoring of pesticide residues in food and drinking water, and for research into the side-effects of pesticide use. In addition, the choices related to the allocation of budget components for agricultural research and extension determine the nature and effectiveness of extension messages and can make a significant difference. Finally, there are financial instruments to provide incentives or disincentives for certain practices in crop production. Examples include pesticide subsidies, taxes or import tariffs, but also financial incentives for the development and use of alternative pest management approaches and products, and support for the local manufacture of such products.

The manner in which governments make choices is not only influenced by available information and budgets, but also by national and international pressures. National concerns about health, environment, food safety and food security are important factors that are further influenced by consumer pressure, media attention and lobbying of civil society groups or specific interest groups. External pressures range from obligations under international conventions to the specific production requirements of export markets. For developing countries, requirements of development assistance agencies and banks related to pest management practices may also play a role.

NATIONAL POLICIES

National policies on pest and pesticide management evolve in response to growing knowledge about the adverse side-effects of pesticides, and the availability of economically viable alternative approaches and products. Clearly, pesticide use and pesticide management practices are affected by pesticide legislation.

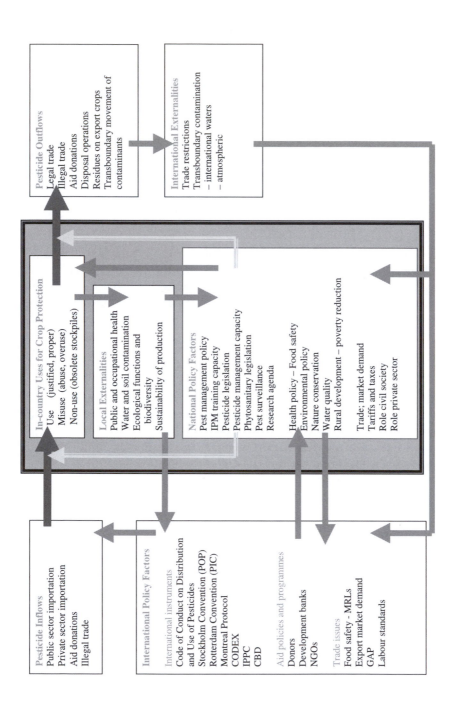

Figure 15.1 *IPM policy framework, as developed by the Global IPM Facility, depicting interactions between pesticide use, side-effects of pesticide use, and national and international policy responses*

In turn, pesticide legislation is shaped by broader policies on public health and environmental protection. Public health policies may address pesticide residues in food and drinking water, and risks associated with the storage, transport and disposal of pesticides. Occupational health policies and labour standards may restrict the use of pesticides hazardous to farmers and plantation workers. Environmental policies on water quality, nature conservation and biodiversity tend to affect the selection and use of pesticides. International regulations and the requirements of export markets also play an important role.

By the end of the 1990s, most industrialized and developing countries had established systems to regulate the availability and use of pesticides. Regulatory systems in Organisation for Economic Co-operation and Development (OECD) countries are continuously being improved as knowledge increases about the side-effects of pesticides and their costs. The European Union (EU) is overhauling its list of registered pesticides, which is leading to hundreds of products being removed from the market. The US is re-evaluating a range of commonly used products, taking into account cumulative effects and impacts on vulnerable groups, such as children. Several notorious products have been further restricted and acceptable daily intake levels are being revised downwards. OECD countries are jointly working on better risk assessment criteria, and some Scandinavian countries have introduced comparative risk assessment systems to reduce yet further the risks of registered pesticides.

Most developing countries are still facing many constraints to the effective enforcement of their regulatory systems. Available financial and human resources for the control of pesticides in developing countries are very small compared with those in industrialized countries. A group of Sahelian countries successfully addressed some of these constraints by pooling their resources in a common pesticide registration scheme. But few developing countries have the primary health care and occupational health systems necessary to monitor, detect and treat pesticide poisoning, nor the agricultural training and extension services to ensure proper advice on pest management and pesticide application. Pesticide residue levels in food and the environment can be high, but a lack of pollution monitoring and good published data often reduces the sense of urgency.

Despite these constraints, many things are changing. The number of countries phasing out extremely and highly hazardous pesticides is steadily growing. Pesticides that are classified as persistent organic pollutants under the Stockholm Convention are now hardly used in agriculture. The monitoring of pesticide residues on produce for domestic markets is gradually becoming a more common practice. And the tightening of export market requirements related to food safety concerns is having a distinct impact on countries that export agricultural produce, particularly those that export fruit and vegetables.

However, in most developing countries, farmer knowledge about pesticides and available alternatives is generally still remarkably limited. Short-term cost considerations remain an important factor in poor farmers' choices of pesticides. Cheap products that present a high risk to users, the public or the environment often continue to be used when less hazardous alternatives are more expensive. Direct savings from such choices are often offset by health costs to the user, or less visible costs to public health or the environment. Training to demonstrate a potential to save money by cutting back pesticide use without loss of production or income, and training to demonstrate the direct relationship between health costs and pesticide use, are therefore the most successful avenues at the farmers' level to achieve change in pest management practices. At national policy level, a combination is required of promoting alternatives and banning, restricting or taxing the use of hazardous products that continue to cause farmer poisoning and involve high health and environmental costs to the society.

This is gradually happening. More and more countries, both industrialized and developing, are phasing out extremely and highly hazardous pesticides, while encouraging less hazardous pest management approaches and products. Integrated pest management (IPM) and biological control (BC) programmes are increasingly recognized and promoted as viable alternatives. Several countries have announced ambitious pesticide reduction programmes, while others have set IPM targets or declared IPM as the preferred approach to pest management.

International Instruments and Policies

A broad range of international instruments has been developed in response to the international health and environmental concerns about pesticide use. Through the ratification of international conventions, governments accept obligations to incorporate the objectives of international policies into their national policies. Examples of instruments directly relevant to the use of pesticides are the Rotterdam Convention on the Prior Informed Consent (PIC) Procedure for Certain Hazardous Chemicals and Pesticides in International Trade, and the Stockholm Convention on Persistent Organic Pollutants, which both became legally binding in 2004, and the Montreal Protocol on the phasing out of methyl bromide. Illegal trade in undesirable pesticides, however, remains a problem for many countries.

Another important instrument is the UN Food and Agriculture Organization (FAO) International Code of Conduct on the Distribution and Use of Pesticides, which sets voluntary standards for the management of pesticides. For many countries, the Code of Conduct has provided guidance for the development of pesticide legislation, and the major pesticide companies have agreed to abide by the Code of Conduct. The Code of Conduct was agreed upon by FAO member states in 1985 and since then has been updated twice. The second update was in 2002 and reflects the growing role of IPM as an alternative to reliance on pesticide use. The Code of Conduct and these other instruments are detailed in Chapters 1 and 4.

Other important international policies that affected pest management are the acceptance of the Polluter Pays Principle by the OECD, Agenda 21 being established at the United Nations Conference on Environment and Development (the 1992 Earth Summit in Rio), and more recently the outcome of the 2002 World Summit on Sustainable Development (WSSD) and recommendations of the Intergovernmental Forum on Chemical Safety (IFCS).

Development Assistance Agency Policies

Most development assistance agencies and lending institutions, such as the World Bank, have pest management policies or environmental policies that influence their pest management assistance to developing countries (Box 15.1). Generally, such policies now promote IPM and involve criteria for the selection of pesticides that tend to exclude WHO Hazard Class I pesticides and restrict the use of WHO Hazard Class II pesticides. In addition, many agencies have supported programmes to develop and promote IPM, and to strengthen capacity for the control of pesticides.

Box 15.1 *World Bank Safeguard Policy on Pest Management*

The World Bank Safeguard Policy on Pest Management (OP 4.09) applies to all World Bank financed projects that involve procurement of pesticides or indirectly increase pesticide use.
  The objectives of the safeguard policy are to:

- Promote the use of biological or environmental control methods and reduce reliance on synthetic chemical pesticides. In Bank-financed agricultural operations, pest populations are normally controlled through IPM approaches. In Bank-financed public health projects, the Bank supports controlling pests primarily through environmental methods.
- Ensure that health and environmental hazards associated with pesticides are minimized. The procurement of pesticides in a Bank-financed project is contingent on an assessment of the nature and degree of associated risk, taking into account the proposed use and the intended user.
- As necessary, strengthen capacity of the country's regulatory framework and institutions to promote and support safe, effective and environmentally sound pest management.

Both an internal World Bank survey (2003) and a survey by the Pesticide Action Network (2002) found that compliance was still weak and that there are several conflicting interests within the Bank. The Bank's safeguard policy unit, however, is working to improve compliance.

Source: Sorby et al, 2003; Tozun, 2001

IPM, as promoted by FAO, is in most cases linked to the empowerment of farming communities, who gain better control over crop production through knowledge, insight and skills. Agroecological knowledge gives farmers the confidence not to select the sprayer if something unexpected happens in their fields. Instead, they analyse the situation, taking into account pest dynamics, natural control processes, available alternatives and the costs/benefits of different pest management options. Resultant savings on pesticides and, often, higher yields, generally lead to higher profits. Development assistance agencies are therefore increasingly placing IPM in a broader context of poverty reduction, improved livelihoods, community empowerment and environmental protection.

A critical issue is the question of obsolete pesticides. Many developing countries and countries with economies in transition have large stockpiles of obsolete pesticides, often scattered over numerous sites. These are old pesticide stocks that can no longer be used and are now classified as hazardous waste. Often such stocks are in a deplorable state and are a hazard to human health and the environment. The removal and destruction of such stockpiles is difficult and expensive. Stockpiles from African countries, for instance, need to be repackaged and shipped to Europe for destruction in dedicated hazardous waste incinerators, as there are no appropriate hazardous waste incineration facilities in Africa.

A number of countries have obsolete stockpiles in excess of 1000 tonnes. A large proportion of obsolete pesticides in Africa results from ill-planned pesticide donations. This problem has received much attention during the last decade and resulted in pressure on development assistance agencies to be more careful with pesticide donations and to contribute to the costs of removing the leftovers of past donations. Likewise, the pesticide industry has come under pressure from public agencies and NGOs to contribute financially to the removal and destruction of old stockpiles because of supplier roles in past donations.

MARKET RESPONSES

The late 1990s saw an interesting development in which some food companies started to overtake the slow process of government pesticide policy reforms. Under pressure from consumers, groups of companies started to impose pesticide use restrictions on their suppliers that went further than existing governmental regulatory requirements to ensure food safety. The British Co-operative Group, for instance, banned certain pesticides from the production of fresh and frozen fruits and vegetables sold through its chain. These included pesticides that were a source of public concern, but had not yet fully passed the review process that eventually was expected to result in regulatory measures. Other companies that obtain their produce from developing countries imposed selection of pesticides or crop production protocols on their suppliers to avoid residue issues or to support a desirable image of sustainable production.

In some cases, food companies are actively engaged in providing assistance to producers in developing countries to develop and introduce more sustainable production methods. There are several such examples from the tomato paste, fruit, chocolate, coffee, tea and oil palm industries. In these cases, the dependency of northern food companies on agricultural commodities from developing countries has provided incentives to develop long-term relationships with producers around the mutual interest of the sustainable production of safe agricultural produce.

Many companies in the food industry now actively pursue internal policies to reduce food safety risks through supply chain management. They demand from their suppliers that certain production protocols are followed, which, among other things, involve more sustainable pest management and more responsible pesticide management. The EUREPGAP initiative of a number of European supermarket chains and their suppliers is an example of this approach, but it has also raised concerns about the access of small producers in developing countries and the advisory role assigned to pesticide companies. A similar initiative is the Sustainable Agriculture Initiative (SAI) of a group of food processing companies, which promotes the sustainable use of production resources to safeguard their long-term economic availability.

A further response to consumer demand is the broad variety of green labels that have been launched, not only in OECD countries, but also in countries such as China and Thailand. There remains, however, debate around the effectiveness of having specific eco-labels and price premiums for food produced under IPM schemes. Clarity and transparency has been found to be important in enhancing consumer confidence in brands and eco-labels. The establishment of certification schemes for such labels continues to provide many challenges.

During the 1990s, the pesticide industry has gone through a process of consolidation that reduced and concentrated the number of companies through a series of mergers. The newly formed companies often rationalized their combined product portfolios, taking into consideration the sharpening of health and environmental criteria for pesticide registration. NGOs have become more successful in exercising pressure on pesticide companies, helped by the general desire of large companies to be recognized as part of social and environmentally responsible leagues, such as a listing in the Footsie4Good, a stock-market index for companies with recognized social and environmentally responsible track records, or the UN's Global Compact. In addition, there have been several Class Action Law suits against pesticide companies that fuelled the debate about liabilities and accountability.

Newly developed products tend to be less hazardous. Companies with significant interests in WHO Hazard Class I products, however, often remain reluctant to withdraw these from the market. Industry-organized 'safe use' programmes play a role in defending the continued use of such products. In the US, several controversial products have been voluntarily withdrawn from the market ahead of re-evaluation against tighter environmental and health criteria. Voluntary withdrawal based on economic or marketing reasons precludes the entry of these products into the Prior Informed Consent procedure under the Rotterdam Convention. This makes it easier for the companies concerned to continue to sell these products in developing countries. There appears to remain a gap between public statements of the pesticide industry and actual field practices, particularly related to the marketing of pesticides in developing countries.

At the same time, there has been an emergence of many small- and medium-scale enterprises offering alternatives, such as biological control agents, soaps, mineral oils and pheromone traps. This has mainly been in the OECD countries, but increasingly also in developing countries. Constraints on market development include cumbersome registration requirements that can be relatively heavy when no distinction is made between such products and pesticides, the absence of broad international marketing networks and advertisement budgets that are dwarfed by the amounts that pesticide companies spend on pesticide promotion.

The role of genetically modified organisms (GMOs) in reducing pesticide use remains ambiguous as both positive and negative evidence about benefits continue to emerge (Nuffield Council on Bioethics, 2004). Herbicide tolerant crops were initially heralded as the way forward to reduce herbicide use, but a review of achievements in the US over the period 1995–2003 showed that herbicide use on herbicide-tolerant crops has now raised some doubts (Benbrook, 2003). Initial claims of pesticide reductions in *Bt*-cotton were followed by mixed reports about actual pesticide use. Nevertheless, there is a potential role for certain biotech-based pest or disease resistant varieties, provided that it is demonstrated that health and environmental risks are negligible, and that there are no market impediments to the GMO concerned. Like any other varieties, their long-term effectiveness will depend on the manner in which these crops are being managed. IPM-based production systems tend to extend the benefits of new varieties. Certain pest and disease resistant GMOs might thus become useful instruments within an IPM strategy, but are unlikely to provide silver bullet solutions to the overuse of pesticides.

THE EMERGENCE OF A NEW AGENDA

As in most processes of change, some stakeholders are moving faster than others. Frontrunners are testing and validating new directions that are eventually mainstreamed in broader agendas. A close look at the front group thus provides a picture of what could be achieved on a broad scale within the next decade. The main emerging trends are:

1   reduced reliance on pesticides;
2   phase-out of pesticides hazardous to human health and the environment;
3   phase-in of alternative approaches and products.

Reduced Reliance on Pesticides

In many systems, pesticide use remains an automatic and primary response to pest problems. Farmers have often become accustomed to applying pesticides in response to signs of crop damage without understanding the ecology of the crop–pest complex and the economics of their intervention.

Both in industrialized and developing countries, there is a growing awareness that current pesticide use levels are unnecessarily high, uneconomic and risky, while available alternatives remain under-utilized. Thorough agroecosystem analysis of pesticide dependent production systems almost invariably reveals significant potential for pesticide reduction. In many cases, pest problems can be reduced through the better management of fields. Nutrient and water management, and agronomic practices to prevent pest development or to encourage natural mortality of pests are part of this. If the use of crop protection inputs is required, then there often are alternatives that can be considered before resorting to pesticide use. Pest management policies should thus encourage pest management practices to be reviewed within a broader agroecological perspective.

Support for IPM to reduce a reliance on pesticides is called for by the FAO Code of Conduct on the Distribution and Use of Pesticides, the World Bank Operational Policy on Pest management, OECD donor policies, Agenda 21, the Convention on Biodiversity, many national policies on pest management and environmental protection, and supply chain requirements of a growing number of companies in the food industry.

Phase-out of Hazardous Pesticides

There is a growing consensus that extremely and highly hazardous pesticides that fall into WHO Acute Toxicity Hazard Class I should be phased out, particularly in developing countries where users generally lack the knowledge and means to reduce the risks to acceptable levels. Generally, pesticide applications involving Class I pesticides can be replaced by alternative pest management approaches or less hazardous products without production risk. There is no compelling case for the continued regular use of Class I pesticides.

Further, there are several other groups of undesirable pesticides:

1   Moderately hazardous pesticides that fall under WHO Hazard Class II. Particularly the higher toxicity range within this Class, which contains several products that are notoriously problematic. Endosulfan, lindane and paraquat are examples that have been discussed in earlier chapters. In many cases, use of problematic Class II pesticides can be replaced by alternative pest management approaches or less hazardous products.
2   Pesticides with chronic health hazards. These include products with carcinogenic and endocrine disrupting properties and products that may cause birth defects or suppress the auto-immune system. Growing attention on this group will increasingly be reflected in pesticide registration decisions and results of re-evaluations. When there are indications that a product might be carcinogenic, it is normally first classified as probably or likely to be carcinogenic, pending further research to scientifically establish its carcinogenity. There seems little justification in permitting continued use of likely or probable carcinogenics if non-carcinogenic alternatives are available.
3   Environmentally persistent pesticides. Persistent pesticides and those with persistent breakdown compounds that continue to cause contamination problems after application as they spread through the ecosystem and food chain. The remaining uses of the most notorious persistent pesticides are being phased out under the Stockholm Convention. Tightening of pesticide residue requirements in both domestic and international markets is providing further pressure against use of other persistent pesticides. The phasing out of lindane from cocoa production is an example (see Chapter 12). Elevated pesticide residue levels in ground water, resulting from prolonged

intensive use, is increasingly becoming a factor leading to decisions to restrict or phase out certain products. The privatization of drinking water supply in some countries is raising questions about liability for pesticide residues.

4   Products that disrupt ecosystems. Broad-spectrum pesticides that affect beneficial organisms and wildlife, and compounds highly toxic to pollinators, fish or birds.

Support for the phasing out of these products comes from a broad range of international and national environmental and health policies, labour standards, pest management policies, and policies that support sustainable development and biodiversity conservation. Problems related to illegal imports of undesirable pesticides are leading to calls for harmonized approaches among groups of countries. Pesticide residue concerns related to trade requirements are increasingly playing a role in the phasing out of certain pesticides. Some food companies have already black-listed some products from use in their supply chain ahead of regulatory decisions.

Phasing out mechanisms include the banning of products, discontinuation of registration of products, or restriction of their use to specific circumstances where risks can be properly contained. Another mechanism that would contribute to the phasing out of undesirable pesticides is taxing these products to internalize the environmental and health costs to society in the prices paid by farmers.

Comparative risk assessment, as developed and introduced in some Scandinavian countries, is providing a new approach to favour less risky products. Broader use of this mechanism would expedite the phasing out of the groups of problematic pesticides described above. In most registration schemes, pesticides are reviewed against set standards. All products that meet these standards can, in principle, be permitted. Comparative risk assessment goes a step further and favours the least risky pesticides within a group of products with comparable uses.

Phase-in of Alternative Approaches and Products

There is growing evidence to show that alternative approaches and products can be very effective at managing pests without the adverse side-effects associated with a reliance on pesticides. Millions of farmers are successfully implementing IPM in many countries and on many different types of farms.

A growing number of both developing and industrialized countries have declared IPM the crop protection approach of choice. This development is also reflected in the policies and guidelines of international organizations like the FAO, OECD and the World Bank.

Impressive achievements are being made by national IPM programmes in developing countries, which use the farmer field school approach. This is a participatory, community-driven and agroecologically-based approach that helps farmers gain the necessary understanding to make sound pest management decisions (see Chapter 8). Over 2 million rice, vegetable and cotton farmers participating in these programmes have managed to cut pesticide use, while increasing farm incomes (see Chapters 3 and 8). The approach, which was developed in Asia, is now being adapted and introduced in other parts of the world.

Supporting research into IPM and biological control has advanced fast and is leading to many new technologies and practices. Several developing countries have gained important experience in biocontrol, although broad application often still needs to be promoted.

In industrialized countries, the number of small- and medium-sized companies offering biocontrol products has grown rapidly. The availability of alternatives such as traps, pheromones and biocontrol products that can play a role in IPM strategies, or outright substitution of pesticide use, are becoming more widely available in a more firmly established sector. Incentives would further support this development and encourage expansion of this sector to other countries. Some developing countries already have interesting programmes to encourage village level production of NPV or neem. The former Soviet Union had an effective and well-organized system for village and district level production of the natural enemies of cotton pests. Local production of alternative products in developing countries may create employment and reduce the use of foreign currency for the importation of pesticides.

Although impressive achievements have been made, there still remains enormous scope for mainstreaming proven alternatives. This process can be accelerated by targeted policy measures that enable and encourage the uptake of alternative pest management approaches and the use of alternative products. Phase-in can work if all the key actors realize that this new agenda can benefit them all, either by providing real benefits, or by avoiding costs or future risks.

CONCLUDING COMMENTS

IPM programmes have demonstrated that current levels of pesticide use in many circumstances are not necessary and, frequently, are even counter-productive. Excessive and otherwise inappropriate pesticide use is an unnecessary burden on farmers' health and income, on public health and on the environment. Many alternative pest management approaches and products are becoming available. The examples marshalled for this book, from both developing and industrialized countries, are showing that there is great potential to clean up agriculture. What is needed now is a strong political will, backed up by consumer awareness and market responses. Then the road is open to detoxify agriculture.

References

Abawi G. S. and Widmer T. L. 2000. Impact of soil health management practices on soilborne pathogens, nematodes and root diseases of vegetable crops. *Applied Soil Ecology* 15, 37–47

Abramovitz J. 1997. Valuing nature's services. In: Brown L., Flavin C. and French H. (eds). *State of the World*. Worldwatch Institute, Washington, DC

Adipala E., Nampala P., Karungi J. and Isubikalu P. 2000. A review on options for management of cowpea pests: Experiences from Uganda. *Integrated Pest Management Reviews* 5, 185–196

Agra CEAS. 2002. Integrated crop management systems in the EU. Amended final report for EC DG Environment, Agra CEAS Consulting, Brussels, 141pp

Agrow. 1996. *World Crop Protection News* 27 September, PJB Publications, Richmond

Agrow. 2001. World news in brief. *AGROW* 379, 29 June 2001, p19

Agrow. 2003. World agrochemical market shrinks. *AGROW* 427, 4 July 2003, p17

Agrow. 2003a. Global agrochemical sales flat in 2002. *AGROW* 419, 28 February, p17, PJB Publications, Richmond

Agrow. 2003b. Insecticide phase-out hits Chinese market. *AGROW* 424, 16 May, PJB Publications, Richmond

Agrow. 2003c. *The World Market for Crop Protection Products in Rice*. PJB Publications, Richmond

Alavanja M. C. R., Sandler D. P., McMaster S. B., Zahm S. H., McDonnell C. J., Lynch C. F., Pennybacker M., Rothman N., Dosemeci M., Bond A. E. and Blair A. 1996. The agricultural health study. *Environmental Health Perspectives* 104, 362–369

Alavanja M. C. R., Samanic C., Dosemeci M., Lubin J., Tarone R., Lynch C. F., Knott C., Thomas K., Hoppin J. A., Barker J., Coble J., Sandler D. P. and Blair A. 2003. Use of agricultural pesticides and prostate cancer risk in the agricultural health study cohort. *American Journal of Epidemiology* 157 (9), 800–814

Allen Y., Matthoessen P., Haworth S., Thain J. E. and Feist S. 1999. Survey of estrogenic activity in UK estuarine and coastal waters and its effects on gonadal development of the flounder. *Environmental Toxicology and Chemistry* 18, 1791–1800

Altieri M. A. and Letourneau D. K. 1984. Vegetation diversity and insect pest outbreaks. *CRC Critical Review of Plant Science* 2, 131–169

Altieri M. A. and Nicholls C. I. 1997. Indigenous and modern approaches to IPM in Latin American. *ILEIA Newsletter* 13 (4), 6

Altin G. N., Paris T., Linquist B., Phengchang S., Chongyikangutor K., Singh A., Singh V. N., Dwivedi J. L., Pandey S., Cenas P., Laza M., Sinha P. K., Mandal N. P. and Suwarno. 2002. Integrating conventional and participatory crop improvement in rainfed rice. *Euphytica* 122, 463–475

Anderson A. 1986. It's raining pesticides in Hokkaido. *Nature* 370, 478

Andow D. A. 1991. Vegetational diversity and arthropod population response. *Annual Review of Entomology* 36, 561–586

Anonymous. 2001. Growing sustainable cocoa. *Pest Management Notes* no 12, 5pp http://www.pan-uk.org/Internat/IPMinDC/pmn12.pdf

Antle J. M. and Wagenet R. J. 1995. Why scientists should talk to economists. *Agronomy Journal* 87, 1033–1040

Antle J. M., Capalbo S. M. and Crissman C. C. 1998a. Tradeoffs in policy analysis: Conceptual foundations and disciplinary integration. In: C. C. Crissman, J. M. Antle and S. M. Capalbo (eds) *Economic, Environmental and Health Tradeoffs in Agriculture: Pesticides and the Sustainability of Andean Potato Production*. International Potato Center and Boston: Kluwer Academic Press, Lima. pp21–40

Antle J. M., Capalbo S. M., Cole D. C., Crissman C. C. and Wagenet R. J. 1998b. Integrated simulation model and analysis of economic, environmental and health tradeoffs in the Carchi potato–pasture production system. In: C. C. Crissman, J. M. Antle, and S. M. Capalbo (eds) *Quantifying Tradeoffs in the Environment, Health and Sustainable Agriculture: Pesticide Use in the Andes*. International Potato Center and Boston: Kluwer Academic Press, Lima. pp243–268

Antle J. M., Cole D. C. and Crissman C. C. 1998c. Further evidence on pesticides, productivity and farmer health: Potato production in Ecuador. *Agricultural Economics* 18, 199–207

Appert J. and Ranaivosoa H. 1971. Un nouveau succes de la lutte biologique a Madagascar: Controle des foreurs de la tige de mais par un parasite introduit *Pediobius furvus Gahan*. *Agronomie Tropicale* 26: 327–331

Aragon A., Aragon C., Thorn A. 2001. Pests, peasants, and pesticides on the Northern Nicaraguan Pacific Plain. *International Journal of Occupational and Environmental Health* 7 (4), 295–302

Archer D. W. and Shogren J. F. 2001. Risk-indexed herbicide taxes to reduce ground and surface water pollution: an integrated ecological economics evaluation. *Ecological Economics* 38, 227–250

Atkin J., Leisinger K. M. (eds) 2000. *Safe and Effective Use of Crop Protection Products in Developing Countries*. Novartis Foundation for Sustainable Development, Switzerland

Avery D. 1995. *Saving the Planet with Pesticides and Plastic*. The Hudson Institute, Indianapolis

Awmack C. S. and Leather S. R. 2002. Host plant quality and fecundity in herbivorous insects. *Annual Review of Entomology* 47, 817–844

Ayer H. and Conklin N. 1990. Economics of agricultural chemicals – flawed methodology and a conflict of interest quagmire. *Choices* Fourth Quarter, 24–30

Azaroff L. S. 1999. Biomakers of exposure to organophosphorous insecticides among farmers' families in rural El Salvador: Factors associated with exposure. *Environmental Research*, Section A 80, 138–147

Azaroff L. S. and Neas L. M. 1999. Acute health effects associated with nonoccupational pesticide exposure in rural El Salvador. *Environmental Research*, Section A 80, 158–164

Azidah A. A., Fitton M. G. and Quicke D. L. J. 2000. Identification of the Diadegma species (Hymenoptera: Ichneumonidae, Campopleginae) attacking the diamondback moth, Plutella xylostella (Lepidoptera: Plutellidae). *Bulletin of Entomological Research* 90, 375–389.

Bairoch P. 1997. New estimates on agricultural productivity and yields of developed countries, 1800–1990. In: A. Bhaduri and R. Skarstein (eds) *Economic Development and Agricultural Productivity*. Edward Elgar, Cheltenham pp45–57

Baker B. 1996. After a long wait, an Environmental Farm Bill passes muster. *Bioscience* 46, 486

Baker B. P., Benbrook C. M., Groth E. and Lutz Benbrook K. 2002. Pesticide residues in conventional, integrated pest management (IPM)-grown and organic foods: insights from three US data sets. *Food Additives & Contaminants* 19 (5), 427–446

Barbosa P. and Benrey B. 1998. The influence of plants in insect parasitoids: implications for conservation biological control. In: Barbosa P. *Conservation Biological Control*. Academic Press, San Diego, California. pp55–82

Barker K. 2003. Presentation at the OECD Risk Reduction Strategy Group seminar on compliance, held in Paris, 10 March. OECD, Paris

Barrera V. and Escudero L. 2004. Encontrando salidas para reducir los costos de producción y la exposición a plaguicidas: Experiencias de la intervencion en la provincia de Carchi, Ecuador. Instituto Nacinal Autonoma de Investigación Agropecuaria (INIAP) and CropLife Ecuador. 69pp

Barrera V. H., Norton G. and Ortiz O. 1998. *Manejo de las principales plagas y enfermedades de la papa por los agricultores en la provincia del Carchi, Ecuador*. Quito, Ecuador: INIAP (National Agricultural Research Institute). 65pp

Barrera V., Escudero L., Norton G. and Sherwood S. 2001. Validación y difusión de modelos de manejo integrado de plagas y enfermedades en el cultivo de papa: Una experiencia de capacitación participativa en la provincia de Carchi, Ecuador. Revista INIAP. 16, 26–28

Barrion A. T. and Litsinger J. A. 1994. Taxonomy of rice insect pests and their arthropod parasites and predators, In: E. A. Heinrichs (ed) *Biology and Management of Rice Insects*, Wiley Eastern Limited. pp13–362

Barsky O. 1984. *Acumulación Campesina en el Ecuador: Los productores de papa del Carchi. Colección de Investigaciones*, No 1. Quito, Ecuador: Facultad Latinoamericana de Ciencias Sociales. 136pp

Barzman M. and Desilles S. 2002. Diversifying rice-based farming systems and empowering farmers in Bangladesh using the farmer field school approach. In Uphoff N. (ed) *Agroecological Innovations*. Earthscan, London

Bateman R. P. 2001. Consultancy on rational pesticide use on sugar cane in Trinidad, with special reference to delivery systems for Metarhizium anisopliae. TCP/TRI project 8921 report. 29pp

Baumol W. J. and Oates W. E. 1988. *The Theory of Environmental Policy*. Cambridge University Press, Cambridge

Bawden R. 2000. The Importance of Praxis in Changing Forestry Practice (prelim. title). Invited Keynote Address for 'Changing Learning and Education in Forestry: A Workshop in Educational Reform', held at Sa Pa, Vietnam, 16–19 April

Bawden R. J. and Packam R. 1993. Systems praxis in the education of the agricultural systems practitioner. Richmond (NSW): University of Western Sidney-Hawkesbury. Paper presented at the 1991 Annual Meeting of the International Society for the Systems Sciences. Östersund, Sweden. *Systems Practice* 6, 7–19

Beanland L., Phelan L. and Saliminen S. 2003. Micronutrient interactions on soybean growth and the developmental performance of three insect herbivores. *Environmental Entomology* 32 (3), 641–651

Beck U. 1992. *Risk Society: Towards a New Modernity*. London: Sage Publications (First published as Risikogesellschaft: Auf dem Weg in eine andere Moderne. Frankfurt am Main, Suhrkamp Verlag 1986)

Benbrook C. M. 2003. *Impacts of Genetically Engineered Crops on Pesticide Use in the United States: The First Eight Years*. Benbrook; Northwest Science and Environmental Policy Center, Ames, Iowa

Bentley J. W. 1992a. Learning about biological pest control. *ILEIA Newsletter*, 8 (4), 16–17.

Bentley J. W. 1992b. The epistemology of plant protection: Honduran campesino knowledge of pests and natural enemies. In: Gibson, R.W. and Sweetmore, A. (eds) *Proceedings of a Seminar on Crop Protection for Resource-Poor Farmers*, Natural Resources Institute, Chatham, UK

Bentley J. W. 2000. The mothers, fathers and midwives of invention. In: Stoll, G. *Natural Crop Protection in the Tropics: Letting Information Come to Life*. Margraf Verlag, Germany. pp281–289.

Bentley J. W., Rodríguez G. and González A. 1994. Science and people: Honduran campesinos and natural pest control inventions. *Agriculture and Human Values*, 11 (2&3), 178–182.

Berger P. L. and Luckman T. 1967. *The Social Construction of Reality. A Treatise in the Sociology of Knowledge*. Doubleday, Garden City: and Anchor Books, Middlesex

Bergman A. 1999. Health condition of the Baltic gray seal during two decaedes. *Acta Pathology Microbiology Immunology Scandinavia* 107, 270–282

Bergman A. and Olsson M. 1985. Pathology of Baltic ringed seal and Gray seal females with special reference to adrenocortical hyperplasia: Is environmental pollution the cause of a widely distributed syndrome? *Finnish Game Research* 44, 47–92

Berkman L. and Kawachi I. 2000. *Social Epidemiology*. Oxford University Press, London. 391pp

Bhatia A. and Munkwold G. P. 2002. Relationships of environmental factors and cultural factors with severity of gray leaf spot in Maize. *Plant Disease* 86, 1127–1133

Bhowmik P. C. and Iderjit. 2003. Challenges and opportunities in implementing allelopathy for natural weed management. *Crop Protection* 22, 661–671

Bichel Committee. 1999. *Report of the Main Committee on Assessing the Overall Consequences of a Partial or Total Phasing-out of Pesticide Use*. 125pp

Biever K. D. 1996. Development and use of a biological control IPM system for insect pests of crucifers. In: Sivapragasam A., Kole W. H., Hassan A. K. and Lim G. S. (eds) *The Management of Diamondback Moth and Other Crucifer Pests*. Proceedings of the third International workshop, Kuala Lumpur, Malaysia, 29 October–1 November. pp257–261

Bills P. S., Mota-Sanchez D. and Whalon M. 2003. Background to the resistance database. Michigan State University. At www.cips.msu.eud/resistance

Blank S. 1998. *The End of Agriculture in the American Portfolio*. Greenwood Publishing Group, New York

Blondell J. 1997. Epidemiology of pesticide poisonings in the United States, with special reference to occupational cases. *Occupational Medicine* 12, 209–220

Blowfield M., Gallat S., Malins A., Maynard B., Nelson V. and Robinson D. 1999. *Ethical Trade and Sustainable Rural Livelihoods*. Natural Resources Institute, University of Greenwich, Chatham, UK. 32pp

BNI. 2001. Wisconsin measures aid branding. *Biocontrol News & Information* 22 (4) 83N–84N.

Boa E., Bentley J. B. and Stonehouse J. 2000. *Sustainable Neighbours: How Ecuador Farmers Manage Cocoa and Other Trees*. Final Technical Report for USDA/CABI Alternative Crops Programme, CABI Bioscience

Bond W. and Grundy A. C. 2001. Non-chemical weed management in organic farming systems. *Weed Research* 41, 383–405

Bottrell D. G. 1984. Social problems in pest management in the tropics. *Insect Science Applications* 4 (1/2), 179–184

Bowden J. 1976. Stem borer ecology and strategy for control. *Annals of Applied Biology* 84, 107–111

Bowman G. 1998. *Steel in the Field: A Farmer's Guide to Weed Management Tools*. Sustainable Agriculture Network, Beltsville, Maryland

Bradshaw N. J., Parham C. J. and Croxford A. C. 1996. The awareness, use and promotion of integrated control techniques of pests, diseases and weeds in British agriculture and horticulture. In: *Brighton Crop Protection Conference Pests & Diseases 1996*, British Crop Protection Council, Farnham, UK, pp591–596

Brandt V. A., Moon S., Ehlers J., Methner M. M. and Struttmann T. 2001. Exposure to endosulfan in farmers: two case studies. *American Journal of Industrial Medicine* 39, 643–649

Braun A. R., Thiele G. and Fernández M. 2000. Farmer Field Schools and Local Agricultural Research Committees: complementary platforms for integrated decision-making in sustainable agriculture. *Agricultural Research & Extension Network* Paper no.105, Overseas Development Institute, London

Brethour C. and Weerskink A. 2001. An economic evaluation of the environmental benefits from pesticide reduction. *Agricultural Economics* 25, 219–226

Bromley D. W. 1994. The language of loss: or how to protect the status quo. *Choices*, Third Quarter, 31–32.

Bryant R. 1999. Agrochemicals in Perspective: Analysis of the Worldwide Demand for Agrochemical Active Ingredients, Synopsis of talk presented at the Fine Chemicals Conference 99. www.agranova.co.uk/resource.htm

Buadu E. J., Gounou S., Cardwell K. F., Mochiah B., Botchey M., Darkwa E. and Schulthess F. 2002. Distribution and relative importance of insect pests and diseases in Southern Ghana. *African Plant Protection* 8, 3–11

Buckle A. P. 1988. Integrated management of rice rats in Indonesia. *FAO Plant Protection Bulletin* 36, 111–118

Buckle A. P. and Smith R. H. 1994. *Rodent Pests and their Control*. CAB International, Wallingford, UK

Buffin D. 2001a. Retailers ban suspect pesticides. *Pesticides News* 53, 3

Buffin D. 2001b. Food retailers aim to restrict pesticide use. *Pesticides News* 54, 3

Buffin D., Williamson S. and Dinham B. 2002. Dying for a bit of chocolate. *The Ecologist* 32 (2), 48–50

Bugg R. L. and Pickett C. H. 1998. Introduction: Enhancing biological control – habitat management to promote natural enemies of agricultural pests, 1–23. In: Pickett C. H. and Bugg R. L. *Enhancing Biological Control: Habitat Management to Promote Natural Enemies of Agricultural Pests*. The Regents of the University of California, Berkeley, CA

Bugg R. L., Ehler L. E. and Wilson L. T. 1987. Effect of common knotweed (*Polygonum aviculare*) on abundance and efficiency of insect predators of crop pests. *Hilgardia* 55, 1–51

Bugg R. L., Anderson J. H., Thomsen C. D. and Chandler J. 1998. Farmscaping in California: Managing hedgerows, roadside and wetland plantings, and wild plants for biointensive pest management, In: Pickett C. H. and Bugg R. L. *Enhancing Biological Control: Habitat Management to Promote Natural Enemies of Agricultural Pests*. University of California Press, Berkeley, CA pp339–374

Burgess H. D. (ed) 1981. *Microbial Control and Plant Diseases, 1970–1980*. Academic Press, London

Busch J. W. and Phelan P. L. 1999. Mixture models of soybean growth and herbivore performance in response to nitrogen-sulphur-phosphorous interactions. *Ecological Entomology* 24, 132–145

Buttel F. 1993. Socioeconomic impacts and social implications of reducing pesticide and agricultural chemical use in the United States. In: Pimentel D. and Lehman H. (eds) *The Pesticide Question – Environment, Economics, and Ethics*. Chapman and Hall, New York, London

Buurma J. S., Smit A. B., van der Linden A. M. A. and Luttik R. 2000. *Integrated Management: The Way Ahead*. LEI, The Hague

Byard J. L. 1999. The impact of rice pesticides on the aquatic ecosystems of the Sacramento River and delta (California). *Review of Environmental Contamination and Toxicology* 159, 95–100

Cade T. J., Lincer J. L., White C. M., Roseneau D. G. and Swartz L. G. 1971. DDE residues and eggshell thinning changes in Alaskan falcons and hawks. *Science* 172, 955–957

Campbell A. 1994. Community first: Landcare in Australia. *Gatekeeper Series* no 42. International Institute of Environment and Development, London, UK

Campbell L. H. and Cooke A. S. (eds) 1995. *The Indirect Effects of Pesticides on Birds*. JNCC, Peterborough

Capra F. 1996. *The Web of Life. A New Synthesis of Mind and Matter*. London: Harper Collins Publishers (Flamingo)

Capra F. 2002. *The Hidden Connections. A Science for Sustainable Living*. Harper Collins, London

Cardwell K., Schulthess F., Ndemah R. and Ngoko Z. 1997. A systems approach to assess crop health and maize yield losses due to pests and diseases in Cameroon. *Agriculture Ecosystems and the Environment* 65, 33–47

Carpenter D. O., Chew F. T., Damstra T., Lam L. H., Landrigan P. J., Makalinao I., Peralta G. L., and Suk W. A. 2000. Environmental threats to the health of children: the Asian perspective. *Environmental Health Perspectives* 108, 989–992

Carruthers K., Prithiviraj B., Fe Q., Cloutier D., Martin R. C. and Smith D. L. 1999. Intercropping of corn with soybean, lupin and forages: silage yield and quality. *Journal of Agronomy and Crop Science* 74, 66–74

Carson R. 1963. *Silent Spring*. Penguin Books, Harmondsworth

Carson R. T. 2000. Contingent valuation: A user's guide. *Environmental Science and Technology* 34, 1413–1418.

Castillo, G. T. 1998. A social harvest reaped from a promise of springtime: User-responsive participatory agricultural research in Asia. In: N. Röling and M. Wagemakers (eds) *Facilitating Sustainable Agriculture. Participatory Learning and Adaptive Management of Environmental Uncertainty*. Cambridge University Press, Cambridge. pp191–214

Castillo J. et al *vs* E. I. DuPont de Nemours & Co Inc et al 2003. Supreme Court of Florida (2003), no SC00–490, 10 July

Castro Ramírez A., Cruz López J. A., Ramírez Salinas C. and Gómez Méndez J. A. 1999. Cambio de prácticas agrícolas y biodiversidad en el cultivo de maíz en la región altos de Chiapas. In: *Memorias del Seminario Internacional sobre Agrodiversidad Campesina, realizado 12–14 mayo 1999*. Facultad de Geografia, Universidad Autonoma del Estado de México. [Change in agricultural practices and biodiversity in maize cultivation in the Chiapas highlands, in Spanish]

Castro-Gutierrez N., McConnell R., Andersson K., Pancheco-Anton F., Hogstedt C. 1998. Respiratory symptoms, spirometry and chronic occupational paraquat exposure. *Scandinavian Journal of Work and Environmental Health* 23, 421–427

CATIE-INTA. 1999. *Final Report on CATIE/INTA IPM Project February 1995–July 1998*. Unpublished report, CATIE/INTA IPM Project, Managua, Nicaragua

Caton B. P., Foin T. C. and Hill J. E. 1999. A plant growth model for integrated weed management in direct-seeded rice. III. Interspecific competition for light. *Field Crops Research* 63, 47–61

Cavigelli M. A., Demming S. R., Probyn L. K. and Mutch D. R. 2000. *Michigan Field Crop Pest Ecology and Management*. Michigan State University. East Lansing, Michigan

CDFA. passim. *Summary of Illnesses and Injuries reported by California Physicians as Potentially Related to Pesticides*. 1972–2002. Calfornia Department of Food and Agriculture, Sacramento

CEDAC. 2000. *Pesticide Pollution in the Tonle Sap Catchment. Project Progress Report* (Sept. 1999–Aug. 2000). CEDAC, Phnom Penh, Cambodia

Chabi-Olaye A., Schulthess F., Nolte C. and Borgemeister C. 2002. Effect of maize/cassava/cowpea and soybean intercropping on the population dynamics of *Busseola fusca* (Fuller) and their natural enemies in the humid forest zones of Cameroon. Abstracts of the Entomological Society of America Conference, Florida

Champion G. T., May M. J., Bennett S., Brooks D. R., Clark S. J., Daniels R. E., Firbank L. G., Haughton A. J., Hawes C., Heard M. S., Perry J. N., Randle Z., Rossall M. J., Rothery P., Skellern M. P., Scott R. J., Squire G. R. and Thomas M. R. 2003. Crop management and agronomic context of the Farm Scale Evaluations of genetically modified herbicide-tolerant crops. *Philosophical Transactions of the Royal Society of London* 358, 1801–1818

Chancellor T. C. B., Tiongco E. R., Holt J.,Villareal S. and Teng P. S. 1999. The influence of varietal resistance and synchrony on tungro incidence in irrigated rice ecosystem in the Philippines. In: Chancellor T. C. B., Azzam O. and Heong K. L. (eds) *Rice Tungro Disease Management*. Proceedings of the International Workshop on Tungro Disease Management, 9–11 November 1998, IRRI, Los Banos, Laguna, Philippines. Makati City (Philippines): International Rice Research Institute, pp121–127

Chandler J. 2003. Pheromones fall foul of EU pesticide directives. *Pesticides News* 59, 10

Chaney W. 1998. Biological control of aphids in lettuce using in-field insectaries, In: Bugg R. L. and Pickett C. H. (eds) *Enhancing Biological Control: Habitat Management to Promote Natural Enemies of Agricultural Pests*. The Regents of the University of California, Berkeley, CA. pp73–83

Chikoye D., Ekeleme F. and Ambe J. T. 1999. Survey of distribution and farmers' perceptions of speargrass [*Imperata cylindrica* (L.) Raeuschel] in cassava-based systems in West Africa. *International Journal of Pest Management* 45 (4), 305–311

Cimatu F. 1997. War on pesticide adverts. *Pesticides News* 38, 21

CIMMYT. 2001. 1999/2000. World maize facts and trends: Meeting world maize needs: Technological opportunities and priorities for the public sector. Centro Internaciaonal the Mais y Trigo, Mexico

CIP-UPWARD. 2003. *Farmer Field Schools. From IPM to Platforms for Learning and Empowerment*. Report of an International Workshop. International Potato Centre-Users' Perspectives with Agricultural Research and Development, Los Banos, Philippines, 83pp

Claridge M. F., Den Hollander J., and Morgan J. C. 1982. Variation within and between populations of the brown planthopper, Nilaparvata lugens (Stal). In: Knight W. J., Pant N. C., Robertson T. S. and Wilson M. R. (eds) *1st International Workshop on Leafhoppers and Planthoppers of Economic Importance*, Commonwealth Institute of Entomology, London, pp36–318

Clark W. S. 2000. Problems of communication and technology transfer in crop protection: a practitioner's perspective. In: *The BCPC Conference Pests & Diseases 2000*, British Crop Protection Council, Farnham, UK, pp1185–1192

Clarke E. E., Levy L. S., Surgeon A. and Calvert I. A. 1997. The problems associated with pesticide use by irrigation workers in Ghana. *Occupational Medicine (London)* 47 (5), 301–308

Clunies-Ross, T. and Hildyard, N. 1992. *The Politics of Industrial Agriculture*. Earthscan, London

Cochrane W. W. 1958. *Farm Prices, Myth and Reality*. University of Minnesota Press, Minneapolis (Especially Chapter 5: The Agricultural Treadmill, pp85–107)

Cohen J. E., Schoenly K., Heong K. L., Justo H., Arida G., Barrion A. T. and Litsinger J. A. 1994. A food web approach to evaluate the effect of insecticide spraying on insect pest population dynamics in a Philippine irrigated rice ecosystem. *Journal of Applied Ecology* 31, 747–763

Colborn T., vom Saal F. S. and Soto A. M. 1993. Developmental effects of endocrine-disrupting chemicals in wildlife and humans. *Environmental Health Perspectives* 101, 378–384

Cole D. C., McConnell R., Murray D. L. and Pacheco Anton F. 1988. Pesticide illness surveillance: the Nicaraguan experience. *Bulletin of the Pan American Health Organization* 22, 119–32

Cole D. C., Carpio F., Julian J. and León N. 1997a. Dermatitis in Ecuadorian farm workers. *Contact Dermatitis – (Environmental and Occupational Dermatitis)* 37, 1–8

Cole D. C., Carpio F., Julian J., León N., Carbotte R. and De Almeida H. 1997b. Neuro-behavioral outcomes among farm and non-farm rural Ecuadorians. *Neurotoxicol Teratology* 19 (4), 277–286

Cole D. C., Carpio F., Jullian J. and León N. 1998a. Assessment of peripheral nerve function in an Ecuadorean rural population exposed to pesticides. *Journal of Toxicology and Environmental Health* 55 (2), 77–91

Cole D. C., Carpio F., Julian J. and León N. 1998b. Health impacts of pesticide use in Carchi farm populations. In: Crissman C. C., Antle J. M. and Capalbo S. M. (eds) *Economic, Environmental and Health Tradeoffs in Agriculture: Pesticides and the Sustainability of Andean Potato Production*. CIP (International Potato Center), Lima, Peru and Kluwer Academic Publishers, Dordrecht /Boston/London. pp209–230

Cole D. C., Carpio F. and León N. 2000. Economic burden of illness from pesticide poisonings in highland Ecuador. *Pan American Review of Public Health* 8 (3), 196–201

COLEACP. 2003. Harmonisation of regulations: ICDCS, an example to be followed. PIP Info. No. 11, January 2003 COLEACP, 2003. *http://www.coleacp.org/fo_internet/en/pesticides/breves/infopip.html*

Colunga-G M., Gage S. H. and Dyer L. E. 1998. The insect community. In: Cavigelli M. A., Deming S. R., Probyn L. K. and Harwood R .R. *Michigan Field Crop Ecology: Managing Biological Processes for Productivity and Environmental Quality*. Michigan State University Extension Bulletin E-2646, East Lansing, Michigan

Conlong D. E. 2001. Indigenous African parasitoids of Eldana saccharina (Lepidoptera: Pyralidae). *Proceedings of the South African Sugar Technologists Association* 74: 201–211.

Consultative Group on International Agricultural Research. 2003. *Annual Report*. Washington DC

Conway G. R. and Pretty J. N. 1991. *Unwelcome Harvest: Agriculture and Pollution*. Earthscan, London

Co-op. 2001. Green and Pleasant Land. How hungry are we for safe, sustainable food? The Co-operative Group, Manchester, UK. 20pp.

Corbett A. 1998. The importance of movement in the response of natural enemies to habitat manipulation. In: Pickett C. H. and Bugg R. L. *Enhancing Biological Control: Habitat Management to Promote Natural Enemies of Agricultural Pests*. University of California Press, Berkeley, California. pp25–48

Cormack B. and Metcalfe P. 2000. *Energy Use in Organic Farming Systems*. ADAS, Terrington

Corriols M. and Hruska A. 1992. Occupational health in Nicaragua. Presented at the 120th Annual Meeting of the American Public Health Association, November 1992, Washington DC

Costa L. G. 1997. Basic toxicology of pesticides. *Occupational Medicine* 12 (2), 251–268

Costanza R., d'Arge R., de Groot R., Farber S., Grasso M., Hannon B., Limburg K., Naeem S., O'Neil R. V., Paruelo J., Raskin R. G., Sutton P. and van den Belt M. (1997 and 1999). The value of the world's ecosystem services and natural capital. *Nature* 387, 253–260; also in *Ecological Economics* 25 (1), 3–15

Coulibaly S. and Nacro S. 2003. The farmer's voice – welcome support for African cotton growers. *Pesticides News* 61, 11

Crain D. A. and Guilette L. J. 1997. Endocrine disrupting contaminants and reproduction in vertebrate wildlife. *Review of Toxicology* 1, 207–231

Cremlyn R. J. 1991. *Agrochemicals: Preparation and Mode of Action*. Wiley, Chichester

Crisp T. M., Clegg E. D., Cooper R. L., Wood W. P., Anderson D. G., Baetcke K. P., Hoffman J. L., Morrow M. S., Rodier D. J., Schaeffer J. E., Tourt L. W., Zeeman M. G. and Patel Y. M. 1998. Environmental endocrine disruption: an effects assessment and analysis. *Environmental Health Perspectives* 106, Suppl.1, 11–56

Crissman, C. C., Cole D. C. and Carpio F. 1994. Pesticide use and farm worker health in Ecuadorian potato production. *American Journal of Agricultural Economics* 76, 593–597

Crissman C. C., Antle J. M. and Capalbo S. M. (eds) 1998. *Economic, Environmental and Health Tradeoffs in Agriculture: Pesticides and the Sustainability of Andean Potato Production*. International Potato Center, Lima and Kluwer Academic Press, Boston. 280pp

Crissman C. C., Espinosa P., Ducrot C. E. H., Cole D. C. and Carpio F. 1998. The case study site: Physical, health and potato farming systems in Carchi Province. In Crissman C. C., Antle J. M. and Capalbo S. M. (eds) *Economic, Environmental and Health Tradeoffs in Agriculture: Pesticides and the Sustainability of Andean Potato Production*. International Potato Center, Lima and Kluwer Academic Press, Boston. pp85–120

Crissman, C. C., Yanggen D., Antle J., Cole D., Stoorvogel J., Barrera V. H., Espinosa P. and Bowen W. 2003. Relaciones de intercambio existentes entre agricultura, medio ambiente y salud humana con el uso de plaguicidas. In: Yanggen D., Crissman C. C. and Espinosa P. (eds) 2003. *Plaguicidas: Impactos en producción, salud y medioambiente en Carchi, Ecuador*. CIP, INIAP, Ediciones Abya Yala, Quito, Ecuador. pp146–162

CropLife International. 2003. *Acutely Toxic Pesticides: Risk assessment, risk management and risk reduction in Developing Countries and Economies in Transition*. www.croplife.org

Curtis J. 1998. *Fields of Change: A New Crop of American Farmers Finds Alternatives to Pesticides*. Natural Resources Defense Council

Cuyno L. C. M., Norton G. W. and Rola A. 2001. Economic analysis of environmental benefits of integrated pest management. A Philippine case study. *Agricultural Economics* 25, 227–233

Daily G. (ed.) 1997. *Nature's Services: Societal Dependence on Natural Ecosystems*. Island Press, Washington DC

Dale D. 1994. Insect pests of the rice plant – their biology and ecology. In: Heinrichs E. A. (ed) *Biology and Management of Rice Insects*, Wiley Eastern Limited. pp363–486

Dalvie M. A., White N., Raine R., Myers J. E., London L., Thompson M. and Christiani D. C. 1999. Long-term respiratory health effects of the herbicide, paraquat, among workers in the Western Cape. *Occupational and Environmental Medicine* 56 (6), 391–396

Daly H. V., Doyen J. T. and Purcell III. A. H. 1998. *Introduction to Insect Biology and Diversity*. 2nd ed. Oxford University Press, New York

Damasio A. 2003a. Virtue in mind. *New Scientist*, 180 (2420), 48–51

Damasio A. 2003b. *Looking for Spinoza. Joy, Sorrow and the Feeling Brain*. Harcourt Inc, Orlando FL

Daniels J. L., Olshan A. F. and Savitz D. A. 1997. Pesticides and childhood cancers. *Environmental Health Perspectives* 105 (10), 1068–1077

Datta S., Hansen L., McConnell L., Baker J., Lenoir J. and Seiber J. N. 1998. Pesticides and PCB contaminants in fish and tadpoles from the Kaweah River Basin, California. *Bulletin of Environmental Contamination and Toxicology* 60, 829–836

Davidson C., Shaffer H. B. and Jennings M. R. 2001. Declines of the Californian red-legged frog: climate, UV-B, habitat and pesticides hypotheses. *Ecological Applications* 11, 464–479

Defra. 2003. *Endocrine Disruption in the Marine Environment (EDMAR)*. Defra, London de Freitas, H. 1999. Transforming microcatchments in Santa Caterina, Brazil. In: Hinch-cliffe F., Thompson J., Pretty J., Guijt I. and Shah P. (eds) *Fertile Ground: The Impacts of Participatory Watershed Development*. IT Publications, London

de Freitas H. 2000. *Soil Management and Conservation for Small Farms*. FAO, Rome

De Groot, A. B. A. 1995. The functioning and sustainability of village crop protection brigades in Niger. *International Journal of Pest Management* 41 (4) 243–248

de Jong F. M. W. and de Snoo G. R. 2002. A comparison of the environmental impact of pesticide use in integrated and conventional potato cultivation in the Netherlands. *Agriculture, Ecosystems & Environment* 91 5–13

De Souza Silva J., Cheaz J. and Calderon J. 2000. Building capacity for strategic mange-ment of institutional change in agricultural science organisations in Latin America: A summary of the project and progress to date. San José (Costa Rica) ISNAR at IICA, Proyecto Neuvo Paradigma

Denis J. 2003. Integrated production in fruit arboriculture: a winning choice. In: Proceed-ings of the Mouvement pour les Droits et Respect des Génerations Futures, PAN Europe and Organic Farming Association of Picardy colloquium on *Alternatives to Reduce or Eliminate the Use of Synthetic Pesticides in Agriculture: Organic Farming, Integrated Crop Management*, held 30 May, Beauvais, France, MDRGF, St Deniscourt, France

Denké D., Schulthess F., Bonato O. and Smith H. 2000. Effet de la fumure en potassium sur le développement, la survie et la fécondité de *Sesamia calamistis* Hampson (Lepid-optera: Noctuidae) et de *Eldana saccharina* Walker (Lepidoptera: Pyralidae). *Insect Science Applications* 20, 151–156

Dennehey T. J., Roelofs W. L., Taschenberg E. F. and Taft T. N. 1990. Mating disruption for control of the grape berry both in New York vineyards, In: Ridgway R. L., Silverstein R. M. and Inscoe M. N. *Behavior-Modifying Chemicals for Insect Management: Applications of Pheromones and Other Attractants*. 1990. Marcel Dekker Inc., New York. pp223–240

Dent D. R. 1991. *Insect Pest Management*. CAB International, Wallingford, UK

Dent D. R. 2000. *Insect Pest Management* 2nd ed. CABI Publishing, Wallingford, UK

Dewar A. 2000. *Agricultural Product Stewardship*, PJB Publications, Richmond, UK

Diallo A., Dieng B. and Everts J. W. 2003. Less pests, more profit, safer vegetables. *Pesticides News* 61, 8–9

Diana S. G., Resetarits W. J., Schaeffer D. J., Beckmen K. B. and Beasley V. R. 2000. Effects of atrazine on amphibian growth and survival in artificial aquatic communities. *Environmental Toxicology and Chemistry* 19, 2961–2967

Dilts R. 1998. Facilitating the emergence of local institutions: Reflection from the experience of the Community IPM Programme in Indonesia. In: *Asian Productivity Organisation Role of Institutions in Rural Community Development: Report of the APO Study Meeting in Colombo, 21–29 Sep 1998*. pp50–65. Available at http://www.community ipm.org/docs/Dilts%20APO%201998.doc. [accessed 8 Feb 2004]

Dinham B. (ed). 1993. *The Pesticide Hazard: A Global Health and Environmental Audit*. Zed Books, London

Dinham B. 2003a. The perils of paraquat: sales targeted at developing countries. *Pesticides News* 60, 4–7

Dinham B. 2003b. Growing vegetables in developing countries for local urban popula-tions and export markets: problems confronting small-scale producers. *Pest Manage-ment Science* 59, 575–582

Dinham B. and Malik S. 2003. Pesticides and human rights. *International Journal of Occupa-tional and Environmental Health* 9, 40–52

Dixon J. and Gulliver A. 2001. *Farming Systems and Poverty: Improving Farmers' Livelihoods in a Changing World*. FAO and World Bank, Rome and Washington

Dolan C., Humphrey J. and Harris-Pascal C. 1999. Horticulture commodity chains: the impact of the UK market on the African fresh vegetable industry. IDS Working paper no. 96, Institute of Development Studies, University of Sussex, Brighton, 39pp

Doutt R. L. and Nakata J. 1973. The Rubus leafhopper and its egg parasitoid: An endemic biotic system useful in grape pest management. *Environmental Entomology* 2, 381–386

Drummond R. O., George, J. E. and Kunz S. E. 1988. *Control of Arthropod Pests of Livestock: A Review of Technology*. CR Press Inc., Boca Raton, Florida

DTI. 2001. *Digest of Energy Statistics*. Department of Trade and Industry, London

Du P. V., Lan N. T. P., Kim P. V., Oanh P. H., Chau N. V., and Chien H. V. 2001. Sheath blight management with antagonistic bacteria in the Mekong Delta. In Mew T. W., Borromeo E. and Hardy B. (eds) 2000. Exploiting biodiversity for sustainable pest management. Proceedings of the Impact Symposium on Exploiting Biodiversity for Sustainable Pest Management, 21–23 August 2000, Kunming, China. International Rice Research Institute, Makati City (Philippines), 241pp

Ebenebe A. A., van den Berg J. and van der Linde T. C. 2001. Farm management practices and farmers' perceptions of stalkborers of maize and sorghum in Lesotho. *International Journal of Pest Management* 47 (1), 41–48

EC. 2002. Annual EU-wide Pesticides Residues Monitoring Report 2001. *http://europa.eu. int/comm/food/fs/inspections/fnaoi/reports/annual_eu/index_en.html*

EC. 2002a. 320 pesticides to be withdrawn in July 2003. European Commission, Brussels

Eddleston M. 2000. Patterns and problems of deliberate self-poisoning in the developing world. *Quarterly Journal of Medicine* 93, 715–731

Ehler L. E. 1998. Conservation biological control: Past, present, and future, In: Barbosa P. *Conservation Biological Control*. Academic Press, San Diego. pp1–8

Eisemon T. O. and Nyamete A. 1990. School literacy and agricultural modernization in Kenya. *Comparative Education Review* 34, 161–176

EJF. 2001. *Death in Small Doses: Cambodia's Pesticides Problems and Solutions*. Environmental Justice Foundation, London, UK. 37pp

Ekström G. and Bergkvist P. 2001. Persistence pays – lower risks from pesticides in Sweden. *Pesticides News* 54, 10–11

Endrödy-Younga S. 1968. The stem-borer *Sesamia botanephaga* Tams and Bowden (Lepidoptera: Noctuidae) and the maize crop in Central Ashanti, Ghana. Ghana *Journal of Agricultural Sciences* 1, 103–131.

ENDS. 1999. Industry glimpses new challenges as endocrine science advances. *ENDS* 290

ENDS. 2002. Evidence of hormone disruption in humans weak, says WHO. *ENDS* 328, 11

Environment Agency. 2002. *Agriculture and Natural Resources: Benefits, Costs and Potential Solutions*. Bristol

EPA. 1999. 1998–1999 Pesticide Market Estimates: Usage. US Environmental Protection Agency. Available at http://www.epa.gov/oppbead1/pestsales/99pestsales/usage 1999_2.html. [accessed 29 November 2003]

EPA. 2000. *Endocrine Disruptor Screening Program Report to Congress. August 2000*. Environmental Protection Agency, Washington DC

EPA. 2001. *Pesticide Industry Sales and Usage. 1998 and 1999 Market Estimates*. Environmental Protection Agency, Washington DC

Erbaugh J. M., Taylor D., Kyamanywa S., Kibwika P., Odeke V. and Mwanje E. 2001. Baseline II; a follow-up survey of farmer pest management practices and IPM knowledge diffusion. IPM CRSP Working paper 01–2, Virginia Tech, Blacksburg, 56pp

Escalada M. M. and Heong K. L. 2004. A participatory exercise for modifying rice farmers' beliefs and practices in stem borer management. *Crop Protection* 23 (1), 11–17

Eskenazi B., Bradman A. and Castorina R. 1999. Exposures of children to organophosphate pesticides and their potential adverse health effects. *Environmental Health Perspectives* 107 suppl. 3, 409–419

Espinosa P., Crissman C. C., Mera-Orcés V., Paredes M. and Basantes L. 2003. Conocimientos, actitudes y practicas de manejo de plaguicidas por familias productoras de papa en Carchi. In: Yanggen D., Crissman C. C. and Espinosa P. (eds) *Los Plaguicidas. Impactos en producción, salud y medio ambiente en Carchi, Ecuador.* CIP, INIAP, Ediciones Abya-Yala, Quito, Ecuador. pp25–48

ETC Group. 2001. Globalization, Inc. Concentration of corporate power: the unmentioned agenda. ETC Group Communique No 71, http://www.etcgroup.org/documents/com_globilization.pdf

EUREPGAP. 2003a. EUREPGAP News Update, June 2003, p8. www.eurep.org

EUREPGAP. 2003b. EUREPGAP General Regulations version 2.1Jan04, Control Points & Compliance Criteria, version2.0-Jan04, Fruit and Vegetables valid from: 12 September 2003. EUREPGAP, FoodPlusGmbH, Cologne

Eurostat. 2002. *The Use of Plant Protection Products in the European Union: Data 1992–1999.* European Communities, Luxembourg

Eusebio J. E. and Morallo-Rejesus B. 1996. Integrated pest management of diamondback moth: the Philippine highlands' experience. In: Sivapragasam A., Kole W. H., Hassan A. K. and Lim G.S. (eds) *The Management of Diamondback Moth and other Crucifer Pests.* Proceedings of the third International workshop, Kuala Lumpur, Malaysia, 29 Oct–1 Nov. pp253–56

Evans S. P., Knezevic S. Z., Lindquist J. L. and Shapiro C. A. 2003. Influence of nitrogen and duration of weed interference on corn growth and development. *Weed Science* 51, 546–556

Eveleens K. 2004. *The History of IPM in Asia*, FAO, Rome

Evenson R. E., Waggoner P. E. and Ruttan V. W. 1979. Economic benefits from research: an example from agriculture. *Science* 205, 1101–1107

Ewald J. A. and Aebischer N. J. 2000.Trends in pesticide use and efficacy during 26 years of changing agriculture in southern England. *Environmental Monitoring & Assessment* 64, 493–529

Eyre N., Downing T., Hoekstra R., Rennings K. and Tol R. S. J. 1997. *Global Warming Damages.* ExternE Global warming Sub-Task, Final Report, European Commission JOS3-CT95-0002, Brussels

Faeth P., Repetto R., Kroll K., Dai Q. and Helmers G. 1991. *Paying the Farm Bill: US Agricultural Policy and the Transition to Sustainable Agriculture.* World Resources Institute, Washington DC

Fairchild W. L., Swansburg E. O., Arsenault J. T. and Brown S. B. 1999. Does an association between pesticide use and subsequent declines in catch of Atlantic Salmon represent a case of endocrine disruption? *Environmental Health Perspectives* 107, 349–357

Fairhead J. 1991. Methodological notes on exploring indigenous knowledge and management of crop health. *RRA Notes* 14, 39–42, International Institute of Environment and Development, London

Falconer K. 1998. Managing diffuse environmental contamination from agricultural pesticides: an economic perspective on policy options, with particular reference to Europe. *Agriculture, Ecosystems & Environment* 69, 37–54

FAO. 1998. Community IPM: Six Cases from Indonesia, FAO Technical Assistance: Indonesian National IPM Program, FAO Rome 258pp

FAO. 2000a. The Impact of IPM Farmer Field Schools on Farmers' Cultivation Practicies in their Own Fields, A report submitted to the FAO Intercountry Programme for Community IPM in Asia by J. Pincus, Economics Department, University of London, 175pp

FAO. 2000b. *Agriculture: Towards 2015/30*. FAO, Rome

FAO. 2001a. Role of pesticide companies in small farm development in Senegal: a case of policy failure, in *Mid-Term Review of the Global IPM Facility*, FAO, Rome, April–June 2001, ppB2:6–11

FAO. 2001b. Baseline study on the problem of obsolete pesticide stocks. *FAO Pesticide Disposal Series* no 9. FAO, Rome, 36pp

FAO. 2001c. *Production Yearbook 1999*. FAO, Rome

FAO. 2001d. From Farmer Field Schools to Community IPM: Ten Years of IPM Training in Asia. (eds) J. Pontius, R. Dilts and A. Bartlett, Food & Agricultural Organisation Regional Office for Asia, Bangkok

FAO. 2002. International Code of Conduct on the Distribution and Use of Pesticides, Rome, Italy, www.fao.int

FAO. 2003. FAO statistical database. www.faostat.org

FAO/ECOWAS. 2002. News Brief: Mali Council of Ministers on the ratification of the regulations common to CILSS member countries on pesticide approval. *Pesticide Management in West Africa Newsletter* no.3, 30–31, FAO/ECOWAS, FAO Regional Office for Africa, Accra

FAO/WHO. 2001. Amount of poor-quality pesticides sold in developing countries alarmingly high. FAO/WHO press release, 1 February

FAOSTAT. 2002. FAOSTAT agricultural data:pesticides trade. http://apps.fao.org/page/collections?subset=agriculture, accessed May 2002

Farah J. 1994. *Pesticide Policies in Developing Countries: Do They Encourage Excessive Use?* World Bank Publications, Washington DC

Farmcare. 2003. *Focus on Farming Practice. The case for Integrated Farm Management 1993–2002*. Farmcare, http://www.co-opfarmcare.com/index2.htm

Farrow R. S., Goldburg C. B. and Small M. J. 2000. Economic valuation of the environment: a special issue. *Environmental Science and Technology* 34 (8), 1381–1383

Feder G., Murgai R. and Quizon J. B. 2004. Sending farmers back to school: the impact of Farmer Field Schools in Indonesia. *Review of Agricultural Economics* 26 (1), 45–62

Feeny P. 1976. Plant apparency and chemical defenses, 1–40. In: Wallace J. W. and Mansell R. L. (eds) *Biochemical Interaction Between Plants and Insects*. Plenum Press, New York

Fennimore S. A. and Jackson L. E. 2003. Organic amendment and tillage effects on vegetable field weed emergence and seedbanks. *Weed Technology* 17, 42–50

Fenske R. A. 1993. *Fluorescent Tracer Evaluation of Protective Clothing Performance*. Risk Reduction Engineering Laboratory, USEPA, Cincinnati

Fenske R., Wong S., Leffingwell J. and Spear R. 1986. A video imaging technique for assessing dermal exposure II. Fluorescent tracer testing. *American Industrial Hygiene Association Journal* 47, 771–775

Fenske R. A., Lu C., Simcox N. J., Loewenherz C., Touchstone J., Moate T. F., Allen E. H. and Kissel J. C. 2000. Strategies for assessing children's organophosphorus pesticide exposures in agricultural communities. *Journal of Experimental and Analytical Environmental Epidemiology* 10 (6 Pt 2), 662–671

Ferraz H. B., Bertolucci P. H., Pereira J. S., Lima J. G. and Andrade L. A. 1988. Chronic exposure to the fungicide maneb may produce symptoms and signs of CNS manganese intoxication. *Neurology* 38 (4), 550–553

Ferrigno S. 2003. The 'Hope of Koussanar'. *Pesticides News* 62, 7

Fianu F. K. and Ohene-Konadu K. 2000. Impact assessment of ICPM Farmers Field Schools on rice, vegetables, and plantain farming at five sites of project GHA/96/001 'National Poverty Reduction Programme' in Ghana. Report for FAO Regional Office for Africa, Accra, Ghana

FIELD. 2003. http://www.thefieldalliance.org/Index.htm

Firbank L. G., Perry J. N., Squire G. R., Bohan D. A., Brooks D. R., Champion G. T., Clarke S. J., Daniels R. E., Deawr A. M., Haughton A. J., Hawes C., Heard M. S., Hill M. O., May M. J., Osborne J. L., Rothery P., Roy D. B., Scott R. J. and Woiwod I. P. 2003. *The Implications of Spring-Sown Genetically Modified Herbicide-Tolerant Crops for Farmland Biodiversity. A Commentary on the Farm Scale Evaluations of Spring Sown Crops.* CEH, Merlewood

Fisher R. N. and Shaffer H. B. 1996. The decline of amphibians in California's Great Central Valley. *Conservation Biology* 10, 1387–1397

Fleischer G. 2000. Resource costs of pesticide use in Germany – the case of atrazine. *Agrarwirtschaft* 49 (11), 379–387

Fleischer G., Jungbluth F., Waibel H. and Zadoks J. C. 1999. *A Field Practitioner's Guide to Economic Evaluation of IPM.* Pesticide Policy Project, Hannover, Germany. 78pp

Fleming L. E. and Herzstein J. A. 1997. Emerging issues in pesticide health studies. *Occupational Medicine* 12, 387–397

Fliessbach A., Mäder P., Dubois D. and Gunst L. 2000. *Organic Farming Enhances Soil Fertility and Biodiversity*, FiBL Dossier no. 1, Research Institute of Organic Agriculture, Frick, Switzerland

Food and Fertiliser Technology Centre (FFTC). (1998). *Pesticide Problems in Asia: Production and Use.* International Seminar: Production and use of pesticides in Asia, Japan, www.agnet.org/library/article/ac1998e.html#0

Forget G. and Lebel J. 2001. An ecosystem approach to human health. *International Journal of Occupational and Environmental Health* 7 (2), S1–S38

Förster P. 2000. Development process of a small-scale neem processing plant. In: Stoll, G. *Natural Crop Protection in the Tropics: Letting Information Come to Life.* Margraf Verlag, Germany. pp321–326

Foster G. M. 1965. Peasant society and the image of the limited good. *American Anthropologist*, 67, 293–315

Frandsen S. E. and Jacobsen L.-B. 2001. Introduction to micro- and macro-economic models. In OECD workshop on the *Economics of Pesticide Risk Reduction in Agriculture*, Copenhagen, 28–30 Nov. OECD, Paris

Friedrich T. 2000. Pesticide application needs in developing countries. *Aspects of Applied Biology* 57, 193–200

Frolick L. M., Sherwood S., Hemphil A. and Guevara E. 2000. Eco-papas: Through potato conservation towards agroecology. *ILEA Newsletter.* December, 44–45

FSA. 2003. Pesticide residue minimisation Action Plan, Food Standards Agency, UK. http://www.foodstandards.gov.uk/aboutus/ourboard/boardmeetings/board2003/boardmeeting120603

Funes F., Garcia L., Bourque M., Perez N. and Rosset P. 2002. *Sustainable Agriculture and Resistance: Transforming Food Production in Cuba.* Food First Books, Oakland CA

Funtowicz S. O. and Ravetz J. R. 1993. Science for the post-normal age. *Futures* 25 (7), 739–755

Galden B. C., Sandler D. P., Zahm S. H., Kamel F., Rowland A. S. and Alavanja M. C. R. 1998. Exposure opportunities of families of farmer pesticide applicators. *American Journal of Industrial Medicine* 34, 581–587

Gallagher K. D. 1988. Effects of host plant resistance on the microevolution of the rice brown planthopper, *Nilaparvata lugens* (Stal) (Homoptera: Delphacidae), PhD dissertation, University of California, Berkeley

Gallagher, K. D. 2000. Community study programmes for integrated production and pest management: Farmer Field Schools. In: FAO, *Human Resources in Agricultural and Rural Development*, Rome. pp60–67

Gallagher K. D., Kenmore P. E. and Sogawa K. 1994. Judicious use of insecticides deter planthopper outbreaks and extend the life of resistant varieties in Southeast Asian rice, In: Denno R. F. and Perfect T. J. (eds) *Planthoppers; Their Ecology and Management*. Chapman & Hall, New York. pp599–614

Gandal N. and Roccas S. 2002. *Good Neighbours/Bad Citizens: Personal Value Priorities of Economists*. London: Centre for Economic Policy Research, Discussion Paper 3660

Gardner D. 2002. Farmer's view on search for safer alternatives. In: *The Pesticide Challenge*, delegates' pack for the conference to promote safer pest management, held 26 November, London, Pesticide Action Network UK, London. pp25–26

GAWI. 1998. Nos pépins sous la loupe. Initiation à la Production Intégrée des fruits à pépins CDRom, GAWI- FRUITNET, Visé, Belgium.

Gent D. H. and Schwartz H. F. 2003. Validation of potato early blight disease forecast models for Colorado using various sources of meteorological data. *Plant Disease* 87, 78–84

Georghiou G. P. 1986. The magnitude of the problem. In: National Research Council. *Pesticide Resistance, Strategies and Tactics*. National Academy Press, Washington DC

Georgiou S., Langford I. H., Bateman I. J. and Turner R. K. 1998. Determinants of individuals' willingness to pay for perceived reductions in environmental health risks: a case study of bathing water quality. *Environment and Planning* 30 (4), 577–594

Gerber A. and Hoffman V. 1998. The diffusion of eco-farming in Germany. In: Röling, N. and Wagemakers M. A. E. (eds) *Social Learning for Sustainable Agriculture*. Cambridge University Press, Cambridge

Gerken A., Suglo J.-V. and Braun M. 2000. Crop protection policy in Ghana. An economic analysis of current practice and factors influencing pesticide use. Draft study on behalf of the Plant Protection and Regulatory Services Directorate, Ministry of Food and Agriculture with the Integrated Crop Protection Project PPRSD/GTZ and the Pesticide Policy project, University of Hannover/GTZ, Accra, 173pp

Giampietro M. 2003. *Multi-Scale Integrated Analysis of Agroecosystems*. Boca Raton, CRC Press

Gibson K. D., Fischer A. J., Foin T. C. and Hill J. E. 2003. Crop traits related to weed suppression in water-seeded rice (*Oryza sativa L.*). *Weed Science* 51, 87–93

GIFAP (undated). Protección de Cultivos: Proyectos de Uso y Manejo Seguro en America Latina. GIFAP, Guatemala

Gigerenzer G. and Todd P. M. 1999. Fast and Frugal Heuristics: The Adaptive Toolbox. In: Gigerenzer G., Todd P. M. and the ABC Research Group (eds) *Simple Heuristics that Make Us Smart*. Oxford University Press, New York and Oxford. pp3–34

Gliessman S. 2000. *Agroecology: Ecological Processes in Sustainable Agriculture*. Lewis Publishers, Boca Raton

Glotfelty D. E., Seiber J. N. and Lidjedahl C. A. 1987. Pesticides in fog. *Nature* 325, 602–605

Goldman B. J. and Clancy K. L. 1991. A survey of organic produce purchases and related attitudes of food cooperative shoppers. *American Journal of Alternative Agriculture* 6 (2), 89–96

Gómez D., Prins C. and Staver C. 1999. Racionalidad en la toma de decisiones de MIP por pequeños y medianos caficultores de Nicaragua. *Manejo Integrado de Plagas*,

CATIE, Costa Rica. 53, 43–51. [Rational decision-making in IPM by small and medium coffee farmers in Nicaragua. In Spanish]

Goulston G. 2002. *The Agrochemical Market in Latin America: Market Analysis*, Agrow Reports, PJB Publications Ltd, Richmond, UK

Gouveia-Vigeant T. and Tickner J. 2003. Toxic chemicals and childhood cancer: A review of the evidence. Lowell Center for Sustainable Production. University of Massachusetts, Lowell. Available at http://sustainableproduction.org/downloads/Child% 20Canc%20Exec%20Summary.pdf [accessed 8 Feb 2004]

Govindasamy R., Italia J., Thatch D. and Adelaja A. 1998. Consumer Response to IPM-grown Produce. *Journal of Extension* 36 (4) http://www.joe.org/joe/1998 august/ rb2.html

Graf B., Lamb R., Heong K. L. and Fabellar L. 1992. A simulation model for the populations dynamic of rice leaf folders (Lepidoptera) and their interactions with rice, *Journal of Applied Ecology* 29, 558–570

Gray J. 2002. *Straw Dogs, Thoughts on Humans and Other Animals.* London, Granta Books

Grossman J. 1991. Organic potatoes in Wisconsin. *IPM Practitioner*, May–June, 16–17

Gu Y. -H. and Mazzola M. 2003. Modification of fluorescent pseudomonad community and control of apple replant disease induced in a wheat cultivar-specific manner. *Applied Soil Ecology* 24, 57–72

Guba E. G. and Lincoln Y. S. 1994. *Fourth Generation Evaluation.* London, Sage Publications

Guest D. I., Anderson R. D., Foard H. J., Phillips D., Worboys S. and Middleton R. M.1994. Long-term control of *Phytophthora* diseases of cocoa using trunk-injected phosphonate. *Plant Pathology* 43, 479–492

Guillette E., Meza M. M., Aquilar M. G., Soto A. D. and Garcia I. E. 1998. An anthropological approach to the evaluation of preschool children exposed to pesticides in Mexico. *Environmental Health Perspectives* 106 (6), 347–353

Gunderson L. H., Holling C. S. and Light S. S. (eds) 1995. *Barriers and Bridges to the Renewal of Ecosystems and Institutions.* New York, Colombia Press

Gurr G. M., Van Emden H. F. and Wratten S. D. 1998. Habitat manipulation and natural enemy efficiency: implications for the control of pests. In: Barbosa P. *Conservation Biological Control.* Academic Press, San Diego, California. pp155–183

Guthman J. 2000. Raising organic: an agroecological assessment of grower practices in California. *Agriculture and Human Values* 17 (3), 257–266

Gutierrez A. P. and Waibel H. 2003. *Review of the CGIAR System Wide Programme on Integrated Pest Management (SP-IPM).* TAC, Rome

Gutierrez A. P., Kogan M. and Stinner R. E. 2003. Report of the External IPM Review Panel to USAID/SPARE (Strategic Partnership for Agriculture Research and Education). Washington, DC

Hagmann J., Chuma E., Murwira K. and Connolly M. 1998. *Learning Together Through Participatory Extension. A Guide to an Approach Developed in Zimbabwe.* AGRITEX, Harare, Zimbabwe. 59pp

Hall R. W. and Ehler L. E. 1979. Rate of establishment of natural enemies in classical biological control. *Bulletin of the Entomological Society of America* 25 (4), 280–282

Hall R. W., Ehler L. E. and Bisabri-Ershadi B. 1980. Rate of success in classical biological control of arthropods. *Bulletin of the Entomological Society of America* 26 (2), 111–114

Halwart M. 1994. The golden apple snail, Pomacea canaliculata in Asian rice farming systems: present impact and future threat, *International Journal of Pest Management* 40 (2), 199–206

Hamilton N. A. 1995. *Learning to Learn with Farmers: An Adult Learning Extension Project.* PhD thesis, University of Wageningen, the Netherlands

Hammond D. A. and Stinner B. J. 1987. Soybean foliage insects in conservation tillage systems: effects of tillage, previous cropping history, and soil insecticide application. *Environmental Entomology* 16, 524–531

Hanley N., MacMillan D., Wright R. E., Bullock C., Simpson I., Parrison D. and Crabtree R. 1998. Contingent valuation versus choice experiments: estimating the benefits of environmentally sensitive areas in Scotland. *Journal of Agricultural Economics* 49 (1), 1–15

Hanshi J. A. 2001. Use of pesticides and personal protective equipment by applicators in a Kenyan district. *African Newsletter on Occupational Health and Safety* 11 (3), 74–76

Harari R., Forastiere F. and Axelson O. 1997. Unacceptable 'occupational' exposure to toxic agents among children in Ecuador. *American Journal of Industrial Medicine* 32 (3), 185–189

Harries J. E., Sheahan D. A., Jobling S., Matthiessen P., Neall P., Sumpter J. P., Tylor T. and Zaman N. 1997. Estrogenic activity in five UK rivers detected by measurement of vitellogenesis in caged male trout. *Environmental Toxicology and Chemistry* 16, 534–542

Harris F. A. 1997. *Transgenic B.t. cotton in the Mississippi Delta. IBC's 2nd Annual Conference on Biopesticides and Transgenic Plants.* International Business Communications, Southborough MA

Hartwig N. L. and Ammon H. U. 2002. 50th Anniversary-Invited Article, Cover crops and living mulches. *Weed Science* 50, 688–699

Hayes T. 2000. Endocrine disruption in amphibians. In Sparling D. W. (ed) *Ecotoxicology of Amphibians and Reptiles.* Society of Environmental Toxicology and Chemistry

Hayes T. B., Collins A., Lee M., Mendoza M., Noriega N., Stuart A. A. and Vonk A. 2002. Hermaphroditic, demasculinized frogs after exposure to the herbicide atrazine at low ecologically relevant doses. *Proceedings of the National Academy of Sciences* 99, 5476–5480

Hellin J. and Higman S. 2001. Competing in the market: farmers need new skills. *Appropriate Technology* 28 (2), 5–7

Henao S., Finkelman J., Albert L. and de Koning H. W. 1993. *Pesticides and Health in the Americas: Pan American Health Organization, Division of Health and Environment.* Washington DC. Environmental Series No 12

Heong K. L. and Escalada M. M. 1997. Perception change in rice pest management: A case study of farmers' evaluation of conflict information. *Journal of Applied Communications* 81 (2), 3–17

Heong K. L. and Escalada M. M. 1998. Changing rice farmers' pest management practices through participation in a small-scale experiment. *International Journal of Pest Management* 44, 191–197

Heong, K. L. and Escalada M. M. 1999. Quantifying rice farmers' pest management decisions: beliefs and subjective norms in stem borer control. *Crop Protection* 18, 315–322

Heong K. L., Escalada M. M. and Lazaro A. A. 1995. Misuse of pesticides among rice farmers in Leyte, Philippines. In: Pingali P. L. and Rogers P. A. (eds) *Impact of Pesticides on Farmers' Health and the Rice Environment.* Kluwer Academic, Norwell, MA. pp97–108

Heong K. L., Escalada M. M., Huan N. H. and Mai V. 1998. Use of communication media in changing rice farmers' pest management in the Mekong Delta, Vietnam. *Crop Protection,* 17 (5), 413–425

Herren H. R. 2003. Genetically engineered crops and sustainable agriculture. In: Ammann K., Jacot Y. and Braun R. (eds) *Methods for Risk Assessment of Transgenic Plants, IV. Biodiversity and Biotechnology.* Birkhäuser Verlag Basel/Switzerland

Herren H. R. and Neuenschwander P. 1991. Biological control of cassava pests in Africa. *Annual Review of Entomology* 36, 257–283

Herrera M. 1999. Estudio del subsector de la papa en Ecuador. INIAP PNRT-Papa. Quito, Ecuador. 140pp

Heydel L., Benoit M. and Schiavon M. 1999. Reducing atrazine leaching by integrating reduced herbicide use with mechanical weeding in corn (Zea mays). *European Journal of Agronomy* 11, 217–225

Hickey J. 1988. Some recollections about eastern North America's peregrine falcon population crash. In Cade T. J., Enderson J. H., Thelander C. G. and White C. M. (eds) *Peregrine Falcon Populations: Their Management and Recovery*. The Peregrine Fund, Boise, Idaho

Hilbeck A. H. 2002. *Transgenic Crops and Integrated Pest Management*. Swiss Federal Institute of Technology, Geobotanical Institute

Hill R. D. 1977. *Rice in Malaya: A Study in Historical Geography*. Oxford University Press, Kuala Lumpur, 213pp

HIV/AIDS Expert Group. 2001. Recommendation 18, The HIV/AIDS pandemic and its gender implications, Report of the expert group meeting 13–17 November 2000, Windhoek, Namibia

Hoffman M. P., Ridgway R. L., Show E. D. and Matteoni J. 1998. Practical application of mass-reared natural enemies: selected case histories, In: Ridgway R. L., Hoffman M. P., Inscoe M. N. and Glenister C. S. *Mass-Reared Natural Enemies: Application, Regulation, and Needs*. Thomas Say Publications in Entomology, Entomological Society of America, Lanham, MD

Holland M., Forster D., Young K., Haworth A. and Watkiss P. 1999. *Economic Evaluation of Proposals for Emission Ceilings for Atmospheric Pollutants*. Report for DG X1 of the European Commission. AEA Technology, Culham, Oxon

Holling C. S. 1995. What Barriers? What Bridges? In: Gunderson L. H., C. S. Holling and S. S. Light (eds) *Barriers and Bridges to the Renewal of Ecosystems and Institutions*. Colombia Press, New York. pp3–37

Holmes K. and Flood J. 2002. Biological control of pests and diseases of tropical tree crops with special reference to cocoa and coffee. In: J. G. M. Vos and P. Neuenchwander (eds) Proceedings of the West Africa regional cocoa IPM workshop (En & Fr), November 13–15, 2001, Cotonou, Benin, CABI Bioscience, IITA and CPL Press, 204pp

Horgan J. 1996. *The End of Science. Facing the Limits of Knowledge in the Twilight*. London, Abacus

Hruska A. and Corriols M. 2002. The impact of training in integrated pest management among Nicaraguan maize farmers: increased net returns and reduced health risk. *International Journal of Occupational and Environmental Health* 8 (3), 191–200

HSE. 1998a. *Pesticides Incidents Report 1997/8*. Health and Safety Executive, Sudbury

HSE. 1998b. *Pesticide Users and their Health: Results of HSE's 1996/7 Feasibility Study* Internet: http://www.open.gov.uk/hse/hsehome.htm

Huan N. H., Mai V., Escalada M. M. and Heong K. L. 1999. Changes in rice farmers' pest management in the Mekong Delta, Vietnam, *Crop Protection* 18, 557–563

Hubert B., R. Ison and N. Röling. 2000. The 'Problematique' with Respect to Industrialised Country 'Agricultures'. In: LEARN GROUP (ed), *Cow Up A Tree: Knowing and Learning for Change in Agriculture. Case Studies from Industrialised Countries*. INRA Editions, Paris. pp13–30

Hummel R. L., Walgenbach J. F., Hoyt G. D. and Kennedy G. G. 2002a. Effects of production system on vegetable arthropods and their natural enemies. *Agriculture, Ecosystems and Environment* 93, 165–176

Hummel R. L., Walgenbach J. F., Hoyt G. D. and Kennedy G. G. 2002b. Effects of vegetable production system on epigeal arthropod populations. *Agriculture, Ecosystems and Environment* 93, 177–188

Hurst D. K. 1995. *Crisis and Renewal: Meeting the Crisis of Organisational Change.* Harvard Business School Press, Boston

Hurst P. 1999. *The Global Pesticide's Industry's 'Safe Use and Handling' Training Project in Guatemala.* International Labour Organization, Geneva, February

IARC, 1979. IARC Monographs on the evaluation of the carcinogenic risk of chemicals to humans – Some halogenated hydrocarbons, IARC, October

Idris A. B. and Grafius E. 1995. Wildflowers as nectar sources for Diadegma insulare (Hymenoptera: Ichneumonidae), a parasitoid of diamondback moth (Lepidoptera: Yponomeutidae). *Environmental Entomology* 24 (6), 1726–1735

IFOAM. 2003. Organic Trade a Growing Reality. An Overview and Facts on Worldwide Organic Agriculture. http://www.ifoam.org/

Iga M. 1997. Effect of release of the introduced ichneumonid parasitoid, Diadegma semiclausum (Hellen), on the diamondback moth, *Plutella xylostella* (L.) in an experimental cabbage field. *Japanese Journal of Applied Entomology and Zoology* 44, 195–199

Imhoff D. 2003. *Farming with the Wild.* Sierra Club Books, San Francisco, California

Intergovernmental Forum of Chemical Safety (IFCS). 2003. *Acutely toxic pesticides: Initial Input on Extent of Problem and Guidance for Risk Management.* Fourth Session – Forum IV. Bangkok, Thailand 1–7 November 2003. Geneva: World Health Organization. (IFCS/FORUM-IV/10w)

International Labour Organisation. 1999. *Integration of Biological Control and Host Plant Resistance.* TCA Wageninigen/CAB International, UK

International Programme on Chemical Safety (IPCS) (2001). *The WHO recommended classification of pesticides by hazard and guidelines to classification 2000–01.* Geneva: United Nations Environment Programme, International Labour Organization, World Health Organization. (WHO/PCS/01.4)

International Rice Research Institute (IRRI), 1979, *Brown Planthopper: Threat to Rice Production in Asia,* Los Baños, Philippines

International Rice Research Institute (IRRI). 2003. *Trends in Pesticide Use: Cause for Hope.* www.knowledgebank.irri.org/pestUseRev/The_pesticide_explosion.htm, viewed 17 September 2003

IOBC. 1999. Integrated production: Principles and technical guidelines. *International Organization for Biological Control Bulletin* 22 (4)

IPCS. 2002. The WHO recommended classification of pesticides by hazard. International Programme on Chemical Safety. WHO/PCS/01.5, Geneva

Irwin M. E. 1999. Implications of movement in developing and deploying integrated pest management strategies. *Agricultural and Forest Meteorology* 97 (4), 235–248

Jarvis P. and Smith A. 2003. *Agrow's Ag-Biotech Top 20.* Agrow Reports, PJB Publications, Richmond

Jeger M. J. 2000. Bottlenecks in IPM. *Crop Protection* 19, 787–792

Jehae M. 2003. Moving towards pesticide use reduction in the Netherlands. *Pesticides News* 60, 17

Jensen J. E. and Petersen P. H. 2001. The Danish Pesticide Action Plan II: obstacles and opportunities to meet the goals. In: *The BCPC Conference Weeds 2001,* British Crop Protection Council, Farnham, UK pp 449–454

Jensen S. and Jansson B. 1976. Anthopogenic substances in seals from the Baltic: methylsulfone metabolites of PCB and DDE. *Ambio* 5, 257–260

Jervis M. S., Kidd M. A. C., Fitton M. D., Huddleson T. and Dawah H. A. 1993. Flower visiting by hymenopteran parasitoids. *Journal of Natural History* 27, 287–294

Jeyaratnam J. 1982. Health hazards awareness of pesticide applicators about pesticides. In: Van Heemstra E. A. H. and Tordoir W. F. (eds) *Education and Safe Handling in Pesticide Application*. Elsevier, Amsterdam. pp23–30

Jeyaratnam J. 1990. Acute pesticide poisoning: a major global health problem. *World Health Statistics Quarterly* 43 (3), 139–144

Jeyaratnam J., Lun K. C. and Phoon W. O. 1987. Blood cholinesterase levels among agricultural workers in four Asian countries. *Toxicology Letters* 33, 195–201

Jiggins J., Röling N. and van Slobbe E. 2002. Social Learning Theory in Relation to Water Management at the Catchment Scale. The Netherlands Team Theoretical Framework. Paper for the European funded SLIM project. Wageningen: University. Unpublished paper. http://slim-open.ac.uk

Jiggins J., Soon L. G. and Eveleens K. G. (eds) (2004). *Farmers, FAO and Field Schools*. FAO/GIF, Rome

Jobin B., DesGranges J.-L. and Boutin C. 1996. Population trends in selected species of farmland birds in relation to recent developments in agriculture in the St Lawrence Valley. *Agriculture, Ecosystems and the Environment* 57, 103–116

Johnson D. A. and Gilmore E. C. 1980. Breeding for resistance to pathogens in wheat. In Harris, M. K. (ed) *Biology & Breeding for Resistance in Arthropods & Pathogens in Agricultural Plants*. Texas A & M University, pp263–275

Jordan V. W. L., Leake A. R., Ogilvy S. and Higginbotham S. 2000. The economics of integrated farming systems in the UK. *Aspects of Applied Biology*, 62, 239–245

Kaergård N. 2001. Highlights from the Bichel committee's assessment in Denmark. In OECD workshop on the *Economics of Pesticide Risk Reduction in Agriculture*, Copenhagen, 28–30 Nov. OECD, Paris

Kalkhoven J. T. R. 1993. Survival of populations and the scale of the fragmented agricultural landscape, In: Bunce R. G. H., Ryszkowski L. and Paoletti M. G. *Landscape Ecology and Agroecosystems*. 1993. Lewis Publishers, Boca Raton. pp83–90

Kaneshiro L. N. and Jones M. W. 1996. Tritrophic effects of leaf nitrogen on Liriomyza trifolii (Burgess) and an associated parasitoid Chrysocharis oscinidis (Ashmead) on bean. *Biological Control* 6, 186–192

Katan J. 2000. Physical and cultural methods for the management of soil-borne pathogens. *Crop Protection* 19, 725–731

Kebe I. B., Guessan J. F., Keli J. Z. and Bekon A. K. 2002. Cocoa IPM research and implementation in Côte d'Ivoire. In: Vos J. G. M. and Neuenchwander P. (eds) *Proceedings of the West Africa Regional Cocoa IPM Workshop (En & Fr)*, November 13–15, 2001, Cotonou, Benin. CABI Bioscience, IITA and CPL Press, 204pp

Keifer M. (ed) 1997. Human health effects of pesticides. *Occupational Medicine: State of the Art Reviews* 12 (2), i–xi and 203–401

Keifer M., Rivas F., Moon J. D. and Checkoway H. 1996. Symptoms and cholinesterase activity among rural residents living near cotton fields in Nicaragua. *Occupational Environmental Medicine* 53, 726–729

Keifer M., McConnell R., Pacheco A. F., Daniel W., and Rosenstock L. 1996a. Estimating underreported pesticide poisonings in Nicaragua. *American Journal of Industrial Medicine* 30, 195–201

Keifer M. C., Murray D. I., Amador R., Coriols M., Gutierrez A., Moliere J., Van der Haar R. and Wesseling C. 1997. Solving pesticide problems in Latin America: a model for health-sector empowerment. *New Solutions* 7 (2), 26–31

Kenmore P. E. 1996. Integrated Pest Management in rice, In: Persley G. J. (ed) *Biotechnology and Integrated Pest Management*, CAB International, UK. pp76–97

Kenmore P. E., Carino F. O., Perez C. A., Dyck V. A. and Gutierrez A. P. 1984. Population regulation of the brown planthopper within rice fields in the Philippines. *Journal of Plant Protection in the Tropics* 1 (1), 19–37

Kenmore P. E., Litsinger J. A., Badong J. P., Santiag A. C. and Salac M. M. 1987. Philippine rice farmers and insecticides: thirty years of growing dependency and new options for change. No. 13. In: Tait J. and Napompeth B. (eds) *Management of Pests and Pesticides: Farmers' Perceptions and Practices*. West View Press, London

Kettlitz B. 2003. Consumer perspective, European Consumers' Organisation. In: Report on the seminar on compliance and risk reduction held 10 March 2003, Paris, OECD Pesticides Working Group, Organisation for Economic Co-operation and Development, ADD website

Kfir R., Overholt W. A., Khan Z. R. and Polaszek A. 2001. Biology and management of economically important lepidopteran cereal stem borers in Africa. *Annual Review of Entomology* 47, 701–31

Khan Z. R., Chiliswa P., Ampong-Nyarkko K., Smart L. A., Polaszek A., Wandera J. and Mulaa M. A. 1997a. Utilization of wild gramineous plants for management of cereal stemborers in Africa. *Insect Science Applications* 17, 143–150

Khan Z. R., Ampong-Nyarkko K., Chiliswa P., Hassanali A., Kimani S., Lwande W., Overholt W. A., Pickett J. A., Smart L. E., Wadhams L. J. and Woodcock M. 1997b. Intercropping increases parasitism of pests. *Nature* 388, 631–632

Khan Z. R., Hassanali A., Overholt W., Khamis T. M., Hooper A. M., Pickett J. A., Wadhams L. J. and Woodcock C. M. 2002. Control of witchweed *Striga hermonthica* by intercropping with *Desmodium* spp., and the mechanism defined as allelopathic. *Journal of Chemical Ecology* 28, 1871–1885

Kiff L. F. 1998. Changes in the status of the peregrine in North America. In: Cade T. J., Enderson J. H., Thelander C. G. and White C. M. (eds) *Peregrine Falcon Populations. Their Management and Recovery*. The Peregrine Fund, Boise, Idaho

Kimani M., Mihindo N. and Williamson S. 2000. We too are proud to be researchers. In: Stoll, G. (ed) *Natural Crop Protection in the Tropics: Letting Information Come to Life*. Margraf Verlag, Germany, pp 327–334

Kimani V. N. and Mwanthi M. A. 1995. Agrochemicals exposure and health implications in Githunguri locatoin, Kenya. *East African Medical Journal* 72 (8), 531–535

King E. G., Hopper K. R. and Powell J. E. 1985. Analysis of systems for biological control of crop arthropod pests in the US by augmentation of predators and parasites, In: Hoy M. A. and Herzog D. C. *Biological Control in Agricultural IPM Systems*. Academic Press, Orlando and London. pp201–228

Kishi, M. 2002. Indonesian farmers' perception of pesticides and resultant health problems from exposure. *Int J Occup Environ Health* 8 (3), 175–81

Kishi M. and LaDou J. 2001. Introduction: international pesticide use. *International Journal of Occupational and Environmental Health* 7 (3), 259–265

Kishi M., Hirschhorn N., Djajadisastra M., Satterlee L. N., Strowman S. and Dilts R. 1995. Relationship of pesticide spraying to signs and symptoms in Indonesian farmers. *Scandinavian Journal of Work, Environment, and Health* 21, 124–133

Knapp M., Wagener B. and Navajas M. 2003. Molecular discrimination between the spider mite *Tetranychus evansi* Baker & Pritchard, an important pest of tomatoes in southern Africa, and the closely related species *T. urticae* Koch (Acarina: Tetranychidae). *African Entomology* 11, 300–304

Knutson R. D., Taylor C. R., Penson J. B. and Smith E. S. 1990. *Economic Impacts of Reduced Chemical Use*. Knutson & Associates, College Station, Texas

Kogan M. 1998. Integrated Pest Management: historical perspectives and contemporary developments. *Annual Review of Entomology* 43, 243–270

Kolb D. 1984. *Experiential Learning: Experience as a source of learning and development*. New Jersey, Prentice Hall

Kolpin D. W. and Martin J. D. 2003. *Pesticides in Ground Water of the United States: Summary of Results of the National Water Quality Assessment Program (NAWQA), 1991–2001*. Sacramento, CA

Kolpin D. W., Barbash J. E. and Gilliom R. J. 2002. Atrazine and metolachlor occurrence in shallow ground water of the United States, 1993–1995: Relations to explanatory factors. *Journal of the American Water Resources Association* 38 (10), 301–311

Krafsur E. S. 1998. Sterile insect technique for suppressing and eradicating insect populations: 55 years and counting. *Journal of Agricultural Entomology* 15, 303–317

Krieger R. (ed) 2001. *Handbook of Pesticide Toxicology*. Academic Press, San Diego

Kroma M. M. and Flora C. B. 2003. Greening pesticides: a historical analysis of the social construction of farm chemical advertisements. *Agriculture and Human Values* 20, 21–35

Kuepper G. and Thomas R. 2002. *Bug Vacuums for Organic Crop Protection*. February 2002. ATTRA Pest Management Technical Note

Kuhn T. S. 1970. *The Structure of Scientific Revolutions*. 2nd ed. University of Chicago Press, Chicago

Kunstadter P., Prapamontol T., Sirirojn B. O., Sontirat A., Tansuhaj A. and Khamboonruang C. 2001. Pesticide exposures among Hmong farmers in Thailand. *International Journal of Occupational and Environmental Health* 7, 313–325

Lamondia J. A., Elmer W. H., Elmer, Mervosh T. L. and Cowles R. S. 2002. Integrated management of strawberry pests by rotation and intercropping. *Crop Protection* 21, 837–846

Lander B. F., Knudsen L. E., Gamborg M. O., Jarventaus H. and Norppa H. 2000. Chromosome aberrations in pesticide-exposed greenhouse workers. *Scandinavian Journal of Work, Environment & Health*, 26(5), 436–442

Landis D. A. and Menalled F. D. 1998. Ecological considerations in the conservation of effective parasitoid communities in agricultural systems, In Barbosa P. *Conservation Biological Control*. Academic Press, San Diego. pp101–122

Landis J. N., Sanchez J. E., Bird G. W., Edson C. E., Isaacs R., Lehnert R. H., Schilder A. M. C. and Swinton S. M. 2002. *Fruit Crop Ecology and Management*. Michigan State University, East Lansing, MI

Landrigan P., Claudio L., Markowitz S., Berkowitz G. S., Brenner B. L., Romero H., Wetmur J. G., Matte T. D., Gore A. C., Godbold J. H. and Wolff M. S. 1999. Pesticides and inner-city children: exposures, risks, and prevention. *Environmental Health Perspectives* 107, suppl 3, 431–437

Langewald J. and Cherry A. 2000. Prospects for microbial pest control in West Africa, *Biocontrol News & Information* 21 (20), 51N–56N

Larsen, E. W. 2001. Farmer field schools: Impact of IPM activities in vegetables during the summer 2001 season in Bangladesh. SPPS 64. Department of Agricultural Extension-DANIDA Strengthening Plant Protection Services Project, Bangladesh. 11pp

Latour B. 1998. To modernize or to ecologize? That's the question. In: Castree N. and Willems-Braun B. (eds) *Remaking Reality: Nature at the Millennium*. Routledge, London and New York. pp221–242

Leach, G. 1976. *Energy and Food Production*. IPC Science and Technology Press, Guilford and IIED, London

Leake A. R. 2002. *Biodiversity in Different Farming Systems*. The Allerton Trust

LEISA. 2003. Aprendiendo con las ECAs. LEISA: Revista de Agroecología. Junio. 19 (1), 87

Letourneau, D. K. and Dyer, L. A. 1998. Experimental test in lowland tropical forest shows top-down effects through four trophic levels. *Ecology* 79 (5), 1678–1687

Letourneau, D. K. 1998. Conserving biology: Lessons for conserving natural enemies, In: Barbosa, P. (ed) *Conservation Biological Control*. Academic Press, San Diego, California, pp9–38

Leung L. K.-P., Singleton, G. R., Sudarmaji. 1999. Ecologically-based populations management of the rice-field rat in Indonesia, In: Singleton G. R., Hinds L., Herwig L. and Zhang Z. (eds) *Ecologically-based Rodent Management*, ACIAR, Canberra, Australia. pp305–318

Levy B. S., Levin J. L. and Teitelbaum D. T. (eds) 1999. DBCP-induced sterility and reduced fertility among men in developing countries: a case study of the export of a known hazard. *International Journal of Occupational and Environmental Health* 5 (2), 115–150

Li Wenhua. 2001. *Agro-Ecological Farming Systems in China*. Man and the Biosphere Series Volume 26. UNESCO, Paris

Liebman M. and Davis A. S. 2000. Integration of soil, crop and weed management in low-external-input farming systems. *Weed Research* 40, 27–47

Lighthall D. 1995. Farm structure and chemical use in the corn belt. *Rural Sociology* 60 (3), 505–520

Lim G. S. and Di Y. B. (eds) 1989. *Status and Management of Major Vegetable Pests in the Asia-Pacific Region (With special focus towards Integrated Pest Management)*. RAPA, FAO, Bangkok, Thailand

Lim G. S. and Ooi P. 2003. Farmer Field Schools: From Rice to other Crops, In: Eveleens K., Jiggins J. and Lim G. S. (eds) Farmers, FAO and Field Schools: Bringing IPM to the Grass Roots in Asia, FAO, Rome

Lipson M. 1997. *Searching for the 'O-word': an analysis of the USDA Current Research Information System (CRIS) for Pertinence to Organic Farming*. Organic Farming Research Foundation, Santa Cruz, California

Loevinsohn M. E. 1987. Insecticide use and increased mortality in rural central Luzon, Philippines. Lancet 1, 1359–1362

Loevinsohn M. E. 1993. Improving pesticide regulation in the Third World: the role of an independent hazard auditor. In Forget G., Goodman T. and de Villiers A. (eds) *Impact of Pesticide Use on Health in Developing Countries: Proceedings of a symposium held in Ottawa, Canada, 17–20 September 1990*. Ottawa, Ontario: IDRC. pp166–177

Loevinsohn M. E. 1994, Rice pests and agricultural environments, In: Heinrichs E. A. (ed) *Biology and Management of Rice Insects*. Wiley Eastern Limited. pp487–515

Loevinsohn M., Meijerink G., ISNAR and Salasya B. 1998. Integrated Pest Management in Smallholder Farming Systems in Kenya: Evaluation of a pilot project. KARI. Unpublished report, CABI Bioscience UK Centre, Ascot, UK. 50pp

Lohr L. and Park T. A. 2002. Choice of insect management portfolios by organic farmers: lessons and comparative analysis. *Ecological Economics* 43, 87–99

Lomborg B. 2000 *The Sceptical Environmentalist: Measuring the Real State of the World*. Cambridge University Press, Cambridge

London L. and Bailie R. 2001. Challenges for improving surveillance for pesticide poisoning: policy implications for developing countries. *International Journal of Epidemiology* 30 (3), 564–570

London L. and Myers J. E. 1995. Critical issues for agrochemical safety in South Africa. *American Journal of Industrial Medicine* 27 (1), 1–14

London L. and Rother H. A. 2000. People, pesticides and the environment: who bears the brunt of backward policy in South Africa? *New Solutions* 10 (4),339–350

London L., Cairncrosse E. and Solomons A. 2000. The quality of surface and groundwater in the rural Western Cape with regard to pesticides. Final report to the Water Research Commission on the Project 'An Assessment of the quality of water supplies in the rural Western Cape with regard to agrichemical pollutants.' WRC Report Number K5/795. Water Research Commission, Pretoria

London L., de Grosbois S., Wesseling C., Kisting S., Rother H. A. and Mergler D. 2002. Pesticide usage and health consequences for women in developing countries: Out of sight, out of mind? *International Journal of Occupational and Environmental Health* 8, 46–59

Lopez-Carrillo L., Blair A., Lopez-Cervantes M., Cebrian M., Rueda C., Reyes R., Mohar A. and Bravo J. 1997. Dichlorodiphenyltrichloroethane serum levels and breast cancer risk: a case-control study from Mexico. *Cancer Research* 57 (17), 3728–32.

Loureiro M. L., McCluskey J. J. and Mittelhammer R. C. 2002. Will consumers pay a premium for eco-labeled apples? *The Journal of Consumer Affairs* 36 (2), 203–219

Lowry L. K. and Frank A. L. 1999. Exporting DBCP and other banned pesticides: consideration of ethical issues. *International Journal of Occupational and Environmental Health* 5 (2), 135–141

Lubchenco J. 1998. Entering the century of the environment: a new social contract for science. *Science* 279, 491–496

Lum K. Y., Josoh Mamat Md., Cheah U. B., Castaneda C. P., Rola A. C. and Sinhaseni P. 1993. Pesticide research for public health and safety in Malaysia, the Philippines, and Thailand. In: Forget G., Goodman T. and Villiers A. (eds) *Impact of Pesticide Use on Health in Developing Countries*. Proceedings of a symposium held in Ottawa, Canada. September 17–20. IDRC, Ottawa. pp31–48

Maarleveld M. 2003. *Social Environmental Learning for Sustainable Natural Resource Management: Theory, Practice and Facilitation*. Wageningen: WUR/CIS, published Doctoral Dissertation

MacArthur R. H. and Wilson E. O. 1963. An equilibrium theory of insular zoogeography. *Evolution* 17, 373–387

MacArthur R. H. and Wilson E. O. 1967. *The Theory of Island Biogeography*. Princeton University Press, Princeton, NJ

Macfarlane R. 2003. Pesticide free production: a tool for use reduction. *Pesticides News* 60, 14–15

Macha M., Rwazo A. and Mkalanga H. 2001. Retail sales of pesticides in Tanzania: occupational human health and safety considerations. *African Newsletter on Occupational Health & Safety* 11 (2), 40–42

Mangan J. and Mangan M. S. 1998. A comparison of two IPM training strategies in China: the importance of concepts of the rice ecosystem for sustainable insect pest management. *Agriculture & Human Values* 15, 209–221

Manu M. and Tetteh E. K., 1987. *A Guide to Cocoa Cultivation*. Cocoa Research Institute Ghana. 51pp

Marino P. C. and Landis D. A. 1996. Effect of landscape structure on parasitoid diversity and parasitism in agroecosystems. *Ecological Applications* 6 (1), 276–284

Marks & Spencer. 2003. http://www2.marksandspencer.com/thecompany/ourcommit menttosociety/environment/reports/food_pesticides_prohibited_list_270703.pdf

Maroni M., Colosio C., Ferioli A. and Fait A. 2000. Biological monitoring of pesticide exposure: a review. *Toxicology* 143, 1–123

Mason C. F. 2002. *Biology of Freshwater Pollution*. 4th ed. Addison, Wesley Longman, Harlow

Matteson P. C. 2000. Insect pest management in tropical Asian irrigated rice. *Annual Review of Entomology* 45, 549–574

Matteson P. C., Gallagher K. D. and Kenmore P. E. 1994. Extension of integrated pest management for planthoppers in Asian irrigated rice: Empowering the user. In: Denno R. F. and Perfect T. J. (eds) *Ecology and Management of Planthoppers*. Chapman and Hall, London. pp656–668

Matthews G., Wiles T. and Baleguel P. 2003. A survey of pesticide application in Cameroon. *Crop Protection* 22, 707–714

Maturana H. R. and Varela F. J. 1987 (and rev ed 1992). *The Tree of Knowledge: The Biological Roots of Human Understanding*. Shambala Publications, Boston, MA

McConnell R. and Hruska A. J. 1993. An epidemic of pesticide poisoning in Nicaragua: Implication for prevention in developing countries. *American Journal of Public Health* 83 (11), 1559–1562

McConnell R. and Magnotti, R. 1994. Screening for insecticide overexposure under field conditions: a reevaluation of the tintometric cholinesterase kit. *American Journal of Public Health* 84, 479–481

McConnell R., Cedillo L., Keifer M. and Palomo M. R. 1992. Monitoring organophosphate insecticide-exposed workers for cholinesterase depression. New technology for office or field use. *Journal of Occupational Medicine* 34, 34–37

McConnell R., Delgado-Tellez E., Cuadra R., Torres E., Keifer M., Almendarez J., Miranda J., El-Fawal H. A., Wolff M., Simpson D. and Lundberg I. 1999a. Organophosphate neuropathy due to methamidophos: biochemical and neurophysiological markers. *Archives of Toxicology* 73, 296–300

McConnell R., Pacheco F., Wahlberg K., Klein W., Malespin O., Magnotti R., Akerblom M. and Murray D. 1999b. Subclinical health effects of environmental pesticide contamination in a developing country: cholinesterase depression in children. *Environmental Research*, Section A. 81, 87–91

McDougall L., Magloire L., Hospedales C. J., Tollefson J. E., Ooms M., Singh N. C. and White F. M. 1993. Attitudes and practices of pesticide users in Saint Lucia, West Indies. *Bulletin of the Pan American Health Organization* 27 (1), 43–51

McNeely J. A. and Scherr S. J. 2003. *Ecoagriculture*. Island Press, Washington DC

Meadows D., Meadows D., Randers J. and Behrens W. 1972. *The Limits to Growth*. Universe Books, New York

Meffe G. K., Nielsen L. A., Knight R. L. and Schenborn D. A. 2002. *Ecosystem Management: Adaptive Community-based Conservation*. Island Press, Washington DC, 313pp

Meikel W. G., Degbey P., Oussou R., Holst N., Nansen C. and Markham R. H. 1999. Pesticide use in grainstores: an evaluation based on survey data from Benin republic. PhAction News #1, IITA Benin

Meir C. J. 1990. Farmer Perception of Pesticides used in Crop Production in the Sixaola District of Talmanca, Costa Rica. MSc Thesis, Imperial College Centre for Environmental Technology, University of London, UK

Meir C. J. 2000. Learning and changing: Helping farmers move to natural pest control. In: Stoll, G. *Natural Crop Protection in the Tropics: Letting Information Come to Life*. Margraf Verlag, Germany. pp265–279

Meir C. J. 2004. *Training for Change: Evaluation of Participatory Training in Natural Pest Control for Smallholder Farmers in Central America*. PhD Thesis, Imperial College Centre for Environmental Technology, University of London, UK

Mekonnen Y. and Agonafir T. 2002. Effects of pesticide applications on respiratory health of Ethiopian farm workers. *International Journal of Occupational and Environmental Health* 8 (1), 35–40

Mekonnen Y. and Agonafir T. 2002. Pesticide sprayers' knowledge, attitude and practice of pesticide use on agricultural farms in Ethiopia. *Occupational Medicine* 52, 311–315

Mendonca G. A., Eluf-Neto J., Andrada-Serpa M. J., Carmo P. A., Barreto H. H., Inomata O. N. and Kussumi T. A. 1999. Organochlorines and breast cancer: a case-control study in Brazil. *International Journal of Cancer* 83 (5), 596–600

Mensah R. K. 2002. Development of an integrated pest management programme for cotton. Part I: establishing and utilizing natural enemies. *International Journal of Pest Management* 48 (20), 87–94

Mera-Orces V. 2000. *Agroecosystems Management, Social Practices and Health: A Case Study on Pesticide Use and Gender in the Ecuadorian Highlands.* A Technical Report to the IDRC. Canadian-CGIAR Ecosystem Approaches to Human Health Training Awards with a particular focus on gender. 39pp

Mera-Orces V. 2001. The sociological dimensions of pesticide use and health risks of potato production in Carchi, Ecuador. Paper prepared for the Open Meeting of the Human Dimensions of Global Environmental Change Research Community. Rio de Janeiro, Brazil, 6–8 October. 21pp

Merino R. and Cole D. C. 2003. Presencia de plaguicidas en el trabajo agrícola, en los productos de consumo, y en el hogar. In Yanggen D., Crissman C. C. and Espinosa P. (eds) *Los Plaguicidas. Impactos en producción, salud y medio ambiente en Carchi, Ecuador.* CIP, INIAP, Ediciones Abya-Yala, Quito, Ecuador. pp71–93

Metcalf, R. L. (1986) Coecolutionary adaptations of rootworm beetles (Coleopera: Chrysomelidae) to cucurbitacins. *Journal of Chemical Ecology* 12, 1109–1124

Miklasiewicz T. J. and Hamond R. B. 2001. Density of potato leafhopper (Homoptera: Cicadellidae) in response to soybean-wheat cropping systems. *Environmental Entomology* 30 (2), 204–214

Miller A. 1983. The influence of personal biases on environmental problem-solving. *Journal of Environmental Management*, 17, 133–142

Miller A. 1985. Technological thinking: its impact on environmental management. *Environmental Management* 9 (3), 179–190

Miranda J., Lundberg I., McConnell R., Delgado E., Cuadra R., Torres E., Wesseling C. and Keifer M. 2002a. Onset of grip- and pinch-strength impairment after acute poisonings with organophosphate insecticides. *International Journal of Occupational and Environmental Health* 8 (1), 19–26

Miranda J., McConnell R., Delgado E., Cuadra R., Keifer M., Wesseling C., Torres E. and Lundberg I. 2002b. Tactile vibration thresholds after acute poisonings with organophosphate insecticides. *International Journal of Occupational and Environmental Health* 8 (3), 212–219

Mital A. 2002. *Giving Away the Farm: The 2002 Farm Bill.* Backgrounder series. Food First Publications

Möller K. (ed) 2000. EUREPGAP Implementation 2000 plus, Official Conference, Barcelona, October, 2000. Euro-retailer Produce Working Group, Eurohandelsinstitut. Köln, Germany. 26pp

Morrill, W. L. (1997) Feeding behavior of Leptocorisa oratorius (F.) in rice. *Recent Research Developments in Entomology*, 1, 11–14

Morse, S. and W. Buhler, 1997. *Integrated Pest Management: Ideals and Realities in Developing Countries*, Lynne Rienner, Boulder, 171pp

Moses M., Johnson E., Anger K. W., Burse V. W., Hortsman S. W., Jackson R. J., Lewis R. G., Maddy K. T., McConnell R., Meggs W. J. and Hoar Zahm S. 1993. Environmental equity and pesticide exposure. *Toxicology and Industrial Health* Sep–Oct; 9 (5), 913–959

Motte F., Compton J. A. F., Magrath P. A. and Aftosa A. 1996. Ghana – a participatory approach to research and extension: the experience of the ODA (UK)/MOFA (Ghana) Large Grain Borer project. In: Farrell G., Greathead A. H., Hill M. G. and Kibata K. N. (eds) *Management of Farm Storage Pests in East and Central Africa*. International Institute of Biological Control, Ascot, UK, pp101–108

MSU. 2000. The Database of Arthopods Resistant to Pesticides. Michigan State University. At www.cips.msu.eud/resistance

Mudimu G. D., Chigume S. and Chikanda M. 1995. *Pesticide Use and Policies in Zimbabwe*. Pesticide Policy Project Publication Series no. 2, Institut für Gartenbauökonomie, University of Hanover, Germany. 30pp

Mullen J. D., Norton G. W. and Reaves D. W. 1994. Economic analysis of environmental benefits of integrated pest management. *Journal of Agriculture and Applied Economics* 29, 243–253

Mumford J. D. 1981. A study of sugar beet growers' pest control decisions. *Annals of Applied Biology*. 97, 243–252

Mumford J. D. and Leach A. 1999. Technology transfer in Asia: a case study of Cocoa Pod Borer management demonstration plots in Sulawesi, Indonesia. In: USDA-CABI-ACRI collaborative cocoa–coffee research meeting, held 7–8 December 1999, London, UK. Unpublished compilation of presentations and abstracts, CABI Bioscience, Ascot, UK

Muramoto J., Gliessman S. R., Koike S. T., Schmida D. and Stephens R. 2003. *Maintaining Agroecosystem Health in the Organic Management of a Strawberry/vegetable Rotation System*. Annual Meeting Abstracts, Agronomy Society of America, Crop Science Society of America, and Soil Science Society of America, Denver, Colorado

Murphy B. C., Rosenheim J. A., Granett J., Pickett C. H. and Dowell R. V. 1998. Measuring the impact of a natural enemy refuge: The prune tree/vineyard example, In: Pickett C. H. and Bugg R. L. *Enhancing Biological Control: Habitat Management to Promote Natural Enemies of Agricultural Pests*. University of California Press, Berkeley, CA. pp297–309

Murphy H. 2001. IPM and farmer's health. *Spider Web: A Newsletter About IPM Training in Asia*. November, 2001. Jakarta: FAO Programme for Community IPM in Asia

Murphy, H. H., Sanusi A., Dilts R., Djajadisastra M., Hirschhorn N. and Yuliatingsih S. 1999. Health effects of pesticide use among Indonesian women farmers: Part 1. Exposure and acute health effects. *Journal of Agromedicine* 6, 61–85

Murphy H. H., Hoan N. P., Matteson P. and Abubakar A. L. 2002. Farmers' self-surveillance of pesticide poisoning: a 12-month pilot in northern Vietnam. *International Journal of Occupational and Environmental Health* 8 (3), 201–211

Murray D. L. 1994. *Cultivating Crisis: The Human Cost of Pesticides in Latin America*. University of Texas Press, Austin

Murray D. L. and Taylor P. L. 2000. Claim no easy victories: evaluating the pesticide industry's global safe use campaign. *World Development* 28 (10),1735–1749

Murray D. L. and Taylor P. L. 2001. Safe Use – not so safe. *Pesticides News* 54, 6–7

Murray D. L. and Taylor P. L. 2001. Beyond safe use: challenging the international pesticide industry's hazard reduction strategy, International Institute for Environment and Development, Gatekeeper Series No 103

Murray D., Wesseling C., Keifer M., Corriols M. and Henao S. 2002. Surveillance of pesticide-related illness in the developing world: putting the data to work. *International Journal of Occupational and Environmental Health* 8 (3), 243–248

Mutch D. R., Martin T. E. and Kosola K. R. 2003. Red clover (*Trifolium pratense*) suppression of common ragweed (*Ambrosia artemisiifolia*) in winter wheat (*Triticum aestivum*). *Weed Technology* 17, 181–185

NAS. 1999. *Hormonally Active Agents in the Environment*. National Academy of Sciences. National Academy Press, Washington DC

Nathaniels N. Q. R., Sijaona M. E. R., Shoo J. A. E., Katinila N. and Mwijage A. 2003a. IPM for control of cashew powdery mildew in Tanzania, II: networks of knowledge and influence. *International Journal of Pest Management* 49 (1), 37–48

Nathaniels N., Nag A., Agli C., Adetonah S., Fagbemissi R., Lantokpode B. and Kakpo Z. 2003b. Tracing the effects of Farmer Field Schools within existing knowledge exchange networks: the case of cowpea in Benin. Unpublished paper, Danish Institute of Agricultural Sciences and International Institute of Tropical Agriculture, Flakkebjerg. 31pp

National Institute for Occupational Safety and Health. 2001. Tracking occupational injuries, illnesses, and hazards: *The NIOSH Surveillance Strategic Plan*. DHHS (NIOSH) Publication No. 2001–118, Washington, DC: NIOSH, January 2001. Available at http:/ /www.cdc.gov/niosh/2001-118.html. [accessed 8 Feb 2004]

Nazarko O., Schoofs A., Van Acker R. C., Entz M., Derksen D., Martens G. and Andrews T. 2002. Pesticide-Free Production. *Pesticide Outlook* 13 (3), 116–118

Ndemah R. and Schulthess F. 2002. Yield of maize in relation to natural field infestations and damage by lepidopterous borers in the forest and forest/savannah transition zones of Cameroon. *Insect Science Applications* 22, 183–193

Ndemah R., Gounou S. and Schulthess F. 2002. The role of wild grasses in the management of lepidopterous cereal stemborers on maize in the forest zone of Cameroon and the derived savanna of southern Benin. *Bulletin of Entomological Research and Environmental Entomology* 32, 61–70

Ndemah R., Schulthess F., Korie, S., Borgemeister C., Poehling, H.-M. and Cardwell, K. 2003. Factors affecting infestations of the Stalk Borer *Busseola fusca* (Lepitoptera: Noctuidae) on maize in the forest zone of Cameroon with special reference to Scelionid Egg Parasitoids. *Environmental Entomology*

Nentwig W. 1998. Weedy plant species and their beneficial arthropods: potential for manipulation in field crops, In: Pickett C. H. and Bugg R. L. *Enhancing Biological Control: Habitat Management to Promote Natural Enemies of Agricultural Pests*. University of California Press, Berkeley, California. pp49–71

Neuchâtel Group. 1999. Common framework on agricultural extension. Neuchâtel Initiative, Ministère des Affaires étrangères, Paris, France. 19pp.

New Sunday Times, Malaysia, 'advertorials' (a term used by the Malaysian Oil Palm Industry Association to describe its two full page advertisements), 12 October 2003 and 9 November 2003.

Newton I. 1979. *Population Ecology of Raptors*. Poyser, Berkhamstead

Newton I. 1986. *The Sparrowhawks*. Poyser, Berkhamstead

Ngouajioa M., McGiffen Jr. M. E. and Hutchinson C. M. 2003. Effect of cover crop and management system on weed populations in lettuce. *Crop Protection* 22, 57–64

Ngowi A. V., Maeda D. N., Wesseling C., Partanen T. J., Sanga M. P. and Mbise G. 2001. Pesticide-handling practices in agriculture in Tanzania: observational data from 27 coffee and cotton farms. *International Journal of Occupational and Environmental Health* 7 (4), 326–332

Nicholls C. I., Parrella M. and Altieri M. A. 2001. The effects of a vegetational corridor on the abundance and dispersal of insect biodiversity within a northern California organic vineyard. *Landscape Ecology* 16, 133–146

Nielsen H. 2002. Danish success with pesticide action plans. *Pesticides News* 57, 10

Norgaard, R. 1994. *History in Development Betrayed. The end of the progress and a coevolutionary revisioning of the future*. Routledge Press, London. pp280

Norse D., Li Ji, Jin L. Eshan and ZhangZheng. 2001. *Environmental Costs of Rice Production in China*. Aileen Press, Bethesda

North D. C. 1990. *Institutions, Institutional Change and Economic Performance*. Cambridge University Press, New York

Norton G. A. 1982. Crop Protection Decision Making: An Overview. In: *Proceedings 1982, British Crop Protection Symposium Decision Making in the Practice of Crop Protection*. BCPC, Brighton

Norton G. A. and Mumford J. D. 1993. *Decision Tools for Pest Management*. CAB International, Wallingford, UK. 264pp

Norton G. A., Adamson D., Aitken L. G., Bilston L. J., Foster J., Frank B. and Harper J. K. 1999. Facilitating IPM: The role of participatory workshops. *International Journal of Pest Management*, 45 (2), 85–90

Nuffield Council on Bioethics. 2004. *The Use of Genetically Modified Crops in Developing Countries*. Nuffield Council on Bioethics, London

OBEPAB (Organisation Beninoise pour la Promotion de l'Agriculture Biologique). 2002. Le Coton au Benin: rapport de consultation sur le cotton conventionnel et le cotton biologique au Benin, PAN UK, London

OECD (ed Parris K). 1999. *OECD Agri-Environmental Indicators*. OECD, Paris

OECD. 2001a. *Environmental Outlook for the Chemicals Industry*. OECD, Paris

OECD. 2001b. *Initiatives to Share the Burden of the Testing and Assessment of Endocrine Disrupting Chemicals*. Environment Directorate. ENV/JM(2001)20/REV1, OECD, Paris

OECD/IEA. 1992. *Energy Balances of OECD Countries*. OECD/International Energy Agency, Paris

Office of Technology Assessment. 1995. *Biologically Based Technologies for Pest Control*. September United States Congress

Ohayo-Mitoko G. J. A., Heederik D. J. J., Kromhout H., Omondi B. E. O. and Boleij J. S. M. 1997. Acethylcholinesterase inhibition as an indicator of organophosphate and carbamate poisoning in Kenyan agricultural workers. *International Journal of Occupational and Environmental Health* 3, 210–220

Ohayo-Mitoko G. J. A., Kromhout H., Karumba P. N. and Boleij J. S. M. 1999. Identification of determinants of pesticide exposure among Kenyan agricultural workers using empirical modeling. *Annals of Occupational Hygiene* 43 (8), 519–525

Okoth J. R. Khisa G. S. and Julianus T. 2003. Towards self-financed farmer field schools. *LEISA Magazine* 19 (1), 28–29

Olaya-Contreras P., Rodriguez-Villamil J., Posso-Valencia H. J. and Cortez J. E. 1998. Organochlorine exposure and breast cancer risk in Colombian women. *Cad Saude Publica*. 14 Suppl 3, 125–132

O'Neill R. V., DeAngelis D. L., Waide J. B. and Allen T. F. H. 1986. Monographs in Population Biology, 23. A Hierarchical Concept of Ecosystems, 253. In: O'Neill R. V., DeAngelis D. L., Waide J. B. and Allen T. F. H. *Monographs in Population Biology, 23: A Hierarchical Concept of Ecosystems*. Princeton University Press, Princeton, NJ

Ooi P. A. C. 1988. *Ecology and Surveillance of Nilaparvata lugens (Stal) – Implications for its Management in Malaysia*, PhD dissertation, University of Malaya. 275pp

Ooi P. 1992. Role of parasitoids in managing diamondback moth in the Cameron Highlands, Malaysia. In: Talekar, N. S. (ed) *Diamondback Moth and Other Crucifer Pests*. Proceedings of the second international workshop, Tainan, Taiwan, 10–14 December 1990. AVRDC publication no 92–368, pp255–262

Ooi P. A. C. 1998. Beyond the Farmer Field School: IPM and empowerment in Indonesia. *Gatekeeper Series No. 78*, International Institute for Environment and Development, London, UK

Ooi P. A. C. and Shepard B. M. 1994. Predators and parasitoids of rice insect pests, In: Heinrichs E. A. (ed) *Biology and Management of Rice Insects*, Wiley Eastern Limited. pp585–612

Ooi P. A. C., Warsiyah, Nanang Budiyanto, and Nguyen Van Son. 2001. Farmer scientists in IPM: a case of technology diffusion. In: Mew T. W., Borromeo E. and Hardy B. (eds) *Exploiting Biodiversity for Sustainable Pest Management*, Proceedings of the Impact Symposium on Exploiting Biodiversity for Sustainable Pest Management, 21–23 August 2000, Kunming, China. Makati City (Philippines), International Rice Research Institute, Los Banos. pp207–215

Oostindie H., Van der Ploeg J. D. and Rentink H. 2002. Farmers' Experiences with and views on rural development practices and processes: outcomes of a transnational European survey. In: Van der Ploeg J. D., Long A. and Banks J. (eds) *Living Country-sides: Rural Development Processes in Europe, the State of the Art*. Doetinchem: Elsevier Bedrijfsinformatie, pp214–230

Opoku I. Y., Akrofi A. Y., Appiah A. A. and Leuterbacher M. C. 1998. Trunk injection of Potassium Phosphonate for the control of black pod disease of cocoa. *Tropical Science* 38 (3), 179–185

Organic Alliance. 2001. Education Still Job #1 for Organic Alliance. January 26, 2001. Press Release. Organic Alliance

Osburn S. 2000. Research report: Do pesticides cause lymphoma? Chevy Chase, MD: Lymphoma Foundation of America. Available at http://www.lymphomahelp.org/rr_2000.pdf

OSTP. 1996. The health and ecological effects of endocrine disrupting chemicals. Office of Science and Technology Policy, Executive Office of the President. Washington DC

Ostrom E. 1998. Coping with Tragedies of the Commons. Paper for 1998 Annual Meeting of the Association for Politics and Life Sciences, Boston, 3–6 September

Overholt W. A., Ngi-Song A. J., Omwega C. O., Kimani-Njogu S. W., Mbapila J., Sallam M. N. and Ofomata V. 1997. A review of the introduction and establishment of *Cotesia flavipes* Cameron (Hymenoptera: Braconidae) in East Africa for biological control of cereal stemborers. *Insect Science Applications* 17, 79–88

Paarlberg R. 2002. Governance and food security in an age of globalisation. 2020Vision Discussion paper no.36, International Food Policy Research Institute, Washington

Padi B., Ackonor J. B., Abitey M. A., Owusu E. B., Fofie A. and Asante E. 2000. Report on the insecticide use and residues in cocoa beans in Ghana. Internal report submitted to the Ghana Cocoa Board. 26pp

Padi B., Ackonor J. B. and Opoku I. Y. 2002a. Cocoa IPM research and implementation in Ghana. In: *Proceedings of the West Africa Regional Cocoa IPM workshop* (English and French), 13–15 November, 2001, Cotonou, Benin (eds J. G. M. Vos and P. Neuenchwander). CABI Bioscience, IITA and CPL Press. 204pp

Padi B., Oduor G. and Hall D. 2002b. Development of mycopesticides and pheromones for cocoa mirids in Ghana. DFID Crop Protection Programme Final technical report. 45pp

Padilla D., Staver C., Monterroso D., Guharay F., Mendoza R., Aguilar A., Monterrey J. and Mendez E. 1999. La implementación participativa del MIP en diferentes zonas cafetaleras de Nicaragua. In: Leyva E. R. and Escobar J. J. (eds) *Memorias del XXII Congreso Nacional de Control Biológico, Montecillo, Mexico, 28–29 Octubre 1999*. Sociedad Mexicana de Control Biologico, Montecillo, Mexico. pp262–265. [Participatory IPM implementation in different coffee growing zones of Nicaragua,. In Spanish.]

Page S. L. J. 2001. Promoting the survival of rural mothers with HIV: A development strategy for southern Africa. *SAfAIDS News*. 9 (1), 2–8

Pain S. 1993. 'Rigid' cultures caught out by climate change. *New Scientist*, 5 March

Paine R. T. 1974. Intertidal community structure: experimental studies on the relationship between a dominant competitor and its principal predator. *Oecologia* 15, 93–120

PAN Africa. 2000. New cases of intoxication in Senegal. *Pesticides & Alternatives* 12, 2–5

PAN Europe. 2003. Worst ever level of residues in grapes in discount supermarkets. *PAN Europe Newsletter* 15, 3

PAN Germany. 2001. From law to field. Pesticide use reduction in agriculture – from pesticide residue analyses to action. http://www.pan-germany.org/download/law-field-sum.pdf

Paredes M. 1995. *Evaluación del Impacto de la Capacitación en Enfermedades de Plantas para Pequeños Agricultores en Tres Comunidades de Honduras*. Ingeniero Agrónomo thesis, Escuela Agrícola Panamericana (Zamorano), Honduras

Paredes M. 2001. We are like the fingers of the same hand: Peasants' heterogeneity at the interface with technology and project intervention in Carchi, Ecuador. MSc thesis. Wageningen, The Netherlands: Wageningen University. 150pp

Parella M. P., Heintz K. M. and Nunney L. 1992. Biological control through augmentative releases of natural enemies: a strategy whose time has come. *American Entomologist* 38 (3), 172–179

Parker W. E. 2002. What impact is ICM having on pest and disease management in field vegetables? ADAS, Woodthorne, Wolverhampton

Parris K. 2002. A framework for analysis of sustainable agriculture: an OECD approach applied to soil and water management. In Proceedings of an OECD Cooperative Research Program Workshop *Agricultural Production and Ecosystems Management*. 12–16 November, Ballina, Australia. OECD, Paris

Payne J., Scholze M. and Kortenkamp A. 2001. Mixtures of four organochlorines enhance human breast cancer cell proliferation. *Environmental Health Perspectives* Apr;109 (4), 391–397

Paz-y-Mino, C., Bustamente G., Sanchez M. E. and Leone P. E. 2002. Cytogenetic monitoring in a population occupationally exposed to pesticides in Ecuador. *Environmental Health Perspectives* 110, 1077–1080

Peakall D. B., Reynolds L. M. and French M. C. 1976. DDE in eggs of the peregrine falcon. *Bird Study* 23, 183–186

Pearce D. and Tinch R. 1998. The true price of pesticides. In: Vorley W. and Keeney D. (eds) *Bugs in the System*. Earthscan, London

Pearce D .W., Cline W. R., Achanta A. N., Fankhauser S., Pachauri R. K., Tol R. S. J. and Vellinga P. 1996. The social costs of climate change: greenhouse damage and benefits of control. In: Bruce et al (eds) *Climate Change 1995: Economic and Social Dimensions of Climate Change*. Cambridge University Press, Cambridge

Penagos H. 2002. Contact dermatitis caused by pesticides among banana plantation workers in Panama. *International Journal of Occupational and Environmental Health* 8 (1), 14–18

Penagos H., Jimenez V., Fallas V., O'Malley M. and Maibach H. I. 1996. Chlorothalonil, a possible cause of erythema dyschromicum perstans (ashy dermatitis). *Contact Dermatitis* 35 (4), 214–218

Perrow C. 1986. *Complex Organisations: A Critical Essay*. Random House, New York

Persley G. J. and MacIntyre L. R. 2002. *Agricultural Biotechnology: Country Case Studies – A Decade of Development*, The Doyle Foundation, Glasgow. ppxix + 228

Peterson D. L. and Parker V. T. 1998. *Ecological Scale: Theory and Applications*. Columbia University Press. New York

Petersen P., Tardin J. M. and Marochi F. 2000. Participatory development of non-tillage systems without herbicides for family farming: the experience of the center-south region of Paraná. *Environmental Development and Sustainability* 1, 235–252

Phillips M. and McDougall J. 1999. Crop Protection's Changing Face. *Farm Chemicals International*, WOW 2000 Global, November

Phillips M. and McDougall J. 2003. The cost of new agrochemical product discovery, development and registration in 1995 and 2000. Final report, a consultancy study for CropLife America and the European Crop Protection Association, Pathhead UK, April

Pimentel D., Acguay H., Biltonen M., Rice P., Silva M., Nelson J., Lipner V., Giordano S., Harowitz A. and D'Amore M. 1992. Environmental and economic cost of pesticide use. *Bioscience* 42 (10), 750–760

Pimentel D., Acquay H., Biltonen M., Rice P., Silva M., Nelson J., Lipner V., Giordano S., Horowitz A. and D'Amore M. 1993. Assessment of environmental and economic impacts of pesticide use. In: Pimentel, D. and Lehman H. (eds) *The Pesticide Question: Environment, Economics, and Ethics*. New York, London. pp47–84

Pingali P. L. and Roger P. A. 1995. *Impact of Pesticides on Farmers' Health and the Rice Environment*. Kluwer Academic Press, Dordrecht

Pinkerton J. N., Ivors K. L., Miller M. L. and Moore L. W. 2000. Effect of soil solarization and cover crops on populations of selected soilborne plant pathogens in western Oregon. *Plant Disease* 84, 952–960

Plog B. A., Niland J., Quinlan P. J. and Plogg H. (eds) 1996. *Fundamentals of Industrial Hygiene*. 4th ed. Ithaca, NY. National Safety Council. 1011pp

Poelking A. 1992. Diamondback moth in the Philippines and its control with *Diadegma semiclausum*. In: Talekar, N. S. (ed) *Diamondback Moth and Other Crucifer Pests*. Proceedings of the second International workshop, Tainan, Taiwan, 10–14 December 1990. AVRDC publication No. 92–368, pp271–278

Polaszek A. 1998. African cereal stem borers; economic importance, taxonomy, natural enemies and control. CAB International in association with the ACP-EU Technical Centre for Agricultural and Rural Co-operation (CTA). 530pp

Pollard E. 1971. Hedges. VI. Habitat diversity and crop pests: A study of Brevicoryne brassica and its syrphid predators. *Journal of Applied Ecology* 8 (3), 751–780

Pollard E. 1973. Hedges: VII. Woodland relic hedges in Huntington and Petersborough. *Journal of Ecology* 61 (2), 343–352

Ponting C. 1991. *A Green History of the World: The Environment and the Collapse of Great Civilizations*. Penguin Books, London. 432pp

Pontius J., Dilts R. and Bartlett A. 2002. *From Farmer Field Schools to Community IPM: Ten Years of IPM Training in Asia*. Bangkok: FAO, Regional Office for Asia and the Pacific

Popper K. R. 1972. *Objective Knowledge: An Evolutionary Approach*. Oxford University Press, Oxford

Potts G. R. 1996. *The Partridge: Pesticides, Predation and Conservation*. Chapman and Hall, London

Powell W. 1991, 1994. Neither Market nor Hierarchy: Network Forms of Organisation. In: Thompson G., Frances J., Levavcic R. and Mitchell J. (eds) *Markets and Hierarchies and Networks: The Co-ordination of Social Life*. Sage, London. pp256–277

Power M. E. 1990. Effect of fish in river food webs. *Science* 250, 411–415

Praneetvatakul S. and Waibel H. 2002. A socio-economic analysis of Farmer Field Schools implemented by the National Program on Integrated Pest Management of Thailand. Paper presented at the CYMMIT impact assessment conference 4–7 February 2002, San Jose, Costa Rica, 23pp, Pesticide Policy Project website http://www.ifgb.uni-hannover.de/ppp/

PRC. 2000. Annual report of the Pesticides Residues Committee 2000, Pesticides Safety Directorate, UK, *www.pesticides.gov.uk*

Pretty J. N. 1995. *Regenerating Agriculture: Policies and Practice for Sustainability and Self-Reliance*. Earthscan, London; National Academy Press, Washington

Pretty J. N. 1998. *The Living Land: Agriculture, Food and Community Regeneration in Rural Europe*. Earthscan, London, 336pp

Pretty J. N. 1998. Supportive policies and practices for scaling up sustainable agriculture. In: Röling, N. and M. A. E. Wagemakers (eds) *Social Learning for Sustainable Agriculture*. Cambridge University Press, Cambridge. pp23–45

Pretty J. N. 2000. Can sustainable agriculture feed Africa? *Environmental Development and Sustainability* 1, 253–274

Pretty J. N. 2001. The rapid emergence of genetic modification in world agriculture: contested risks and benefits. *Environmental Conservation* 28 (3), 248–262

Pretty J. N. 2002. *Agri-Culture: Reconnecting People, Land and Nature*. Earthscan, London

Pretty J. N. 2003. Social capital and the collective management of resources. *Science* 302, 1912–1915

Pretty J. N. and Chambers R. 1994. Towards a learning paradigm: new professionalism and institutions for agriculture. In: Scoones, I. and Thompson, J. (eds) *Beyond Farmer First: Rural People's Knowledge, Agricultural Research and Extension Practice*. Intermediate Technology Publications, London. pp182–202

Pretty J. N. and Hine R. 2000. The promising spread of sustainable agriculture in Asia. *National Resources Forum* 24, 107–126

Pretty J. N. and Hine R. 2001. Reducing food poverty with sustainable agriculture: a summary of new evidence. Centre for Environment & Society, University of Essex, Colchester

Pretty J. N. and Ward H. 2001. Social capital and the environment. *World Development* 29 (2), 209–227

Pretty J. N., Brett C., Gee D., Hine R., Mason C. F., Morison J. I. L., Raven H., Rayment M. and van der Bijl G. 2000. An assessment of the total external costs of UK agriculture. *Agricultural Systems* 65 (2), 113–136

Pretty J. N., Brett C., Gee D., Hine R. E., Mason C. F., Morison J. I. L., Rayment M., van der Bijl G. and Dobbs T. 2001. Policy challenges and priorities for internalising the externalities of agriculture. *Journal of Environmental Planning and Management* 44 (2), 263–283

Pretty J. N., Ball A. S., Li Xiaoyun and Ravindranath N. H. 2002. The role of sustainable agriculture and renewable resource management in reducing greenhouse gas emissions and increasing sinks in China and India. *Philosophical Transactions of the Royal Society Series A* 360, 1741–1761

Pretty J. N., Mason C. F., Nedwell D. B. and Hine R. E. 2003a. Environmental costs of freshwater eutrophication in England and Wales. *Environmental Science and Technology* 37 (2), 201–208

Pretty J. N., Morison J. I. L. and Hine R. E. 2003b. Reducing food poverty by increasing agricultural sustainability in developing countries. *Agricultural Ecosystems and the Environment* 95 (1), 217–234

Price P. W. 1976. Colonization of crops by arthropods: non-equilibrium communities in soybean fields. *Environmental Entomology* 5 (4), 605–611

Price P. W. 1997. *Insect Ecology*. 3rd ed. John Wiley & Sons, New York

Price P. W. and Waldbauer G. P. 1994. Ecological aspects of pest management, In: Metalf R. L. and Luckman W. H. *Introduction to Insect Pest Management*. Wiley-Interscience, New York. pp33–65

PROFEPA. 2001. Denuncia PROFEPA Envenenamiento de Peces, Grupo Reforma Servicio Informativo (http://www.mural.com/occidente/articulo/092807/)

Prokopy R. J. 2003. Two decades of bottom-up, ecologically based pest management in a small commercial apple orchard in Massachusetts. *Agriculture, Ecosystems and Environment* 94, 299–309

Protected Harvest. 2001. New Eco–label greatly reduces growers' use of toxic pesticides, Protect Harvest press release, 5th October 2001, and PH website and Accomplishments and Directions 2003 piece

Pumisacho M. and Sherwood S. (eds) 2000. *Herramientas de Aprendizaje para Facilitadores. Manejo Integrado del Cultivo de Papa.* Quito, Ecuador: INIAP and CIP. 181pp

Quijano R. 2002. *Endosulfan Poisoning in Kasargod, Kerala, India.* Report of a Fact Finding Mission, Pesticide Action Network Asia and the Pacific, Malaysia, At URL panap@panap.net

Quizon J., Feder G. and Murgai R. 2000. *A Note on the Sustainability of the Farmer Field School Approach to Agricultural Extension.* Washington: The World Bank, Development Economics Group

Rastogi S. K., Gupta B. N., Husain T., Mathur N. and Garg N. 1989. Study of respiratory impairment among pesticide sprayers in Mango plantations. *American Journal of Industrial Medicine* 16 (5), 529–538

Ratcliffe D. A. 1958. Broken eggs in peregrine eyries. *British Birds* 51, 23–26

Ratcliffe D. A. 1970. Changes attributable to pesticides in egg breakage frequency and eggshell thickness in some British birds. *Journal of Animal Ecology* 45, 831–849

Ratcliffe D. A. 1980. *The Peregrine Falcon.* Poyser, Berkhamstead

RCEP. 1996. *Sustainable Use of Soil.* 19th report of the Royal Commission on Environmental Pollution. Cmnd 3165, HMSO, London

Reardon T. and Barrett C. B. 2000. Agroindustrialization, globalization, and international development. An overview of issues, patterns and determinants. *Agricultural Economics* 23, 195–205

Reganold J. P., Glover J. D., Andrews P. K. and Hinman H. R. 2001. Sustainability of three apple production systems. *Nature* 410, 926–930

Reis S. R., Kund G. S., Carson W. G., Phillips P. A. and Trumble J. T. 1999. Economics of reducing insecticide use on celery through low-input pest management strategies. *Agriculture, Ecosystems & Environment* 73, 185–197

Renner K. A. 2000. Weed ecology and management, In: Cavigelli M. A., Deming S. R., Probyn L. K. and Mutch D. R. *Michigan Field Crop Pest Ecology and Management.* Michigan State University Extension Bulletin E-2704, East Lansing, Michigan. pp51–68

Repetto R. and Baliga S. S. 1996. *Pesticides and the Immune System: the Public Health Risks.* World Resources Institute, Washington DC

Repetto R. and Baliga S. 1997a. *Pesticides and the Immune System: The Public Health Risks.* World Resources Institute, Washington DC

Repetto R. and Baliga S. 1997b. Pesticides and immunosuppression: the risks to public health. *Health Policy Planning* 12 (2), 97–106

Restrepo M., Munoz N., Day N. E., Parra J. E., de Romero L. and Nguyen-Dinh X. 1990a. Prevalence of adverse reproductive outcomes in a population occupationally exposed to pesticides in Colombia. *Scandinavian Journal of Work Environment and Health* 16, 232–238

Restrepo M., Munoz N., Day N., Parra J. E., Hernandez C., Blettner M. and Giraldo A. 1990b. Birth defects among children born to a population occupationally exposed to pesticides in Colombia. *Scandinavian Journal of Work Environment and Health* 16, 239–246

Reus J. A. W. A. and Leendertse P. C. 2000. The environmental yardstick for pesticides: a practical indicator used in the Netherlands. *Crop Protection* 19, 637–641

Rhoades D. F. and Gates R. G. 1976. Toward a general theory of plant anti-herbivore chemistry, In: Wallace J. W. and Mansell R. L. *Biochemical Interaction Between Plants and Insects*. Planum Press, New York. pp168–213

Ribaudo M. O., Horan R. D. and Smith M. E. (1999). *Economics of Water Quality Protection from Nonpoint Sources: Theory and Practice*. Agricultural Economic Report 782. Economic Research Service, US Department of Agriculture, Washington, DC

Rice R. E. and Kirsch P. 1990. Mating disruption of Oriental Fruit Moth in the United States, In: Ridgway R. L., Silverstein R. M. and Inscoe M. N. *Behavior-modifying Chemicals for Insect Management: Applications of Pheromones and Other Attractants*. Marcel Dekker, Inc, New York. pp193–211

Richter E. D. and Safi J. 1997. Pesticide use, exposure, and risk: a joint Israel–Palestinian Perspective. *Environmental Research* 73, 211–218

Richter E. D., Chuwers P., Levy Y., Gordon M., Grauer F., Mazordouk J., Levy S., Barron S. and Gruener N. 1992. Health effects from exposure to organophosphates in workers and residents in Israel. *Israel Journal of Medical Science* 28, 584–598

Ricketts T. H. 2001. The matrix matters: Effective isolation in fragmented landscapes. *The American Naturalist* 158 (1), 87–99

Riddell I. 2001. Monitor farms and farmer discussion groups in New Zealand. Their role in improving physical and financial performance through collective learning. Occasional report, Scottish Agricultural College, St Boswells, UK

Ridley M. 1995. *The Origins of Virtue*. Penguin Books, Harmondsworth

Rigby D., Woodhouse P., Young T. and Burton M. 2001. Constructing a farm level indicator of sustainable agricultural practice. *Ecological Economics* 39, 463–478

Risch S. J., Andow D. and Altieri, M. A. 1983. Agroecosystem diversity and pest control: data, tentative conclusions, and new research directions. *Environmental Entomology* 12, 625–629.

Rita P., Reddy P. P. and Reddy S. V. 1987. Monitoring of workers occupationally exposed to pesticides in grape gardens of Andhra Pradesh. *Environmental Research* 44, 1–5

Rogers E. M. 1995. *Diffusion of Innovations*. 4th ed. Free Press, New York

Rogers J. 1989. *Adults Learning*. Open University Press, Milton Keynes

Rojas A., Ojeda M. E. and Barraza X. 2000. [Congenital malformations and pesticide exposure]. *Revista Medica de Chile* 128 (4), 399–404 [In Spanish]

Rola A. and Pingali P. 1993. *Pesticides, Rice Productivity, and Farmers' Health: An Economic Assessment*. IRRI, Los Baños, Philippines

Rola A. C., Provido Z. S., Olanday M. O., Paraguas F. J., Sirue A. S., Espadon M. A. and Hupeda S. P. 1997. *Socio-economic Evaluation of Farmer Decision-Making among IPM Farmer Field School Graduates*. SEAMEO Regional Center for Graduate Study & Research in Agriculture, Los Baños, Laguna, Philippines. 128pp

Röling, N. 2000. Gateway to the global garden: Beta-gamma science for dealing with ecological rationality. Eight annual Hopper Lecture. University of Guelph, Canada, 24 October. (available at: www.uoguelph.ca/cip). 47pp

Röling N. 2002a. Beyond the Aggregation of Individual Preferences. Moving from multiple to distributed cognition in resource dilemmas. In: Leeuwis, C. and Pyburn R. (eds) *Wheelbarrows Full of Frogs: Social Learning in Natural Resource Management*. Koninklijke Van Gorcum, Assen. pp25–28

Röling N. 2002b. Is There Life After Agricultural Science? Public Lecture held at the occasion of his retirement, 27 June in the Assembly Hall of Wageningen University. Also presented at Workshop on New Directions in Agro-ecology Research and

Education. Sponsored by the University of Wisconsin Agro-ecology Initiative. 29–31 May 2002

Röling N. 2003. From causes to reasons: the human dimension of agricultural sustainability. *International Journal of Agricultural Sustainability* 1 (1), 73–88

Röling N. and de Jong F. 1998. Learning: Shifting paradigms in education and extension studies. *Journal of Agricultural Education and Extension* 5 (3), 143–161

Röling N. and Wagemakers M. A. E. (eds) 1998. *Facilitating Sustainable Agriculture*. Cambridge University Press, Cambridge

Rombach, M. C. and K. D. Gallagher, 1994, The brown planthopper: Promises, problems and prospects, In: Heinrichs E. A. (ed) *Biology and Management of Rice Insects*, Wiley Eastern Limited. pp693–711

Romieu I., Hernandez-Avila M., Lazcano-Ponce E., Weber J. P. and Dewailly E. 2000. Breast cancer, lactation history, and serum organochlorines. *American Journal of Epidemiology* 152 (4), 363–370

Root R. B. 1973. Organization of a plant-arthropod association in simple and diverse habitats: the fauna of collards (Brassica oleracea). *Ecological Monographs* 43, 95–124

Rosenheim J. A., Kaya H. K., Ehler L. E., Marois J. J. and Jaffee B. A. 1995. Intraguild predation among biological-control agents: Theory and evidence. *Biological Control* 5 (3), 303–335

Rosenstock L., Keifer M., Daniell W., McConnell R., Claypoole K., and the Pesticide Health Effects Study Group. 1991. Chronic central nervous system effects of acute organophosphate pesticide intoxication. *Lancet* 338, 223–227

Rosenthal E. 2003. The tragedy of Tauccamarca: A human rights perspective on the pesticide poisoning deaths of 24 children in the Peruvian Andes. *International Journal of Occupational and Environmental Health* 9, 53–58.

Rossett P. M. and Benjamen M. (eds) 1994. *The Greening of the Revolution: Cuba's Experiment with Organic Agriculture*. Ocean Press, Melbourne

Rowley J. 1992. Different constraints on individual and community action. In: Gibson R.W. and Sweetmore A. (eds) *Proceedings of a Seminar on Crop Protection for Resource-Poor Farmers*, Natural Resources Institute, Chatham, UK

Rubia E. G., Heong K. L., Zalucki M., Gonzales B., and Norton G. A. 1996, Mechanisms of compensation of rice plants to yellow stem borer *Scirpophaga incertulas* (Walker) injury. *Crop Protection* 15, 335–340

Rueda A. 1995. What do farmers really want from IPM? In: *IPM Implementation Workshop for Central America and the Caribbean. Workshop Proceedings, Costa Rica, 1994*. Natural Resources Institute, Chatham, UK. 116pp

Rupa D. S., Reddy P. P. and Reddi O. S. 1991. Reproductive performance in population exposed to pesticides in cotton fields in India. *Environmental Research* 55 (2), 123–128

Russell D. A., Kranthi K. R., Surulivelu T., Jadhav D. R., Regupathy A. and Singh J. 2000. Developing and implementing insecticide resistance management practices in cotton ICM programmes in India. In: *British Crop Protection Council Proceedings of the International BCPC Conference on Pests and Diseases 2000*, held in Brighton, UK 13–16 November, BCPC, Farnham, UK. pp205–212

Sæthre M.-G., Ørpen H. M. and Hofsvang T. 1999. Action programmes for pesticide risk reduction and pesticide use in different crops in Norway. *Crop Protection* 18 (2), 7–215

SAI. 2003. Sustainable Agriculture Initiative Platform, *http://www.saiplatform.org/*

SAN. 2000. A Whole Farm Approach to Managing Pests. USDA Sustainable Agriculture Research and Education Program. Available at www.sare.org/htdocs/pubs/. [accessed 11/15/03]

Schafer K. (ed) 1998. *Learning from the BIOS Approach*. Community Alliance with Family Farmers and World Resources Institute, California. 37pp

Schettler T. 2001. Toxic threats to neurologic development of children. *Environmental Health Perspectives* 109 Suppl 6, 813–816

Schettler T., Stein J., Valenti M., Reich F. and Wallinga D. 2000. In harm's way: toxic threats to child development. Boston, MA: Greater Boston Physicians for Social Responsibility. Available at http://www.igc.org/psr/ihw-report.htm. [accessed 8 Feb 2004]

Schmitz P. M. 2001. Overview of cost–benefit assessment. In OECD workshop on the *Economics of Pesticide Risk Reduction in Agriculture*, Copenhagen, 28–30 Nov. OECD, Paris

Schulthess F., Bosque-Pérez N. A., Chabi-Olaye A., Gounou S., Ndemah R. and Goergen, G. 1997. Exchanging natural enemies species of lepidopterous cereal stemborers between African regions. *Insect Science Applications* 17, 97–108

Schulthess F., Chabi-Olaye A. and Goergen G. 2001. Seasonal fluctuations of noctuid stemborer egg parasitism in southern Benin with special reference to Sesamia calamistis Hampson (Lepidoptera: Noctuidae) and *Telenomus* spp. (Hymenoptera: Scelionidae) on maize. *Biocontrol, Science and Technology* 11, 765–777

Schulthess F., Chabi-Olaye A. and Gounou S. 2004. Multi-trophic level interactions in a cassava-maize relay cropping system in the humid tropics of West Africa. *Bulletin of Entomological Research* 94 (3), 261–272

Scialabba N. and Hattam C. 2002. *Organic Agriculture, Environment and Food Security*. FAO, Rome

Seif A., Varela A. M., Michalik S. and Löhr B. 2001. A guide to IPM in French beans production. ICIPE Science Press, Nairobi

Sétamou M. and Schulthess F. 1995. The influence of egg parasitoids belonging to the *Telenomus busseolae* (Hym.: Scelionidae) species complex on *Sesamia calamistis* (Lepidoptera: Noctuidae) populations in maize fields in southern Benin. *Biocontrol Science and Technology* 5, 69–81

Sétamou M., Schulthess F., Bosque-Pérez N. A. and Thomas-Odjo A. 1993. Effect of nitrogen and silica on the bionomics of *Sesamia calamistis* Hampson (Lepidoptera: Noctuidae). *Bulletin of Entomological Research* 83, 405–411

Sétamou M., Schulthess R., Bosque-Pérez N. A. and Thomas-Odjo A. 1995. The effect of stem and cob borers on maize subjected to different nitrogen treatments with special reference to *Sesamia calamistis* Hampson (Lepidoptera: Noctuidae). *Entomologica Experimentalis et Applicata* 77, 205–210

Settle W. H., Ariawan H., Tri Astuti E., Cahyana W., Hakim A. L., Hindayana D., Sri Lestari A. and Pajarningsih. 1996. Managing tropical rice pests through conservation of generalist natural enemies and alternative prey. *Ecology* 77 (7), 1975–1988

Settle W. H. and Whitten M. J. 2000. Plenary Lecture: The Role of Small Scale Farmers in Strengthening the Link between Sustainable Agriculture and Biodiversity. In: *The XXIst Congress of Entomology*, Iguassu, Brazil, August 2000

Shanower T. G., Schulthess F. and Bosque-Pérez N. A. 1993. The effect of larval diet on the growth and development of Sesamia calamistis Hampson (Lepidoptera: Noctuidae) and *Eldana saccharina* Walker (Lepidoptera: Pyralidae). *Insect Science Applications* 14, 681–685

Sharpe C. R., Franco E. L., de Camargo B., Lopes L. F., Barreto J. H., Johnsson R. R. and Mauad M. A. 1995. Parental exposures to pesticides and risk of Wilms' tumor in Brazil. *American Journal of Epidemiology* 141 (3), 210–217

Shennan C. and Bode C. A. 2002. Integrating wetland habitat with agriculture, In: Jackson L. L. and Jackson D. *The Farm as a Natural Habitat*. Island Press, Washington. pp189–204

Shennan C., Cecchettini C. L., Goldman G. B. and Zalom F. G. 2001. Profiles of California farmers by degree of IPM use as indicated by self-descriptions in a phone survey. *Agriculture, Ecosystems and Environment* 84, 267–275

Shepard B. M. and Ooi P. A. C. 1991. Techniques for evaluating predators and parasitoids in rice. In: Heinrichs E. A. and Miller T. A. (eds) *Rice Insects: Management Strategies*, Springer-Verlag, New York. pp197–214

Shepard, B. M. and Shepard E. F. 1997. IPM Research, Development and Training Activities for Palawija Crops in Indonesia. Final Report, 1 October 1995–15 October. Clemson University – Institut Pertanian Bogor. Clemson University, S Carolina, US

Shepard B. M., Justo H. D., Rubia E. G. and Estano D. B. 1990. Response of the rice plant to damage by the rice whorl maggot, Hydriella philippina Ferino (Diptera: Ephydridae). *Journal of Plant Protection in the Tropics* 7, 173–177

Shepard B. M., Shepard E. F., Carner G. R., Hammig M. D., Rauf A. and Turnipseed S. G. 2001. Integrated pest management reduces pesticides and production costs of vegetables and soybean in Indonesia: Field studies with local farmers. *Journal of Agromedicine* 7 (3), 31–66

Sherwood S. G. 1995. Mastering Mystery: Learning to Manage Plant Diseases with Farmers of Honduras and Nicaragua. Master of Professional Studies thesis, Cornell University, US. 169pp

Sherwood S. and Bentley J. 1995. Rural farmers explore causes of plant disease. *ILEIA Newsletter* 11 (1), 20–22

Sherwood, S. and Pumisacho. M. In press. Guia Metodológica de Escuelas de Campo de Agricultores. INIAP-CIP-FAO-WN, World Neighbours, Jardines de San Juan, Guatemala

Sherwood, S., Crissman C. and Cole D. 2002. Pesticide Exposure and Poisonings in the Northern Andes: A Call for International Action. Pesticide Action Network United Kingdom. Spring edition. 10pp

Sherwood S. G., Cole D. C. and Paredes M. 2003. Estrategias de intervención para reducir los riesgos causados por plaguicidas. In: Yanggen D., Crissman C. C. and Espinosa P. (eds) *Los Plaguicidas. Impactos en producción, salud y medio ambiente en Carchi, Ecuador*. CIP, INIAP, Ediciones Abya-Yala, Quito, Ecuador. pp163–185

Sibanda T., Dobson H. M., Cooper J. F., Manyangarirwa W. and Chiimba W. 2000. Pest management challenges for smallholder vegetable farmers in Zimbabwe. *Crop Protection* 19, 807–815

Sissoko M. 1994. Lutte Intégrée contre les ennemis du mil dans la zone de Banamba au Mali. Sahel PV Info., No. 60, 15–19

Sivayoganathan C., Gnanachandran S., Lewis J. and Fernando M. 1995. Protective measure use and symptoms among agropesticide applicators in Sri Lanka. *Social Science and Medicine* 40 (4), 431–436

Smith, B. D. 1995. *The Emergence of Agriculture*. Scientific American Library, New York. 232pp

Smith C. M. 1989. *Plant Resistence to Insects*. John Wiley & Sons, New York

Smith C. 2001. Pesticide exports from U.S. ports, 1997–2000. *International Journal of Occupational and Environmental Health* 7 (4), 266–274

Snapp S. S. and Shennan C. 1994. Salinity effects on root growth and senescence in tomato and the consequences for severity of Phytophthora root rot infection. *Journal of the American Society for Horticultural Science* 119 (3), 458–463

Sodavy P., Sitha M., Nugent R. and Murphy H. 2000. Farmers' Awareness and Perceptions of the effect of Pesticides on their Health. FAO-IPM, Phnom Penh, Cambodia (unpublished)

Sogawa K., Kilin D., and Bhagiawati A. H. 1984. Characterization of the brown plant-hopper population on IR42 in North Sumatra, Indonesia. *International Rice Research Newsletter* 9 (1), 25

Sokuthea N. 2002. *The Empowerment of Farmer Life Schools*. Report of the Project 'Mobilization and Empowerment of Rural Communities Along the Asian Highway (Route 5) in Cambodia to Reduce HIV Vulnerability' (UNDP-FAO/RAS/97/202). Available at http://www.communityipm.org/docs/Farmer_Life_Schools.doc. [accessed 8 Feb 2004]

Solomon G. M. and Motts L. 1998. *Trouble on the Farm: Growing Up With Pesticides in Agricultural Communities*. San Francisco, CA: Natural Resources Defense Council. Available at http://www.nrdc.org/health/kids/farm/farminx.asp. [accessed 8 Feb 2004]

Solomon G. M., Ogunseitan O. A., and Kirsch J. 2000. *Pesticides and Human Health: A Resource for Health Care Professionals*. San Francisco, CA: Physicians for Social Responsibility, Los Angeles and Californians for Pesticide Reform

Sorby K., Fleischer G. and Pehu E. 2003. *Integrated Pest Management in Development, Review of Trends and Implementation Strategies*. ARD working paper 5, World Bank, Washington DC

Sotomayor D., Allen L. H., Chen Z., Dickson D. W. and Hewlett T. 1999. Anaerobic soil management practices and solarization for nematode control in Florida. *Nematropica* 29, 153–170

Spencer J. and Dent D. R. 1991. Walking speed as a variable in knapsack sprayer operation: perception of speed and the effect of training. *Tropical Pest Management* 37, 321–323

Steiner R., McLaughlin L., Faeth P. and Janke R. 1995. Incorporating externality costs in productivity measures: a case study using US agriculture. In Barbett V., Payne R. and Steiner R. (eds). *Agricultural Sustainability: Environmental and Statistical Considerations*. John Wiley, New York. pp209–230

Steinmetz R., Young P. C. M., Caperell-Grant A., Gize E. A., Madhukar B. V., Ben-Jonathan N. and Bigsby R. M. 1996. Novel estrogenic action of the pesticide residue ß hexachlorocyclohexane in human breast cancer cells. *Cancer Research* 56, 5403–5409

Stern V. M., van den Bosch, R. and Leigh T. F. 1964. Strip cutting of alfafa for *Lygus* bug control. *California Agriculture* 18 (4–6)

Stern, V. M., Smith, R. F., van den Bosch, R. and Hagen, R. 1959. The integrated control concept. *Hilgardia* 29, 81–101

Stevens C., Khan V. A., Rodriguez-Kabana R., Ploper L. D., Backman P. A., Collins D. J., Brown J. E., Wilson M. A. and Igwegbe E. C. K. 2003. Integration of soil solarization with chemical, biological and cultural control for the management of soilborne diseases of vegetables. *Plant and Soil* 253, 493–506

Stoorvogel J. J., Antle J. M., Crissman C. C. and Bowen W. 2004. The Tradeoff Analysis Model: Integrated Bio-physical and Economic Modeling of Agricultural Production Systems. *Agricultural Systems* 80 (1), 43–66

Suárez F., Naveso M. and de Juana E. 1997. Farming in the drylands of Spain: birds of the pseudosteppes. In: Pain D. J. and Pienkowski M. W. (eds) *Farming and Birds in Europe*. Academic Press Ltd, London

Sullivan P. G. 2001. *Principles of Sustainable Weed Management for Croplands*. Agronomy Assistance Guide Series. ATTRA publications

Syngenta Group. 2002a. Annual Report 2002. www.syngenta.com

Syngenta. 2002b. Summary of rebuttal to 'Paraquat – Syngenta's controversial herbicide' www.syngenta.com

Szmedra P. 1999. The health impacts of pesticide use on sugarcane farmers in Fiji. *Asia Pacific Journal of Public Health* 11 (2), 82–88

Tafuri J. and Roberts J. 1987. Organophosphate poisoning. *Annals of Emergency Medicine* 16 (2), 193–202

Taha T. E. and Gray R. H. 1993. Agricultural pesticide exposure and perinatal mortality in central Sudan. *Bulletin of the World Health Organization* 71 (3–4), 317–321

Tait E. J. 1983. Pest control decision-making on brassica crops. *Advanced Applied Biology* 8, 121–188

Talekar N. S, Yang J. C. and Lee S. T. 1992. Introduction of *Diadegma semiclausum* to control diamondback moth in Taiwan. In: Talekar, N. S. (ed) *Diamondback Moth and Other Crucifer Pests*. Proceedings of the second International workshop, Tainan, Taiwan, 10–14 December 1990. AVRDC publication no 92–368, pp263–270

Tamm L. 2001. Organic agriculture: development and state of the art: Pesticides in perspective, *Journal of Environmental Monitoring* 3 92N–96N

Tegtmeier E. M. and Duffy M. D. 2004. External costs of agricultural production in the United States. *International Journal of Agricultural Sustainability*, 2 (1), 155–175

Tekelenburg A. 2002. Cactus Pear and Cochineal in Cochabamba. The development of a cross-epistemological management toolkit for interactive design of farm innovation. Wageningen: University, published doctoral dissertation

Teng P. S. 1994. The epidemiological basis for blast management. In: Zeigler R. S., Leong S. A. and Teng P. S. (eds) *Rice Blast Disease*. CAB International, Wallingford,UK. pp409–433

ter Weel P. and van der Wulp H. 1999. *Participatory Integrated Pest Management: Policy and Best Practice*. The Hague, Netherlands; Ministry of Foreign Affairs. 67pp

Thies C. and Tscharntke T. 1999. Landscape structure and biological control in agroecosystems. *Science* 285, 893–895

Thiruchelvam M., Richfield E. K., Baggs R. B., Tank A. W. and Cory-Slechta D. A. 2000. The nigrostriatal dopaminergic system as a preferential target of repeated exposures to combined paraquat and maneb: implications for Parkinson's disease. *Journal of Neuroscience* 20 (24), 9207–9214

Thomas M. and Waage J. K. 1996. *Integration of Biological Control and Host-plant Resistance Breeding: A Scientific and Literature Review*. Ocean Press, Sydney

Thrupp L. A. 1991. Sterilization of workers from pesticide exposure: the causes and consequences of DBCP-induced damage in Costa Rica and beyond. *International Journal of Health Services* 21 (4), 731–757

Thrupp L. A. 2002. Fruits of Progress. *Growing Sustainable Farming and Food Systems*. World Resources Institute, Washington, US. 80pp

Tinoco-Ojanguren R. and Halperin D. C. 1998. Poverty, production, and health: inhibition of erythrocyte cholinesterase via occupational exposure to organophosphate insecticides in Chiapas, Mexico. *Archives of Environmental Health* 53 (1), 29–35

Tobin R. J. 1994. *Bilateral Donor Agencies and the Environment: Pest and Pesticide Management*. US Agency for International Development, Washington, DC

Tobin R. J. 1996. Pest management, the environment, and Japanese foreign assistance. *Food Policy* 21, 211–228

Tomlin C. D. S. 2000. *The Pesticide Manual*. 12th ed. British Crop Protection Council, Brighton

Tompkins D. K., Wright A. T. and Fowler D. B. 1992. Foliar disease development in no-till winter wheat: influence of agronomic practices on powdery mildew development. *Canadian Journal of Plant Science* 72, 965–972

Ton P. 2002. *Organic Cotton Production in sub-Saharan Africa: The Need for Scaling Up*. PAN UK, London

Ton P., Tovignan S., Vodouhe S. D. (2000). Endosulfan deaths and poisonings in Benin. *Pesticides News*, 47, 12–14

Tovignan S., Vodouhe S. D. and Dinham B. 2001. Cotton pesticides cause more deaths in Benin. *Pesticides News* 52, 12–14

Tozun N. 2001. New Policy, Old Patterns: A Survey of IPM in World Bank Projects. *The Global Pesticide Campaigner* 11 (11)

Trimble R. M. 1993. Efficacy of mating disruption for controlling the grape berry moth, *Endopiza viteana* (Clemens) (Lepidoptera: Tortricidae): a case study over three consecutive growing seasons. *The Canadian Entomologist* 125 (1), 1–9

Tyagarajan K. 2002. China Agrochemicals Market: The Challenges Ahead, *Agrolinks*, CropLife Asia, March

Crop Protection Association UK (1986–2001) *Annual Reports*, Wood Mackenzie (company analyst), Edinburgh

Unilever. 2003a. Sustainable Agriculture: the good stories. Sustainable Agriculture Programme, Unilever, 20pp. www.unilever.com

Unilever. 2003b. *Sustainable Vining Peas: Good Agricultural Practice Guidelines*. Sustainable Agriculture Programme, Unilever, 25pp

Uphoff N. (ed) 2002. *Agroecological Innovations*. Earthscan, London

Uphoff, N. Esman M. J. and Krishna A. 1998. *Reasons for Success: Learning From Instructive Experiences in Rural Development*. Kumarian Press, West Hartford, Connecticut. 240pp

Urquhart P. 1999. IPM and the citrus industry in South Africa. *Gatekeeper Series* No. 86. International Institute of Environment and Development, London, UK. 20pp

Useem M., Setti L. and Pincus J. 1992. The science of Javanese management: organizational alignment in an Indonesian development programme. *Public Administration and Development* 12, 447–471

USEPA. 1992. Ethylene bisdithiocarbamates (EBDCs); Notice of intent to cancel and conclusion of Special Review. *Federal Register* 57 (41), 7434–7539

Vaeck M., Botterman J., Reynaerts A., Block M. de, Hofte H. and Leemans J. 1987. *Plant Gene Systems and Their Biology*. Proceedings of a CIBA-Geigy-UCLA symposium held at Tamarron, Colorado, USA, 2–8 Feb. pp171–181

van Bruggen A. H. C. and Grunwald N. J. 1994. The need for a dual hierarchical approach to study plant disease suppression. *Applied Soil Ecology* 1, 91–95

Van Bruggen A. H. C. and Termorshuizen A. J. 2003. Integrated approaches to root disease management in organic farming systems. *Australasian Plant Pathology* 32 (2), 141–156

Van Bruggen A. H. C., Neher D. A., and Ferris H. 1998. Root disease caused by fungi and nematodes in conventional and alternative farming systems. Abstr., Session 5.1, 7th International Congress of Plant Pathology, Edinburgh 9–16 August

Van de Fliert E. 1993. Integrated Pest Management: Farmer Field Schools Generate Sustainable Practices. PhD thesis. University of Wageningen, The Netherlands. 304pp

Van de Fliert E. and Proost J. (eds) 1999. *Women and IPM: Crop Protection Practices and Strategies*. Royal Tropical Insititute (KIT) Amsterdam, the Netherlands/Intermediate Technology Publications, London UK

Van de Fliert E., Thiele G., Campilan D., Ortiz O., Olanya M. and Sherwood S. 2002. Development and linkages of farmer field school and other platforms for participatory research and learning. Paper presented at the *International Learning Workshop on FFS: Emerging Issues and Challenges*, 21–25 October, Yogyakarta, Indonesia. http://www.eseap.cipotato.org/UPWARD/Index.htm

Van de Vrie M., McMurtry J. A. and Huffacker C. B. 1972. Ecology of tetranychid mites and their natural enemies: A review. III. Biology, ecology and pest status, and host plant relations of tetranychids. *Hilgardia* 41, 343–432

van den Berg H. 2001. Facilitating scientific method as follow-up for FFS graduates. FAO Community IPM Programme, Manila, http://www.communityipm.org/downloads.html

van den Berg H. and Lestar A. S. 2001. Improving local cultivation of soybean in Indonesia through farmers' experiments. *Experimental Agriculture* 37, 183–193

van den Berg, H. and Soehardi (2000) The influence of the rice bug Leptocorisa oratorius on rice yield. *Journal of Applied Ecology*, 37, 959–970

Van den Berg J., Nur A. F. and Polaszek A. 1998. Cultural control. In: Polaszek A. (ed) *Cereal Stem-borers in Africa: Economic Importance, Taxonomy, Natural Enemies and Control*. International Institute of Entomology. CAB International. pp333–347

van den Berg H., Senerath H. and Amarasinghe L. 2003. Farmer Field Schools in Sri Lanka: assessing the impact. *Pesticides News* 61, 14–16.

Van der Hoek W., Konradsen F., Athukorala K. and Wanigadewa T. 1998. Pesticide poisoning: a major health problem in Sri Lanka. *Social Science and Medicine* 46 (4–5), 495–504

van der Ploeg J. D. 1994. Styles of farming: an introductory note on concepts and methodology. In: van der Ploeg J. D. and Long A. (eds) *Born from Within: Practice and Perspectives of Endogenous Rural Development*. Assen: Van Gorcum. pp7–31

van Duuren B. 2003. Report on Consultancy on the Initial Survey for Impact Assessment in Cambodia. FAO Regional Vegetable IPM Programme, Phnom Penh, Cambodia

van Emden H. F. and Peakall D. B. 1996. *Beyond Silent Spring: IPM and Chemical Safety*. Chapman and Hall, London

van Huis A. and Meerman F. 1997. Can we make IPM work for resource-poor farmers in sub-Saharan Africa? *International Journal of Pest Management* 43 (4), 313–320

Van Schoubroeck F. 1999. *Learning to Fight a Fly: Developing Citrus IPM in Bhutan*. Wageningen: Wageningen University. Published Doctoral Dissertation

Van Slobbe E. 2002. *Waterbeheer Tussen Crisis en Vernieuwing: Een Studie naar Vernieuwingsprocessen in de Sturing van Regionaal Waterbeheer*. Wageningen: University, Published Doctoral Dissertation (in Dutch)

van Wendel de Joode B. N., De Graaf I. A. M., Wesseling C. and Kromhout H. 1996. Paraquat exposure of knapsack spray operators on banana plantations in Costa Rica. *International Journal of Occupational and Environmental Health* 2 (4), 294–304

van Wendel de Joode B. N., Wesseling C., Kromhout H., Monge P., Garcia M. and Mergler D. 2001. Chronic nervous-system effects of long-term occupational exposure to DDT. *Lancet* 357 (9261), 1014–1016

Verma K. 1998. Cotton, pesticides and suicides. *Global Pesticide Campaigner* 8 (2), 3–5

Verschueren C. 2001. Director General, CropLife International, press release on the launch of CropLife International www.croplife.org

Vetterli W., Perkins R., Clay J. and Guttenstein E. 2002. *Organic Farming and Nature Conservation*. Paper presented at the OECD workshop on Organic Agriculture, held 23–26 September 2002, Washington, DC

Visser I., Cawley S. and Röling N. 1998. A co-learning approach to extension: Soil nitrogen workshops in Queensland, Australia. *Journal of Agricultural Education and Extension* 5 (3), 179–191

Vorgetts J. 2001. Africa Emergency Locust and Grasshopper Assistance (AELGA) Project. In: *5th FAO Consultation on Obsolete, Unwanted and Banned Pesticide Stocks*, Rome, 10–11 May, FAO, Rome

Vorley W. and Keeney D. (eds) 1998. *Bugs in the System: Redesigning the Pesticide Industry for Sustainable Agriculture*. Earthscan, London

Vos J. 1998. Development of decision-making tools for vegetable farmers in South East Asia. In: *Ecotoxicology; Pesticides and Beneficial Organisms*. Kluwer Academic Publishers, US. pp404–409

Vos J. G. M. 1999. Visit Report 21 June–5 July 1999, Ghana. (BCCCA/GIPMF funded project 'sustainable cocoa production in West Africa: development of a farmer-participatory integrated crop management project'). CABI Bioscience

Vos J. G. M. 2000. Vietnam Showcase. Training News, *Biocontrol News & Information* 21 (1) http://pest.cabweb.org/Journals/BNI/BNI21-1/TRAIN.HTM

Vos J. G. M. 2001. *IPM Case Study on Disease Management Action Research in Vietnam*. Paper presented at the Regional Trainers' workshop on Vegetable IPM Curriculum Development, held at Chiangmai, Thailand, 18–24 February. IPMForum website http://www.cabi-publishing.org/IPM/case.htm

Vos J. G. M. and Neuenschwander P. 2002. Proceedings of the West Africa regional cocoa IPM workshop (English and French), 13–15 November, 2001, Cotonou, Benin. CABI Bioscience, IITA and CPL Press. 204pp

Vos J. G. M., Dybing E., Greim H. A., ladefoged O., Lambre C., Tarazona J. V., Brandt I. and Vethaak A. D. 2000. Health effects of endocrine-disrupting chemicals on wildlife, with special reference to the European situation. *Critical Reviews in Toxicology* 30, 71–133

Wagener B., Löhr B., Reineke A. and Zebitz C. P. W. 2002. Molecular identification of Diadegma species (Ichneumonidae) parasitising Diamondback Moth *Plutella xylostella* (Plutellidae) in eastern and southern Africa. In: International Symposium: Improving biocontrol of Plutella xylostella. CIRAD, Montpellier, 21–24 October

Waibel H. 1987. Farmers' practices and recommended economic threshold levels in irrigated rice in the Philippines. No. 23 In: Tait, J. and Napompeth, B. (eds) *Management of Pests and Pesticides: Farmers' Perceptions and Practices*. West View Press, London

Waibel, H. and Fleischer G. 1998. Kosten und Nutzen des chemischen Pflanzenschutzes in der deutschen Landwirtschaft aus gesamtwirtschaftlicher Sicht (Social costs and benefits of chemical plant protection in German agriculture). Kiel: Vauk Verlag

Waibel H. and Fleischer G. 2001. Experience with cost–benefit studies of pesticides in Germany. In OECD workshop on the *Economics of Pesticide Risk Reduction in Agriculture*, Copenhagen, 28–30 Nov. OECD, Paris

Waibel H., Fleischer G. and Becker H. 1999a. The economic benefits of pesticides: A case study from Germany. *Agrarwirtschaft* 48 H. 6, S. 219–230

Waibel H., Fleischer G. and Becker H. 1999b. Cost–benefit analysis of pesticides. *Agrarwirtschaft* 48 H. 8/9, 338–342

Walde S. J. 1995. How quality of host plants affects a predator-prey interaction in biological control. *Ecology* 76, 1206–1219

Walker S. R. and Medd R. W. 2002. Improved management of *Avena ludoviciana* and *Phalaris paradoxa* with more densely sown wheat and less herbicide. *Weed Research* 42, 257–270

Watkins S. 2003. *The World Market for Crop Protection Products in Rice*, Agrow Reports, PJB Publications Ltd, Richmond, UK

Watterson A. 1988. *Pesticide Users' Health and Safety Handbook: An International Guide*. Aldershot, Gower Technical Press and Van Nostrand Reinhold, New York. 504pp

Watterson A. 2000. Agricultural science and food policy for consumers and workers: recipes for public health successes or disasters? *New Solutions* 10 (4), 317–324

Way M. J. and Heong K. L. 1994. The role of biodiversity in the dynamics and management of insect pests of tropical irrigated rice – a review. *Bulletin of Entomology Research* 84, 567–587

Way, M. O., Grigarick A. A., Litsinger J. A., Palis F. and Pingali P. L. 1991. Economic thresholds and injury levels for insect pests of rice, In: Heinrichs E. A. and Miller T. A. (eds), *Rice Insects: Management Strategies*, Springer-Verlag, New York. pp67–106

Weller D. M., Raaijmakers J. M., McSpadden Gardener B. B. and Thomashow L. S. 2002. Microbial populations responsible for specific soil suppressiveness to plant pathogens. *Annual Review of Phytopathology* 40, 309–348

Wesseling C., Castillo L. and Elinder C. G. 1993. Pesticide poisonings in Costa Rica. *Scandinavian Journal of Work Environment and Health* 19 (4), 227–223

Wesseling C., Ahlbom A., Antich D., Rodriguez A. C. and Castro R. 1996. Cancer in banana plantation workers in Costa Rica. *International Journal of Epidemiology* 25, 1125–1131

Wesseling C., McConnell R., Partanen T. and Hogstedt C. 1997a. Agricultural pesticide use in developing countries: health effects and research needs. *International Journal of Health Services* 27 (2), 273–308

Wesseling C., Hogstedt C., Picado A. and Johansson L. 1997b. Unintentional fatal paraquat poisonings among agricultural workers in Costa Rica: fifteen case reports. *American Journal of Industrial Medicine* 32, 433–441

Wesseling C., Hogstedt C., Fernandez P. and Ahlbom A. 2000. Time trends of occupational pesticide-related injuries in Costa Rica, 1982–92. *International Journal of Occupational and Environmental Health* 7 (1), 1–6

Wesseling C., Aragon A., Castillo L., Corriols M., Chaverri F., de la Cruz E., Keifer M., Monge P., Partanen T. J., Ruepert C. and van Wendel de Joode B. 2001. Hazardous pesticides in Central America. *International Journal of Occupational and Environmental Health* 7 (4), 287–294

Wesseling C., Van Wendel de Joode B., Ruepert C., Leon C., Monge P., Hermosillo H. and Partanen T. 2001. Paraquat in developing countries. *International Journal of Occupational and Environmental Health*, 7,275–286

Wesseling C., Keifer M., Ahlbom A., McConnell R., Moon J. D., Rosenstock L. and Hogstedt C. 2002. Long-term neurobehavioral effects of mild poisonings with organophosphate and n-methyl carbamate pesticides among banana workers. *International Journal of Occupational and Environmental Health* 8 (1), 27–34

Whitten, M. J. and Ketelaar J. W. 2003. Farmer Field Schools: From Crop Protection to Crop Husbandry, In: Eveleens K., Jiggins J. and Lim G. S. (eds) *Farmers, FAO and Field Schools: Bringing IPM to the Grass Roots in Asia*, FAO, Rome

WHO/EEA. 2002. *Children, Health and Environment: A Review of Evidence*. Environmental Issue report no.29, WHO Regional Office for Europe and European Environment Agency, Copenhagen

Whorton D., Krauss R. M., Marshall S. and Milby T. H. 1977. Infertility in male pesticide workers. *Lancet* 17 (8051), 1259–1261

Williams P. R. D. and Hammitt J. K. 2000. A comparison of organic and conventional fresh produce buyers in the Boston area. *Risk Analysis* 20 (5), 735–746

Williamson S. 2002. Challenges for farmer participation in integrated and organic production of agricultural tree crops. *Biocontrol News & Information* 23 (1) 25N–36N

Williamson S. 2003. *The Dependency Syndrome: Pesticide Use by African Smallholders*. Pesticide Action Network UK, London. 126pp

Williamson S. and Ali B. 2000. *Delivery of Biocontrol Technologies to IPM Farmers – Nicaragua*. CABI/UNEP Critical Issues Series, Dent, D. R. and Gopalan, H. N. B. (eds) CABI Bioscience UK Centre, Ascot. 22pp

Williamson S., Little A., Arif Ali M., Kimani M., Meir C. and Oruko L. 2003. Aspects of cotton and vegetable farmers' pest management decision-making in India and Kenya. *International Journal of Pest Management* 49 (3), 187–198

Wilson B. W., Sanborn J. R., O'Malley M. A., Henderson J. D. and Billitti J. R. 1997. Monitoring the pesticide-exposed worker. *Occupational Medicine* 12 (2), 347–363

Wilson C. and Tisdell. 2001. Why farmers continue to use pesticides despite environmental, health and sustainability costs. *Ecological Economics* 39, 449–462

Wilson M. 1997. Biocontrol of aerial plant diseases in agriculture and horticulture: current approaches and future prospects. *Journal of Industrial Microbiology and Biotechnology* 19, 188–191

Winrock International. 1994. Pesticides and the agrichemical industry in Sub-Saharan Africa. Paper prepared for USAID by Environmental and natural resources Policy and Training project, Winrock International Environmental Alliance, Arlington, VA. 117pp

Wolfe M. S. 2002. *The Role of Functional Biodiversity in Managing Pests and Diseases in Organic Production Systems*. IOR Elm Farm Research Centre, Newbury, Berkshire

Wong C. S., Capel P. D. and Nowell L. H. 2000. *Organochlorine Pesticides and PCBs in Stream Sediment and Aquatic Biota – Initial Results from the National Water-Quality Assessment Program, 1992–1995*. US Geological Survey. Water-Resources Investigations Report 00–4053. Sacramento, CA

Workneh F., van Bruggen A. H. C., Drinkwater L. E. and Shennan C. 1993. Variables associated with corky root and Phytophthora root rot of tomatoes in organic and conventional farms. *Phytopathology* 83, 581–589

World Bank. 2000. Project appraisal document on a proposed adaptable program credit in the amount of US$67 million equivalent to the Republic of Ghana in support of the first phase for an Agricultural Services Sub-sector Investment Program, June 2000. Africa Regional Office, World Bank, Washington

World Development Index. 2003. World Bank, Washington, DC. (an on-line database available at: www.worldbank.org [accessed 6 January 2004])

World Health Organization (WHO). 1990. *The Public Health Impact of Pesticides Use in Agriculture*. World Health Organization, Geneva

World Health Organisation (WHO). 2001. *Recommended Classification of Pesticides by Hazard and Guidelines to Classification 2000–2002*. International Programme on Chemical Safety, WHO/PCS/01.5

World Resources Institute. 2000. *World Resources 2000–2001. People and Ecosystems. The Fraying Web of Life*. Washington: World Resources Institute with UNDP, UNEP, and World Bank

WPGVA. 2003. Accomplishments and directions by the WWF/WPGVA/UW Collaboration. Partnership for Environmental Potato Production, University of Wisconsin, Madison, US. At http://ipcm.wisc.edu/bioipm/news

Yanggen, D., Crissman C. C. and Espinosa P. (eds) 2003. *Plaguicidas: Impactos en producción, salud y medioambiente en Carchi, Ecuador*. CIP, INIAP, Ediciones Abya Yala, Quito, Ecuador. 198pp

Youdeowei A. 2000. Proposals for a *Pest Management Plan*. Consultant's report for the Rwanda Agricultural and Rural Market Development Project, World Bank, Washington, US

Yussefi M. and Willer H. (eds) 2003. *The World of Organic Agriculture 2003: Statistics and Future Prospects*. www.soel.de/inhalte/publikationen/s/s_74.pdf

Zahm S. H. and Blair A. 1993. Cancer among migrant and seasonal farmworkers: an epidemiologic review and research agenda. *American Journal of Industrialized Medicine* 24, 753–766

Zahm S. H. and Ward M. H. 1998. Pesticides and childhood cancer. *Environmental Health Perspectives* 106 (Suppl 3)

Zahm S. H., Ward M. H. and Blair A. 1997. Pesticides and cancer. *Occupational Medicine* 12 (2), 269–289

Zalom F. G. 1993. Reorganising to facilitate the development and use of integrated pest management. *Agriculture, Ecosystems and Environment* 46, 245–256

Zalom F. G., Phillips P. A., Toscano N. C. and Udayagiri S. 2001. *UC Pest Management Guidelines: Strawberry-Lygus Bug*. February 2001. University of California Department of Agriculture and Natural Resources. Berkeley, California

Zeddies J., Schaab R. P., Neuenschwander P., and Herren H. R. 2001. Economics of biological control of cassava mealybug in Africa. *Agricultural Economics* 24, 209–219

Zhang J., Cai W.-W. and Lee D. J. 1992. Occupational hazards and pregnancy outcomes. *American Journal of Industrial Medicine* 21, 397–408

Zhou G., Baumgartner J. and Overholt W. A. 2001. Impact of an exotic parasitoid on stemborer (Lepidoptera) populations. *Ecological Applications* 11, 1554–1551

Zhu, Y., Chen, H., Fen, J., Wang, Y., Li, Y., Cxhen, J., Fan, J., Yang, S., Hu, L., Leaung, H., Meng, T. W., Teng, A. S., Wang, Z. and Mundt, C. C. 2000. Genetic diversity and disease control in rice. *Nature* 406, 718–722

Zilberman D. and Millock K. 1997. Financial incentives and pesticide use. *Food Policy* 22 (2), 133–144

Index